Nanoengineering and Smart Materials

Nanoengineering and Smart Materials

Edited by
Peggy Rusk

Willford Press
www.willfordpress.com

Published by Willford Press,
118-35 Queens Blvd., Suite 400,
Forest Hills, NY 11375, USA

ISBN: 978-1-68285-427-3

Cataloging-in-Publication Data

Nanoengineering and smart materials / edited by Peggy Rusk.
p. cm.
Includes bibliographical references and index.
ISBN 978-1-68285-427-3
1. Nanotechnology. 2. Materials science. 3. Nanostructured materials.
I. Rusk, Peggy.
T174.7 .N36 2018
620.5--dc23

For information on all Willford Press publications
visit our website at www.willfordpress.com

Willford Press

Printed in China.

Contents

Preface

I am honored to present to you this unique book which encompasses the most up-to-date data in the field. I was extremely pleased to get this opportunity of editing the work of experts from across the globe. I have also written papers in this field and researched the various aspects revolving around the progress of the discipline. I have tried to unify my knowledge along with that of stalwarts from every corner of the world, to produce a text which not only benefits the readers but also facilitates the growth of the field.

Nanoengineering involves the engineering and manufacturing of materials at the nanoscale level. Traditional properties of materials are fundamentally altered at the molecular level. Nanomaterials are used in the manufacture of sensors, microfluids, thermoelectrics, photovoltaics, biomedical devices, etc. This book is a compilation of chapters that discuss the most vital concepts and emerging trends in this field. It is meant for students who are looking for an elaborate reference text on nanoengineering.

Finally, I would like to thank all the contributing authors for their valuable time and contributions. This book would not have been possible without their efforts. I would also like to thank my friends and family for their constant support.

Editor

1

A layer-wise MITC9 finite element for the free-vibration analysis of plates with piezo-patches

Maria Cinefra[a]*, Stefano Valvano[a] and Erasmo Carrera[a,b,c]

[a]Department of Mechanical and Aerospace Engineering, Politecnico di Torino, Torino, Italy; [b]School of Aerospace, Mechanical and Manufacturing Engineering College, RMIT University, Melbourne, Australia; [c]Department of Mathematics, Faculty of Science, King Abdulaziz University, P.O. Box 80203, Jeddah 21589, Saudi Arabia

The present article considers the free-vibration analysis of plate structures with piezoelectric patches by means of a plate finite element with variable through-the-thickness layer-wise kinematic. The refined models used are derived from Carrera's Unified Formulation (CUF) and they permit the vibration modes along the thickness to be accurately described. The finite-element method is employed and the plate element implemented has nine nodes, and the mixed interpolation of tensorial component (MITC) method is used to contrast the membrane and shear locking phenomenon. The related governing equations are derived from the principle of virtual displacement, extended to the analysis of electromechanical problems. An isotropic plate with piezoelectric patches is analyzed, with clamped-free boundary conditions and subjected to open- and short-circuit configurations. The results, obtained with different theories, are compared with the higher-order type solutions given in the literature. The conclusion is reached that the plate element based on the CUF is more suitable and efficient compared to the classical models in the study of multilayered structures embedding piezo-patches.

Keywords: plate; patches; finite-element method; piezoelectric materials; mixed interpolated; tensorial components; Carrera's Unified Formulation; layer-wise

1. Introduction

Piezoelectric materials have the ability to convert mechanical energy into electrical energy and vice versa. For the last 50 years, the use of piezoelectric components as electromechanical transducers in sensor as well as in actuator applications has been continuously increasing. More recently, piezoelectrics have been considered to be among the most suitable materials for extending the structural capabilities beyond that of purely passive load-carrying one. Vibration and noise suppression, controlled active deformation, and health monitoring are among the most important applications of these "intelligent" structural components.

Finding analytical solutions for general smart structural problems is a very difficult task, and they only exist for a very few specialized and idealized cases. Meanwhile, the

*Corresponding author. Email: maria.cinefra@polito.it

finite-element (FE) method has become the most widely used technique to model various physical processes, including piezoelectricity. The introduction of piezoelectric material into a passive structure naturally leads to a multilayered component, and it has been recognized that classical models are not suitable for an accurate design of such structures, see for example the review article by Noor and Burton [1] and the references cited herein. The strain induced by the piezoelectric actuators has been used in the works of Crawley and Luis [2], Bailey and Hubbard [3], and Robbins and Reddy [4] as an applied strain that contributes to the total strain of the non-active structure. Interlaminar continuity (IC) of the transverse stresses and the related discontinuity of the slopes of the displacement distributions in thickness direction at the layer interfaces are the main effects arising in multilayered structures which cannot be captured by classical formulations based on love first approximation theories (LFATs) (see e.g. [5]). Many refined theories for plates have been proposed in order to meet the modeling requirements – known as C-requirements – posed by these characteristics; further details can be found in the monograph by Reddy [6] and in the paper by Carrera [7]. For curved structures, Koiter [8] recognized the importance of transverse stress effects even for homogeneous plates and recommended the inclusion of such effects whenever a consistent higher-order model has to be proposed. The fundamentals for the modeling of piezoelectric materials have been given in many contributions, in particular in the pioneering works of Mindlin [9], EerNisse [10], Tiersten and Mindlin [11], and in the monograph by Tiersten [12]. The embedding of piezoelectric layers into plates sharpens the requirements of an accurate modeling of the resulting adaptive structure due to the localized electromechanical coupling, see e.g. the review by Saravanos and Heyliger [13]. Therefore, within the framework of two-dimensional approaches, layer-wise (LW) descriptions have often been proposed, either for the electric field only (see e.g. the works of Kapuria [14] and of Ossadzow-David and Touratier [15]) or for both the mechanical and electrical unknowns (e.g. Heyliger et al. [16]). Ballhause et al. [17] have shown that a fourth-order assumption for the displacements leads to the correct closed-form solution. They conclude that the analysis of local responses requires at least a LW description of the displacements, see also [18]. Benjeddou et al. [19] emphasized that a quadratic electric potential through the plate thickness satisfies the electric charge conservation law exactly. An attempt to mathematically substantiate axiomatic two-dimensional piezoelectric shell formulations by the means of asymptotic expansions can be found in the book by Rogacheva [20]. Zhou et al. (2000) [21] have presented coupled FE models based on a third-order theory for the dynamic response of smart composite plates. In [22] and [23], Arau`jo et al. present a new FE model for the parameter estimation and the analysis of active sandwich laminated plates with a viscoelastic core and laminated anisotropic face layers, as well as piezoelectric sensor and actuator layers. The model is formulated using a mixed LW approach, by considering a higher-order shear deformation theory to represent the displacement field of the viscoelastic core and a first-order shear deformation theory (FSDT) for the displacement field of the adjacent laminated anisotropic face layers and exterior piezoelectric layers. An exhaustive overview of the many different modeling approaches and solution techniques for laminated piezoelectric plates is far beyond the scope of this article; more details on this topic can be found in the already cited review by Saravanos and Heyliger, and in the surveys by Gopinathan et al. [24] and by Benjeddou [25]. Over the last few years, interest has been emerging for mixed formulations also involving stresses and dielectric displacements as primary variables, see for example the recent works of Lammering et al. [26] and of Benjeddou et al. [27]. Some of the most recent contributions to the FE analysis of piezoelectric plates that includes an FSDT description of displacements and a LW form of the electric potential have been developed by Sheik et al. [28]. The numerical, membrane, and bending behavior of FEs that are based on FSDTs were analyzed by Auricchio et al. [29] in the framework of a suitable variational formulation. The third-order theory of higher-order theories type has been applied by Thornburg and Chattopadhyay [30] to derive FEs that take into consideration electromechanical coupling. Similar elements have more recently been considered by Shu [31]. The extension of the third-order zigzag multilayered theory to finite analysis of electromechanical problems has been proposed by Oh and Cho [32]. An extension to the piezoelectricity of numerically efficient plate elements based on the mixed interpolation of tensorial component (MITC) formulation has recently been provided by Kogl and Bucalem [33,34]. Some of the most recent contributions to the FE analysis of piezoelectric shells that are based on exact geometry solid-shell element with the first-order seven-parameter equivalent single-layer (ESL) theory have been developed by Kulikov et al. [35], and a piezoelectric solid-shell element with a mixed variational formulation and a geometrically nonlinear theory has been developed by Klinkel et al. [36].

A plate FE is presented in this article for the analysis of plate structures with piezo-patches. It is based on Carrera's Unified Formulation (CUF), which has been developed by Carrera for multilayered structures [37]. Many works have been devoted to the extension of CUF to electromechanical problems, see [38–41]. Among others, Carrera [42] extends principle of virtual displacement (PVD) and RMVT variational statements to piezo-laminated plates, see also [43–46]. The plate geometry is considered and the MITC method [47–52] is used to contrast the shear locking. The governing equations for the free-vibration analysis of plate structures are derived from the PVD, in order to apply the FE method. An isotropic plate with piezoelectric patches is analyzed, and the results, obtained with the different models contained in CUF, are compared with the higher-order type solution given in the literature by Yasin et al. [53].

2. Unified Formulation

The main feature of the Unified Formulation by Carrera [42] [54,55] is the unified manner in which the displacement variables are handled. According to CUF, the displacement field and the potential field are written by means of approximating functions in the thickness direction as follows:

$$\boldsymbol{u}^k(x,y,z) = F_\tau(z)\boldsymbol{u}_\tau^k(x,y)\ , \quad \delta\boldsymbol{u}^k(x,y,z) = F_s(z)\delta\boldsymbol{u}_s^k(x,y)\ , \quad \tau,s = 0,1,\ldots,N\ , \tag{1}$$

$$\Phi^k(x,y,z) = F_\tau(z)\,\Phi_\tau^k(x,y)\ , \quad \delta\,\Phi^k(x,y,z) = F_s(z)\delta\,\Phi_s^k(x,y)\ , \quad \tau,s = 0,1,\ldots,N\ , \tag{2}$$

where (x, y, z) is a Cartesian coordinate reference system, defined in the next section, and the displacement $\boldsymbol{u} = \{u, v, w\}$ and the potential Φ are referred to such system. δ indicates the virtual variation and $k = 1, \ldots, n_l$ identifies the layer where n_l is the total number of layers. In this work, the layer number n_l assumes two values: $n_l = 1$ for the plate and $n_l = 3$ for the plate area in with both sensors and/or actuators are present. F_τ and F_s are the so-called thickness functions depending only on z. $\boldsymbol{u}_s$ and Φ_s are the unknown variables depending on the coordinates x and y. τ and s are sum indexes and N is the order of expansion in the thickness direction assumed for the variables. In the case of ESL models, a Taylor expansion is employed as thickness functions:

$$u = F_0\, u_0 + F_1\, u_1 + \ldots + F_N\, u_N = F_s\, u_s, \qquad s = 0, 1, \ldots, N. \tag{3}$$

$$F_0 = z^0 = 1, \quad F_1 = z^1 = z, \quad \ldots, \quad F_N = z^N. \tag{4}$$

Classical theories, such as the FSDT, can be obtained from an ESL model with $N = 1$, by imposing a constant transverse displacement through the thickness via penalty techniques. The correction of the Poisson's locking is applied to the FSDT model and the ESL model with $N = 1$ according to the procedure explained in the literature [56].

In the case of LW models, the displacement and the potential are defined at the k-layer level:

$$u^k = F_t\, u_t^k + F_b\, u_b^k + F_r\, u_r^k = F_\tau\, u_\tau^k, \qquad \tau = t, b, r, \quad r = 2, \ldots, N. \tag{5}$$

$$\Phi^k = F_t\, \Phi_t^k + F_b\, \Phi_b^k + F_r\, \Phi_r^k = F_\tau\, \Phi_\tau^k, \qquad \tau = t, b, r, \quad r = 2, \ldots, N. \tag{6}$$

$$F_t = \frac{P_0 + P_1}{2}, \quad F_b = \frac{P_0 - P_1}{2}, \quad F_r = P_r - P_{r-2}. \tag{7}$$

in which $P_j = P_j(\zeta_k)$ is the Legendre polynomial of j-order defined in the ζ_k-domain: $P_0 = 1$, $P_1 = \zeta_k$, $P_2 = (3\zeta_k^2 - 1)/2$, $P_3 = (5\zeta_k^3 - 3\zeta_k)/2$, $P_4 = (35\zeta_k^4 - 30\zeta_k^2 + 3)/8$. The top ($t$) and bottom ($b$) values of the displacements and the potential are used as unknown variables, and one can impose the following compatibility conditions:

$$u_t^k = u_b^{k+1}, \quad \Phi_t^k = \Phi_b^{k+1}, \quad k = 1, n_l - 1. \tag{8}$$

The LW models, in respect to the ESLs, allow the zigzag form of the variable distribution in layered structures to be modelled.

3. Plate geometry

In this section, the derivation of a plate FE for the analysis of multilayered structures is presented. The element is based on LW and ESL theories contained in the Unified Formulation. A nine-node plate element is considered. Plates are bi-dimensional structures in which one dimension (in general the thickness in the z-direction) is negligible with respect to the other two in-plane dimensions. In this work, the layer number n_l assumes two values: $n_l = 1$ for the plate, $n_l = 3$ for the plate area in which both sensors and/or actuators are present. Geometry and the reference system are indicated in Figure 1. Geometrical relations can be expressed in matrix form as:

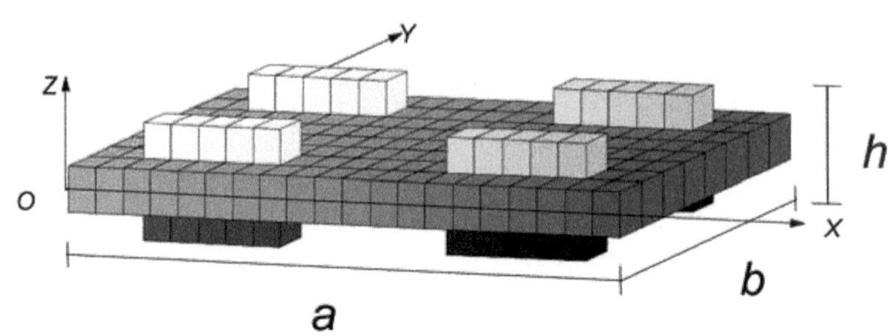

Figure 1. Geometry of the plate.

$$\begin{aligned}\boldsymbol{\sigma}_p &= [\sigma_{xx}, \sigma_{yy}, \gamma_{xy}] = (\boldsymbol{D}_p)\boldsymbol{u},\\ \boldsymbol{\sigma}_n &= [\gamma_{xz}, \gamma_{yz}, \sigma_{zz}] = (\boldsymbol{D}_{np} + \boldsymbol{D}_{nz})\boldsymbol{u},\end{aligned} \tag{9}$$

where the differential operators are defined as follows:

$$\boldsymbol{D}_p = \begin{bmatrix} \partial_x & 0 & 0 \\ 0 & \partial_y & 0 \\ \partial_y & \partial_x & 0 \end{bmatrix} \boldsymbol{D}_{np} = \begin{bmatrix} 0 & 0 & \partial_x \\ 0 & 0 & \partial_y \\ 0 & 0 & 0 \end{bmatrix} \boldsymbol{D}_{nz} = \begin{bmatrix} \partial_z & 0 & 0 \\ 0 & \partial_z & 0 \\ 0 & 0 & \partial_z \end{bmatrix} \tag{10}$$

The geometrical relations between electric field **E** and potential Φ are defined as follows:

$$\begin{aligned}\mathrm{E}_p &= [\mathrm{E}_x, \mathrm{E}_y]^T = -\boldsymbol{D}_{ep}\,\Phi,\\ \mathrm{E}_n &= [\mathrm{E}_z]^T = -\boldsymbol{D}_{en}\,\Phi,\end{aligned} \tag{11}$$

where the differential operators are defined as follows:

$$\boldsymbol{D}_{ep} = \begin{bmatrix} \partial_x \\ \partial_y \end{bmatrix}, \boldsymbol{D}_{en} = [\partial_z].$$

4. Finite-element approximation and MITC method

According to the FE method, the displacement and the potential components are interpolated on the nodes of the element by means of the Lagrangian shape functions *Ni*:

$$\delta\boldsymbol{u}_s = N_i\delta\boldsymbol{q}_{u s_i} \qquad \boldsymbol{u}_\tau = N_j\boldsymbol{q}_{u\tau_j} \qquad \text{with} \quad i,j = 1,\ldots,9 \tag{12}$$

$$\delta\Phi_s = N_i\delta q_{\Phi s_i} \qquad \Phi_\tau = N_j q_{\Phi\tau_j} \qquad \text{with} \quad i,j = 1,\ldots,9 \tag{13}$$

where $\boldsymbol{q}_{u\tau_j} = (\boldsymbol{q}_{u_\tau}, \boldsymbol{q}_{v_\tau}, \boldsymbol{q}_{w_\tau})$ and $\delta\boldsymbol{q}_{us_i} = (\delta\boldsymbol{q}_{u_s}, \delta\boldsymbol{q}_{v_s}, \delta\boldsymbol{q}_{w_s})$ are the nodal displacements and their virtual variations for the mechanical variables, and $q_{\Phi\tau_j} = (q_{\Phi_\tau})$ and $\delta q_{\Phi s_i} = (\delta q_{\Phi_s})$ are the nodal potential and their virtual variation for the electric variable. Substituting in the geometrical relations (9), one has:

$$\begin{aligned}\boldsymbol{\epsilon}_p &= F_\tau\,(\boldsymbol{D}_p)(N_i)\boldsymbol{q}_{\tau_i}\\ \boldsymbol{\epsilon}_n &= F_\tau\,(\boldsymbol{D}_{n\Omega})(N_i)\boldsymbol{q}_{\tau_i} + F_{\tau,z}\boldsymbol{D}_{nz}(N_i)\boldsymbol{q}_{\tau_i}\end{aligned} \tag{14}$$

Considering the local coordinate system (ξ, η), the MITC plate elements [57,58] are formulated by using, instead of the strain components directly computed from the displacements, an interpolation of these within each element using a specific interpolation strategy for each component. The corresponding interpolation points, called tying points, are shown in Figure 2 for the MITC9 plate element.

The interpolating functions are Lagrangian and are calculated by imposing that the function assumes the value 1 in the corresponding tying point and 0 in the others. These are arranged in the following arrays:

$$\begin{aligned}\boldsymbol{N}_{m1} &= [N_{A1}, N_{B1}, N_{C1}, N_{D1}, N_{E1}, N_{F1}]\\ \boldsymbol{N}_{m2} &= [N_{A2}, N_{B2}, N_{C2}, N_{D2}, N_{E2}, N_{F2}]\\ \boldsymbol{N}_{m3} &= [N_P, N_Q, N_R, N_S]\end{aligned} \tag{15}$$

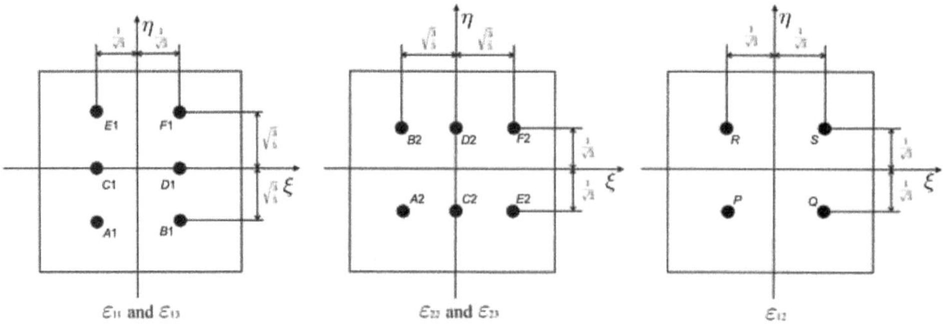

Figure 2. Tying points for the MITC9 shell finite element.

From this point on, the subscripts $m1$, $m2$, and $m3$ indicate quantities calculated in the points ($A1$, $B1$, $C1$, $D1$, $E1$, $F1$), ($A2$, $B2$, $C2$, $D2$, $E2$, $F2$) and (P, Q, R, S), respectively. Therefore, the strain components are interpolated as follows:

$$
\begin{aligned}
\boldsymbol{\epsilon}_{\mathrm{p}} = \begin{bmatrix} \epsilon_{\mathrm{xx}} \\ \epsilon_{\mathrm{yy}} \\ \gamma_{\mathrm{xy}} \end{bmatrix} &= \begin{bmatrix} \mathbf{N}_{\mathrm{m1}} & 0 & 0 \\ 0 & \mathbf{N}_{\mathrm{m2}} & 0 \\ 0 & 0 & \mathbf{N}_{\mathrm{m3}} \end{bmatrix} \begin{bmatrix} \epsilon_{\mathrm{xx}_{\mathrm{m1}}} \\ \epsilon_{\mathrm{yy}_{\mathrm{m2}}} \\ \gamma_{\mathrm{xy}_{\mathrm{m3}}} \end{bmatrix} \\
\boldsymbol{\epsilon}_{\mathrm{n}} = \begin{bmatrix} \gamma_{\mathrm{xz}} \\ \gamma_{\mathrm{yz}} \\ \epsilon_{\mathrm{zz}} \end{bmatrix} &= \begin{bmatrix} \mathbf{N}_{\mathrm{m1}} & 0 & 0 \\ 0 & \mathbf{N}_{\mathrm{m2}} & 0 \\ 0 & 0 & 1 \end{bmatrix} \begin{bmatrix} \gamma_{\mathrm{xz}_{\mathrm{m1}}} \\ \gamma_{\mathrm{yz}_{\mathrm{m2}}} \\ \epsilon_{\mathrm{zz}} \end{bmatrix}
\end{aligned}
\tag{16}
$$

where the strains $\epsilon_{xx_{m1}}, \epsilon_{yy_{m2}}, \epsilon_{xy_{m3}}, \epsilon_{xz_{m1}}, \epsilon_{yz_{m2}}$ are expressed by means of Equation (14) and the shape functions N_i are calculated in the tying points.

5. Constitutive equations

The second step toward the governing equations is the definition of the constitutive equations that permit the stresses and the electric displacements to be expressed by means of the strains and the electric fields. The generalized Hooke's law is considered by employing a linear constitutive model for infinitesimal deformations. A linear coupling of the electric fields is employed to complete stress equations and to describe the electric displacement equations. In the numerical section, an isotropic plate with piezo-patches is analyzed, but the constitutive equations have been developed for composite materials as a future extension of this work. These equations are obtained in material coordinates (1, 2, 3) for each layer k and then rotated in the general reference system (x, y, z). Therefore, the stresses $\boldsymbol{\sigma}_p^k = \left\{\sigma_{xx}^k, \sigma_{yy}^k, \gamma_{xy}^k\right\}$ and $\boldsymbol{\sigma}_n^k = \left\{\gamma_{xz}^k, \gamma_{yz}^k, \sigma_{zz}^k\right\}$ and electric displacements $\boldsymbol{D}_p^k = \left\{\mathcal{D}_x^k, \mathcal{D}_y^k\right\}$ and $\boldsymbol{D}_n^k = \left\{\mathcal{D}_z^k\right\}$ after the rotation are:

$$\boldsymbol{\sigma}_{pC}^k = \boldsymbol{C}_{pp}^k \boldsymbol{\epsilon}_{pG}^k + \boldsymbol{C}_{pn}^k \boldsymbol{\epsilon}_{nG}^k - \boldsymbol{e}_{pp}^{k^T} \mathbf{E}_{pG}^k - \boldsymbol{e}_{np}^{k^T} \mathbf{E}_{nG}^k \tag{17}$$

$$\boldsymbol{\sigma}_{nC}^k = \boldsymbol{C}_{np}^k \epsilon_{pG}^k + \boldsymbol{C}_{nn}^k \epsilon_{nG}^k - \boldsymbol{e}_{pn}^{k^T} \mathbf{E}_{pG}^k - \boldsymbol{e}_{nn}^{k^T} \mathbf{E}_{nG}^k \tag{18}$$

$$\boldsymbol{D}^k_{pC} = \boldsymbol{e}^k_{pp}\epsilon^k_{pG} + \boldsymbol{e}^k_{pn}\epsilon^k_{nG} + \boldsymbol{\varepsilon}^k_{pp}\mathbf{E}^k_{pG} + \boldsymbol{\varepsilon}^k_{pn}\mathbf{E}^k_{nG} \tag{19}$$

$$\boldsymbol{D}^k_{nC} = \boldsymbol{e}^k_{np}\epsilon^k_{pG} + \boldsymbol{e}^k_{nn}\epsilon^k_{nG} + \boldsymbol{\varepsilon}^k_{np}\mathbf{E}^k_{pG} + \boldsymbol{\varepsilon}^k_{nn}\mathbf{E}^k_{nG} \tag{20}$$

where

$$\begin{aligned}
\boldsymbol{C}^k_{pp} &= \begin{bmatrix} C^k_{11} & C^k_{12} & C^k_{16} \\ C^k_{12} & C^k_{22} & C^k_{26} \\ C^k_{16} & C^k_{26} & C^k_{66} \end{bmatrix} \qquad \boldsymbol{C}^k_{pn} = \begin{bmatrix} 0 & 0 & C^k_{13} \\ 0 & 0 & C^k_{23} \\ 0 & 0 & C^k_{36} \end{bmatrix} \\
\boldsymbol{C}^k_{np} &= \begin{bmatrix} 0 & 0 & 0 \\ 0 & 0 & 0 \\ C^k_{13} & C^k_{23} & C^k_{36} \end{bmatrix} \qquad \boldsymbol{C}^k_{nn} = \begin{bmatrix} C^k_{55} & C^k_{45} & 0 \\ C^k_{45} & C^k_{44} & 0 \\ 0 & 0 & C^k_{33} \end{bmatrix}
\end{aligned} \tag{21}$$

$$\begin{aligned}
\boldsymbol{e}^k_{pp} &= \begin{bmatrix} 0 & 0 & 0 \\ 0 & 0 & 0 \end{bmatrix}, \quad \boldsymbol{e}^k_{pn} = \begin{bmatrix} e^k_{15} & e^k_{14} & 0 \\ e^k_{25} & e^k_{24} & 0 \end{bmatrix}, \\
\boldsymbol{e}^k_{np} &= \begin{bmatrix} e^k_{31} & e^k_{32} & e^k_{36} \end{bmatrix}, \quad \boldsymbol{e}^k_{nn} = \begin{bmatrix} 0 & 0 & e^k_{33} \end{bmatrix}.
\end{aligned} \tag{22}$$

$$\boldsymbol{\varepsilon}^k_{pp} = \begin{bmatrix} \varepsilon^k_{11} & \varepsilon^k_{12} \\ \varepsilon^k_{12} & \varepsilon^k_{22} \end{bmatrix}, \quad \boldsymbol{\varepsilon}^k_{pn} = \begin{bmatrix} 0 \\ 0 \end{bmatrix}, \quad \boldsymbol{\varepsilon}^k_{np} = \begin{bmatrix} 0 & 0 \end{bmatrix}, \quad \boldsymbol{\varepsilon}^k_{nn} = \begin{bmatrix} \varepsilon^k_{33} \end{bmatrix}. \tag{23}$$

The material coefficients Cij depend on the Young's moduli E_1, E_2, E_3, the shear moduli G_{12}, G_{13}, G_{23} and Poisson moduli ν_{12}, ν_{13}, ν_{23}, ν_{21}, ν_{31}, ν_{32} that characterize the layer material. The piezoelectric material is characterized by the piezoelectric stiffness coefficients eij and the permittivity coefficients εij. For the problem analyzed in this work, the piezoelectric stiffness coefficients are calculated from the piezoelectric coefficients dij as follows:

$$\mathbf{e}^{k^T} = \mathbf{C}^k \mathbf{d}^{k^T} \tag{24}$$

where

$$\begin{aligned}
\boldsymbol{d}^k_{pp} &= \begin{bmatrix} 0 & 0 & 0 \\ 0 & 0 & 0 \end{bmatrix}, \quad \boldsymbol{d}^k_{pn} = \begin{bmatrix} d^k_{15} & d^k_{14} & 0 \\ d^k_{25} & d^k_{24} & 0 \end{bmatrix}, \\
\boldsymbol{d}^k_{np} &= \begin{bmatrix} d^k_{31} & d^k_{32} & d^k_{36} \end{bmatrix}, \quad \boldsymbol{d}^k_{nn} = \begin{bmatrix} 0 & 0 & d^k_{33} \end{bmatrix}.
\end{aligned} \tag{25}$$

6. Governing equations and FE matrices

This section presents the derivation of the governing FE stiffness matrix based on the PVD in the case of multilayered plate structures subjected to electromechanical loads. The PVD for a multilayered piezoelectric structure reads:

$$\int_V \left(\delta\boldsymbol{\epsilon}_{pG}{}^T \boldsymbol{\sigma}_{pC} + \delta\boldsymbol{\epsilon}_{nG}{}^T \boldsymbol{\sigma}_{nC} - \delta\mathbf{E}^T_{pG}\boldsymbol{D}_{pC} - \delta\mathbf{E}^T_{pG}\boldsymbol{D}_{\mathrm{pC}} \right) dV = \delta L_{\mathrm{e}} - \delta L_{\mathrm{in}} \tag{26}$$

where V is the volume, integration domain of the structure. The first member of the equation represents the variation of the internal work, while the second members are respectively the external work δL_e and the inertial work δL_{in}. In order to refer the integration domains to the midsurface of each layer Ω_k in the plate coordinate system and along the thickness direction A_k, one has to introduce the following integral notation:

$$\int_{\Omega_k}\int_{A_k}\left\{\delta\boldsymbol{\epsilon}_{pG}^{k\ T}\boldsymbol{\sigma}_{pC}^{k}+\delta\boldsymbol{\epsilon}_{nG}^{k\ T}\boldsymbol{\sigma}_{nC}^{k}-\delta\mathbf{E}_{pG}^{k\ T}\boldsymbol{D}_{pC}^{k}-\delta\mathbf{E}_{pG}^{k\ T}\boldsymbol{D}_{pC}^{k}\right\}d\Omega_k dz=\delta L_e-\delta L_{in} \tag{27}$$

where

$$\delta L_{in}=\int_{\Omega_k}\int_{A_k}\{\rho\,\delta\boldsymbol{u}^T\ddot{\boldsymbol{u}}\}\,d\Omega_k dz \tag{28}$$

$\boldsymbol{u}$ are the general unknown displacements, $\ddot{\boldsymbol{u}}$ are the accelerations, and ρ is the material density. $\delta L_e = 0$ in the case of free-vibration analysis.

Substituting the constitutive Equations (17–20), the geometrical relations written via the MITC method (16) and applying the Unified Formulation 1 and the FEM approximation 12, one obtains the following governing equations:

$$\delta\boldsymbol{u}_s^k : \boldsymbol{K}_{uu}^{k\tau s}\boldsymbol{u}_\tau^k+\boldsymbol{K}_{u\Phi}^{k\tau s}\boldsymbol{\Phi}_\tau^k=-\boldsymbol{M}_{uu}^{k\tau s}\ddot{\boldsymbol{u}}_\tau^k \tag{29}$$

$$\delta\boldsymbol{\Phi}_s^k : \boldsymbol{K}_{\Phi u}^{k\tau s}u_\tau^k+\boldsymbol{K}_{\Phi\Phi}^{k\tau s}\boldsymbol{\Phi}_\tau^k=0 \tag{30}$$

In compact form:

$$\delta\boldsymbol{q}_s^k\ :\ \boldsymbol{K}^{k\tau s}\,\boldsymbol{q}_\tau^k\ =\ -\boldsymbol{M}^{k\tau s}\ddot{\boldsymbol{q}}_\tau^k \tag{31}$$

where

$$\boldsymbol{K}^{k\tau s}=\begin{bmatrix}\boldsymbol{K}_{uu} & \boldsymbol{K}_{u\Phi}\\ \boldsymbol{K}_{\Phi u} & \boldsymbol{K}_{\Phi\Phi}\end{bmatrix}^{k\tau s},\quad \boldsymbol{M}^{k\tau s}=\begin{bmatrix}\boldsymbol{M}_{uu} & 0\\ 0 & 0\end{bmatrix}^{k\tau s} \tag{32}$$

The undamped dynamic problem can be written as follows:

$$\boldsymbol{M}^{k\tau s}\ddot{\boldsymbol{q}}_\tau^k+\boldsymbol{K}^{k\tau s}\boldsymbol{q}_\tau^k=0 \tag{33}$$

Introducing harmonic solutions, it is possible to compute the natural frequencies ω_n by solving an eigenvalue problem:

$$(\boldsymbol{K}^{k\tau s}-\omega_n^2\boldsymbol{M}^{k\tau s})\,\boldsymbol{q}_\tau^k=0 \tag{34}$$

where q_τ^k are the eigenvectors. The mechanical part $\boldsymbol{K}_{uu}^{k\tau sij}$ is a 3×3 matrix, the coupling matrices $\boldsymbol{K}_{u\phi}^{k\tau sij}$ and $\boldsymbol{K}_{\phi u}^{k\tau sij}$ have dimensions 3×1 and 1×3, respectively, and the electrical part $\boldsymbol{K}_{\phi\phi}^{k\tau sij}$ is a 1×1 matrix. This is the basic element from which the stiffness matrix of the whole structure is computed. The mechanical part of the mass matrix $\boldsymbol{M}_{uu}^{k\tau sij}$ is a 3 × 3 matrix, the other parts of the global mass matrix are imposed to zero. The global matrix

$K^{k\tau sij}$ and $M^{k\tau sij}$ are called fundamental nucleus, and their explicit expression is given in Appendix 1. For the expansion of the fundamental nucleus on the indexes τ and s and the assembling procedure at multilayer level, the reader can refer to [42]. $q_\tau = (q_{u_\tau}, q_{v_\tau}, q_{w_\tau}, q_{\Phi_\tau})$ is the vector of the nodal mechanical displacements and the nodal electric potential.

7. Numerical results and discussion

In this work, a simple displacement formulation is employed for the analysis of plate structures with piezoelectric patches. The refined theories contained in CUF, coupled with the MITC method, permit a high degree of accuracy to be reached by increasing the order of expansion of displacements in the thickness direction and the number of used elements. In the electromechanical problems, it is necessary to impose the values of the electric potential variable at the top and bottom surfaces of the plate. To obtain this, LW models with Legendre polynomials are particularly suitable. The efficiency of LW models is tested with the FE scheme, and the numerical results are compared with those presented in the literature. In this direction, two kinds of reference problems are considered: an aluminum rectangular plate with eight piezoelectric patches symmetrically placed (four on the top and four on the bottom surfaces) in short- and open-circuit configurations. In this article, some acronyms are used in tables and figures to indicate models employed. The first part indicates the multilayer approach, LW or ESL. The number N indicates the order of expansion used in the thickness direction.

7.1. *Cantilevered plate with piezo-patches*

The structure analyzed by Yasin et al. [53] (see Figure 3) is a cantilevered aluminum plate with eight piezoelectric patches, four on the top surface and symmetrically four on the bottom ones. The plate is clamped on one short edge and free on the other three sides. The physical and geometrical properties of the plate are given in Table 1. A convergence study is

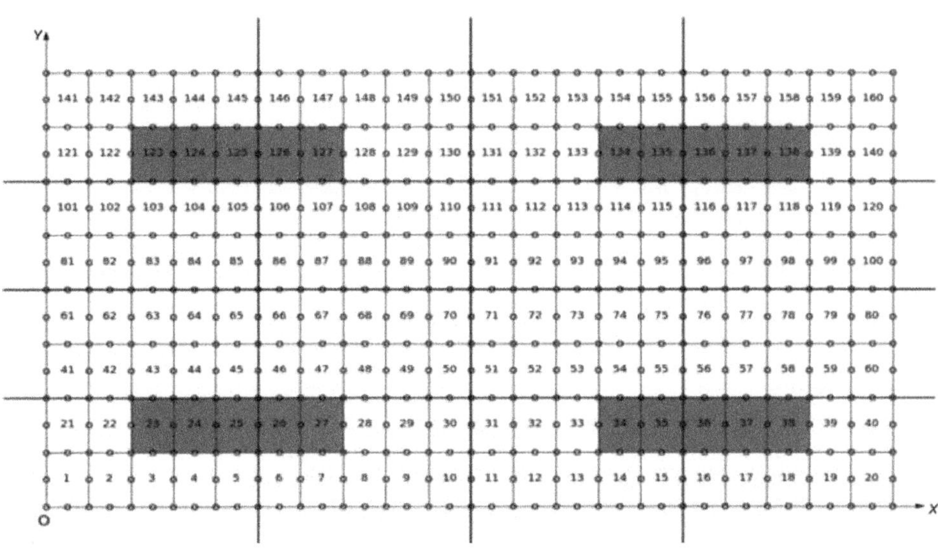

Figure 3. Mesh adopted for the isotropic plate with piezo-patches.

Table 1. Physical and geometrical data of the cantilevered plate with piezo-electric patches.

Material	PZT	Aluminum
a [m]	0.075	0.3
b [m]	0.025	0.2
h [m]	0.63×10^{-3}	0.8×10^{-3}
Density [K g/m^3]	7600.0	2700.0
E_{11} [N/m^2]	63.0×10^9	70.0×10^9
E_{22} [N/m^2]	63.0×10^9	70.0×10^9
E_{33} [N/m^2]	63.0×10^9	70.0×10^9
ν_{12} [−]	0.28	0.32
ν_{13} [−]	0.28	0.32
ν_{23} [−]	0.28	0.32
G_{12} [N/m^2]	24.8×10^9	26.515×10^9
G_{13} [N/m^2]	24.8×10^9	26.515×10^9
G_{23} [N/m^2]	24.8×10^9	26.515×10^9
d_{15} [m/V]	670.0×10^{-12}	0
d_{24} [m/V]	670.0×10^{-12}	0
d_{31} [m/V]	-220.0×10^{-12}	0
d_{32} [m/V]	-220.0×10^{-12}	0
d_{33} [m/V]	374.0×10^{-12}	0
ε_{11} [F/m]	15.3×10^{-9}	30.975×10^{-12}
ε_{22} [F/m]	15.3×10^{-9}	26.55×10^{-12}
ε_{33} [F/m]	15.0×10^{-9}	26.55×10^{-12}

Table 2. Convergence study.

Mesh	10 × 4	20 × 8
1	7.6526	7.6074
2	25.967	25.351
3	46.379	45.790
4	92.789	90.454
5	126.01	123.76
6	153.40	149.00
7	189.47	184.88
8	249.12	235.48
9	316.52	327.03
10	347.66	332.31

Frequencies of a cantilevered plate with piezo-patches. Open-circuit configuration. LW4 model.

presented in Table 2 for the open-circuit configuration, and the LW4 model is employed. Two meshes are considered: 10 × 4 and 20 × 8 elements (this last is represented in Figure 3). Since the results show small differences, the mesh 20 × 8 can be taken as convergence solution, and it will be adopted for the following analysis. First of all, a comparison between the MITC method and the selective reduced integration technique is presented in Table 3 for both short- and open-circuit configurations, and the LW4 model is employed. The table shows that the present MITC9 element is as efficient in contrasting the shear locking phenomenon as classical reduced integration techniques. Moreover, in the work [49], it was demonstrated that the MITC method permits the spurious modes to be reduced/removed.

Table 3. Frequencies of a cantilevered plate with piezo-patches.

	Short circuit		Open circuit	
Mode	MITC	Selective	MITC	Selective
1	7.6059	7.5897	7.6074	7.5914
2	25.346	25.306	25.351	25.313
3	45.787	45.732	45.790	45.737
4	90.446	90.250	90.454	90.274
5	123.76	123.62	123.76	123.64
6	148.93	148.23	149.00	148.28
7	184.69	184.02	184.88	184.09
8	235.50	234.94	235.48	235.01
9	328.22	327.38	327.03	327.46
10	332.49	329.79	332.31	329.96

Comparison between MITC method and selective reduced integration technique. Short- and open-circuit configurations. LW4 model.

7.1.1. Short-circuit mode

For the first case, the piezoelectric patches are short circuited, thereby rendering ineffective the piezoelectric coupling effect that enhances the stiffness of otherwise passive structure. This case includes only the pure structural stiffness of the PZT patches and the modal analysis, made by Yasin et al. [53], involves the solution of the simplified following eigenvalue problem:

$$(\boldsymbol{K}_{uu} - \omega_n^2 \boldsymbol{M})\boldsymbol{q}_n = 0$$

In this work, two methods are used for short-circuit mode: the free-vibration analysis using the mechanical stiffness only, as Yasin et al. [53], and the free-vibration analysis with the complete stiffness matrix applying the penalty technique for the electrical degrees of freedom of the top and bottom surfaces of piezoelectric patches. In general, the results, given in terms of natural frequencies, approach the exact solution by increasing the order of expansion N, see Table 4. One can note that the element does not suffer the locking phenomenon for the examined thin plate ($a/h = 300$). The mechanical solution, obtained by using the higher-order LW models, matches the reference solution. The small differences between them can be explained with the different description of the displacement field that in this work is more refined than the reference solution case. The first four modes are shown in Figures 4–7. The solution with the penalty technique applied to the electrical DOFs shows few differences with the mechanical results, see Table 5. This fact is due to the higher-order modeling of the electric potential along the thickness of the patches.

7.1.2. Open-circuit mode

For the second case, the piezoelectric patches are acting as sensors, so in open-circuit mode no electric potential is imposed. A static condensation of the electrical DOFs is made by Yasin et al. [53]:

Table 4. Frequencies of a cantilevered plate with piezo-patches.

Mode	Ref [53]	LW4	LW3	LW2	LW1
1	7.5236	7.5639	7.5640	7.5661	8.2918
2	25.195	25.312	25.324	25.356	26.324
3	45.542	45.718	45.723	45.736	50.113
4	90.215	90.236	90.277	90.386	96.301
5	123.03	123.66	123.69	123.76	133.99
6	147.62	147.08	147.09	147.18	161.14
7	183.16	182.65	182.69	182.85	194.84
8	234.05	235.40	235.51	235.76	254.69
9	326.93	325.14	325.15	325.27	347.75
10	328.04	326.98	327.13	327.52	364.35
Mode	**ESL4**	**ESL3**	**ESL2**	**ESL1**	**FSDT**
1	7.7235	7.7318	7.7413	7.6713	7.6882
2	25.654	25.689	25.814	25.737	25.781
3	46.004	46.035	46.072	45.830	45.855
4	91.347	91.459	91.864	91.545	91.667
5	124.78	124.87	125.05	124.75	124.87
6	152.86	153.06	153.45	152.02	152.63
7	188.45	188.61	189.26	187.96	188.51
8	240.23	240.33	241.04	239.63	240.24
9	329.53	329.71	331.25	329.07	329.19
10	343.19	346.92	347.63	342.85	345.16

Pure mechanical case.

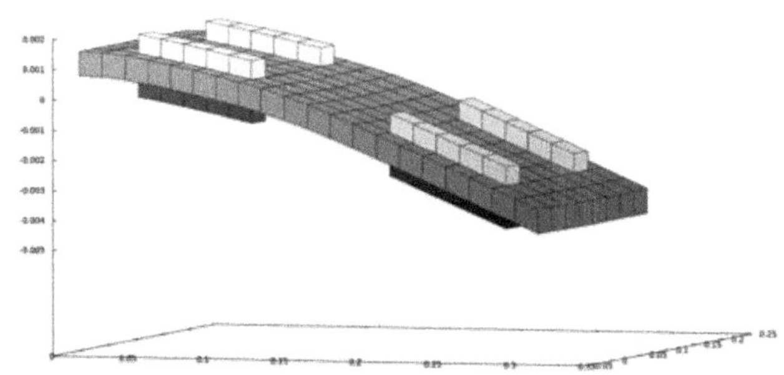

Figure 4. Natural frequency 7.5639 Hz. Mode 1.

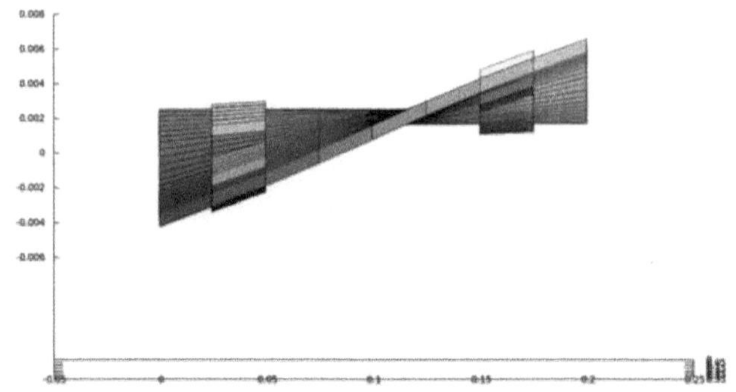

Figure 5. Natural frequency 25.312 Hz. Mode 2.

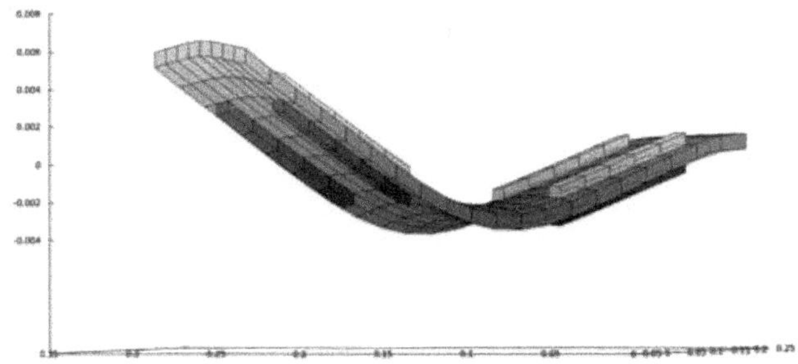

Figure 6. Natural frequency 45.718 Hz. Mode 3.

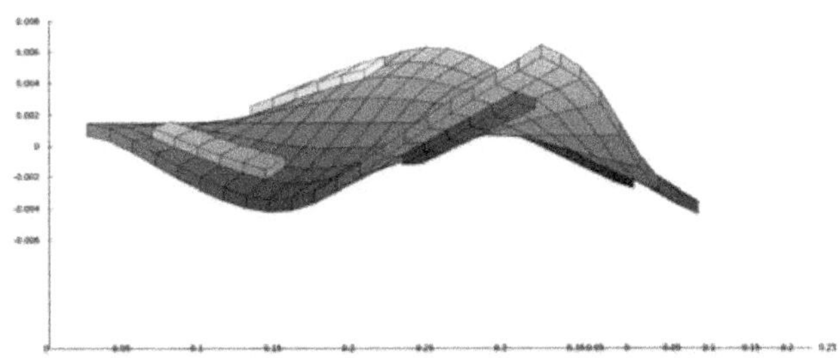

Figure 7. Natural frequency 90.236 Hz. Mode 4.

Table 5. Frequencies of a cantilevered plate with piezo-patches.

Mode	Ref [53]	LW4	LW3	LW2	LW1
1	7.5236	7.6059	7.6061	7.6083	8.3386
2	25.195	25.346	25.359	25.393	26.363
3	45.542	45.787	45.792	45.805	50.190
4	90.215	90.446	90.490	90.607	96.513
5	123.03	123.76	123.79	123.86	134.22
6	147.62	148.93	148.94	149.04	162.99
7	183.16	184.69	184.73	184.91	196.96
8	234.05	235.50	235.62	235.88	254.79
9	326.93	328.22	328.37	328.78	355.13
10	328.04	332.49	332.50	332.64	365.61

Short-circuit case.

$$\left(\left\{\boldsymbol{K}_{uu}-\boldsymbol{K}_{u\Phi}\boldsymbol{K}_{\Phi\Phi}^{-1}\boldsymbol{K}_{\Phi u}^{T}\right\}-\omega_n^2\boldsymbol{M}\right)\boldsymbol{q}_n=0$$

For the open-circuit mode, in this work, free-vibration analysis with complete stiffness matrix is done, see Table 6. In general, the results approach the exact solution by increasing the order of expansion N. One can note that the element does not suffer the locking phenomenon for the examined thin plate (a/h = 300). The higher-order LW models are in good agreement with the reference solution. As in the previous case, there are few differences due to different expansion of the displacement field that in this work is higher-order in respect to the reference solution.

Table 6. Frequencies of a cantilevered plate with piezo-patches.

Mode	Ref [53]	LW4	LW3	LW2	LW1
1	7.5639	7.6074	7.6076	7.6098	8.3402
2	25.221	25.351	25.362	25.396	26.368
3	45.591	45.790	45.795	45.808	50.194
4	90.371	90.454	90.500	90.617	96.526
5	123.10	123.76	123.80	123.87	134.24
6	149.56	149.00	148.98	149.08	163.03
7	185.11	184.88	184.77	184.96	197.01
8	234.05	235.48	235.65	235.91	254.83
9	328.18	327.03	328.40	328.81	355.22
10	335.68	332.31	332.59	332.73	365.65

Open-circuit case.

7.1.3. Benchmark solutions

Table 7 shows the first 10 natural frequencies of the same structures, but considering two different geometries: in the first one, the thickness of the aluminum plate and patches is multiplied by 10; in the second case, it is divided by 10. Both short- and open-circuit configurations are studied and the models LW1 and LW4 are employed for the analysis. Since no reference results are provided in the literature for these cases, the solutions presented can be used as benchmark solutions for future comparisons.

8. Conclusion

This article has presented the free-vibration analysis of an isotropic plate with piezo-patches by means of a plate FE based on the CUF [54,55]. The results have been provided in terms of free-vibration natural frequencies for very thin plates, and the performances of the various models, equivalent-single-layer and LW theories contained in CUF, have been tested. The conclusions that can be drawn are the following:

Table 7. Frequencies of a cantilevered plate with piezo-patches.

$h$new = h * 10				$h$new = h/10				
Short circuit		Open circuit		Short circuit		Open circuit		
Mode	LW4	LW1	LW4	LW1	LW4	LW1	LW4	LW1
1	74.139	81.481	74.157	81.521	0.7766	0.8454	0.7711	0.8425
2	238.45	247.81	238.84	248.44	2.5428	2.6442	2.5381	2.6425
3	447.52	490.01	447.59	490.18	4.5842	5.0236	4.5837	5.0252
4	853.34	908.09	854.49	910.04	9.0567	9.6632	9.0410	9.6725
5	1176.4	1210.9	1177.1	1211.3	12.388	13.435	12.296	13.431
6	1207.0	1277.2	1207.2	1278.2	14.927	16.324	15.329	16.322
7	1411.9	1551.3	1412.8	1552.8	18.509	19.729	19.027	19.791
8	1733.1	1851.8	1734.6	1854.4	23.593	25.520	23.517	25.601
9	2147.1	2328.3	2151.0	2334.1	32.863	35.573	30.072	34.862
10	3023.0	3322.6	3024.8	3327.9	33.325	36.602	31.904	36.237

Different aspect ratios.

(1) the plate element is locking free when the plate is very thin;
(2) the exact solution is achieved by increasing the order of expansion of the variables in the thickness;
(3) the use of LW models gives more accurate results, if the free-vibration behavior of the structure needs to be accurately described;
(4) the analysis of short-circuit mode shows that the higher-order description of the electric potential gives some differences in the frequency results compared to those obtained with pure mechanical analysis;
(5) Future works could be devoted to extending this plate element to shell structure with single or double curvature.

Disclosure statement

No potential conflict of interest was reported by the authors.

References

[1] A.K. Noor and W.S. Burton, *Assessment of computational models for multilayered shells*, Appl. Mech. Rev. 43(4) (1990), pp. 67–97.
[2] E. Crawley and J. De Luis, *Use of piezoelectric actuators as elements of intelligent structures*, AIAA J. 25 (1987), pp. 1373–1385. doi:10.2514/3.9792.
[3] T. Bailey and J. Hubbard, *Distributed piezoelectric polymer active vibration control of a cantilever beam*, AIAA J. 8 (1985), pp. 605–611.
[4] D. Robbins and J. Reddy, Analysis of piezoelectrically actuated beams using a layer-wise displacement theory, Comput. Struct. 41 (1991a), pp. 265–279. doi:10.1016/0045-7949(91)90430-T.
[5] E. Carrera, *Historical review of zig-zag theories for multilayered plates and shells*, Appl. Mech. Rev. 56(3) (2003), pp. 287–308.
[6] J.N. Reddy, *Mechanics of Laminated Composite Plates and Shells: Theory and Analysis*, CRC Press, Boca Raton, 2004.
[7] E. Carrera, *CZ requirements—Models for the two dimensional analysis of multilayered structures*, Composite Struct. 37 (1997), pp. 373–383. doi:10.1016/S0263-8223(98)80005-6.
[8] W.T. Koiter, *A consistent first approximation in the general theory of thin elastic shells*, in *The Theory of Thin Elastic Shells*, W.T. Koiter, ed., IUTAM, Delft, North-Holland, 1959, pp. 12–33.
[9] R.D. Mindlin, *Forced thickness-shear and flexural vibrations of piezoelectric crystal plates*, J. Appl. Phys. 23 (1952), pp. 83–88.
[10] E.P. EerNisse, *Variational method for electroelastic vibration analysis*, IEEE Trans. Sonics Ultra- Sonics 14(4) (1967), pp. 153–159. doi:10.1109/T-SU.1967.29431.
[11] H.F. Tiersten and R.D. Mindlin, *Forced vibrations of piezoelectric crystal plates*, Quart. Appl. Math. 20(2) (1962), pp. 107–119.
[12] H.F. Tiersten, *Linear Piezoelectric Plate Vibrations*, Plenum, New York, 1969.
[13] D.A. Saravanos and P.R. Heyliger, *Mechanics and computational models for laminated piezoelectric beams, plates and shells*, Appl. Mech. Rev. 52(10) (1999), pp. 305–319.
[14] S. Kapuria, *A coupled zig-zag third-order theory for piezoelectric hybrid cross-ply plates*, J. Appl. Mech. 71 (2004), pp. 604–614.
[15] C. Ossadzow-David and M. Touratier, *A multilayered piezoelectric shell theory*, Compos. Sci. Technol. 64 (2004), pp. 2121–2137. doi:10.1016/j.compscitech.2004.03.005.
[16] P. Heyliger, K.C. Pei, and D.A. Saravanos, Layerwise mechanics and finite element model for laminated piezoelectric shells, AIAA J. 34(11) (1996), pp. 2353–2360. doi:10.2514/3.13401.
[17] D. Ballhause, M. D'Ottavio, B. Kröplin, and E. Carrera, *A unified formulation to assess multilayered theories for piezoelectric plates*, Comput. Struct. 83(15–16) (2005), pp. 1217–1235. doi:10.1016/j.compstruc.2004.09.015.

[18] M. D'Ottavio, D. Ballhause, B. Kroeplin, and E. Carrera, *Closed-form solutions for the free-vibration problem of multilayered piezoelectric shells*, Comput. Struct. 84 (2006), pp. 1506–1518. doi:10.1016/j.compstruc.2006.01.030.

[19] A. Benjeddou, J. Deu, and S. Letombe, *Free vibrations of simply-supported piezoelectric adaptive plates: An exact sandwich formulation*, Thin-Walled Struct. 40 (2002), pp. 573–593. doi:10.1016/S0263-8231(02)00013-7.

[20] N.N. Rogacheva, *The Theory of Piezoelectric Shells and Plates*, CRC Press, Boca Raton, 1994

[21] X. Zhou, A. Chattopadhyay, and H. Gu, Dynamic responses of smart composites using a coupled thermo-piezoelectric-mechanical model, AIAA J. 38 (2000), pp. 1939–1948. doi:10.2514/2.848.

[22] A.L. Araújo, C.M.M. Soares, C.A.M. Soares, and J. Herskovits, *Characterisation by inverse techniques of elastic, viscoelastic and piezoelectric properties of anisotropic sandwich adaptive structures*, Appl. Composite Mater. 17(5) (2010), pp. 543–556. doi:10.1007/s10443-010-9142-2.

[23] A.L. Araujo, C.M.M. Soares, and C.A.M. Soares, *Finite element model for hybrid active-passive damping analysis of anisotropic laminated sandwich structures*, J. Sandwich Struct. Mater. 12 (2010), pp. 397–419. doi:10.1177/1099636209104534.

[24] S.V. Gopinathan, V.V. Varadan, and V.K. Varadan, *A review and critique of theories for piezoelectric laminates*, Smart Mater. Struct. 9 (2000), pp. 24–48. doi:10.1088/0964-1726/9/1/304.

[25] A. Benjeddou, *Advances in piezoelectric finite element modeling of adaptive structural elements: A survey*, Comput. Struct. 76 (2000), pp. 347–363. doi:10.1016/S0045-7949(99)00151-0.

[26] R. Lammering and S. Mesecke-Rischmann, *Multi-field variational formulations and related finite elements for piezoelectric shells*, Smart Mater. Struct. 12(6) (2003), pp. 904–913. doi:10.1088/0964-1726/12/6/007.

[27] A. Benjeddou and O. Andrianarison, *Extension of Reissners mixed variational theorem to piezoelectric multilayered composites*, in *Proceedings of the Sixth Hellenic-European Conference on Computer Mathematics and Its Applications*, E.A. Lipitakis, ed., LEA, Athens, Hellas, 2003, pp. 116–124.

[28] A.H. Sheikh, P. Topdar, and S. Halder, An appropriate FE model for through-thickness variation of displacement and potential in thin/moderately thick smart laminates, Composite Struct. 51 (2001), pp. 401–409. doi:10.1016/S0263-8223(00)00156-2.

[29] F. Auricchio, P. Bisegna, and C. Lovadina, *Finite element approximation of piezoelectric plates*, Int. J. Numer. Methods Eng. 50 (2001), pp. 1469–1499. doi:10.1002/(ISSN)1097-0207.

[30] R.P. Thornburgh and A. Chattopadhyay, *Simultaneous modeling of mechanical and electrical response of smart composite structures*, AIAA J. 40(8) (2002), pp. 1603–1610. doi:10.2514/2.1830.

[31] X. Shu, *Free-vibration of laminated piezoelectric composite plates based on an accurate theory*, Composite Struct. 67 (2005), pp. 375–382. doi:10.1016/j.compstruct.2004.01.022.

[32] J. Oh and M. Cho, *A finite element based on cubic zig-zag plate theory for the prediction of thermoelectric-mechanical behaviors*, Int. J. Solids Struct. 41(5–6) (2004), pp. 1357–1375. doi:10.1016/j.ijsolstr.2003.10.019.

[33] M. Kogl and M.L. Bucalem, *Analysis of smart laminates using piezoelectric MITC plate and shell elements*, Comput. Struct. 83 (2005), pp. 1153–1163. doi:10.1016/j.compstruc.2004.08.024.

[34] M. Kogl and M.L. Bucalem, *A family of piezoelectric MITC plate elements*, Comput. Struct. 83 (2005), pp. 1277–1297. doi:10.1016/j.compstruc.2004.04.025.

[35] G.M. Kulikov and S.V. Plotnikova, *Exact geometry piezoelectric solid-shell element based on the 7-parameter model*, Mech. Adv. Mater. Struct. 18 (2011), pp. 133–146. doi:10.1080/15376494.2010.496067.

[36] S. Klinkel and W. Wagner, *A piezoelectric solid shell element based on a mixed variational formulation for geometrically linear and nonlinear applications*, Comput. Struct. 86 (2008), pp. 38–46. doi:10.1016/j.compstruc.2007.05.032.

[37] E. Carrera, *A class of two-dimensional theories for anisotropic multilayered plates analysis*, Accademia Delle Scienze Di Torino, Memorie Sci. Fisiche, 1995-1996 19-20 (1995), pp. 1–39.

[38] A. Robaldo, E. Carrera, and A. Benjeddou, *A unified formulation for finite element analysis of piezo- electric adaptive plates*, Comput. Struct. 84 (2006), pp. 1494–1505. doi:10.1016/j.compstruc.2006.01.029.

[39] E. Carrera, M. Boscolo, and A. Robaldo, *Hierarchic multilayered plate elements for coupled multifield problems of piezoelectric adaptive structures: Formulation and numerical assessment*, Arch. Comput. Methods Eng. 14 (2007), pp. 383–430. doi:10.1007/s11831-007-9012-8.

[40] E. Carrera and A. Robaldo, *Hierarchic finite elements based on a unified formulation for the static analysis of shear actuated multilayered piezoelectric plates*, Multidiscipline Model. Mater. Struct. 6(1) (2010), pp. 45–77. doi:10.1108/15736101011055266.

[41] M. Cinefra, E. Carrera, and S. Valvano, *Variable kinematic shell elements for the analysis of electro-mechanical problems*, SMART 2013 – Modelling and Analysis of Smart Structures, Mech. Adv. Mater. Struct. 22 (2015), pp. 77–106. doi:10.1080/15376494.2014.908042.

[42] E. Carrera, *Theories and finite elements for multilayered plates and shells: A unified compact formulation with numerical assessment and benchmarking*, Arch. Comput. Methods Eng. 10(3) (2003), pp. 215–296. doi:10.1007/BF02736224.

[43] E. Carrera and M. Boscolo, *Classical and mixed finite elements for static and dynamic analysis of piezoelectric plates*, Int. J. Numer. Methods Eng. 70 (2007), pp. 1135–1181. doi:10.1002/(ISSN)1097-0207.

[44] E. Carrera, S. Brischetto, and M. Cinefra, *Variable kinematics and advanced variational statements for free vibrations analysis of piezoelectric plates and shells*, Comput. Model. Eng. Sci. 65(3) (2010), pp. 259–341.

[45] E. Carrera and P. Nali, *Classical and mixed finite plate elements for the analysis of multifield problems and smart layered structures*, Acta Mech. Sinica 23 (2010), pp. 115–121.

[46] E. Carrera and P. Nali, *Multilayered plate elements for the analysis of multifield problems*, Finite Elem. Anal. Des. 46 (2010), pp. 732–742. doi:10.1016/j.finel.2010.04.001.

[47] K.-J. Bathe, P.-S. Lee, and J.-F. Hiller, *Towards improving the MITC9 shell element*, Comput. Struct. 81 (2003), pp. 477–489. doi:10.1016/S0045-7949(02)00483-2.

[48] P. Panasz and K. Wisniewski, *Nine-node shell elements with 6 dofs/node based on two-level approximations. Part I theory and linear tests*, Finite Elem. Anal. Des. 44 (2008), pp. 784–796. doi:10.1016/j.finel.2008.05.002.

[49] E. Carrera, M. Cinefra, and P. Nali, *MITC technique extended to variable kinematic multilayered plate elements*, Composite Struct. 92 (2010), pp. 1888–1895. doi:10.1016/j.compstruct.2010.01.009.

[50] M. Cinefra, C. Chinosi, and L. Della Croce, *MITC9 shell elements based on refined theories for the analysis of isotropic cylindrical structures*, Mech. Adv. Mater. Struct. 20 (2013), pp. 91–100. doi:10.1080/15376494.2011.581417.

[51] M. Cinefra and E. Carrera, *Shell finite elements with different through-the-thickness kinematics for the linear analysis of cylindrical multilayered structures*, Int. J. Numer. Methods Eng. 93 (2013), pp. 160–182. doi:10.1002/nme.4377.

[52] M. Cinefra and S. Valvano, *A variable kinematic doubly-curved MITC9 shell element for the analysis of laminated composites*, Mech. Adv. Mater. Struct. (In Press). ISSN 1537-6494.

[53] M.Y. Yasin, N. Ahmad, and M.N. Alam, *Finite element analysis of actively controlled smart plate with patched actuators and sensors*, Latin American J. of Solids & Struct. 7 (2010), pp. 227–247. doi:10.1590/S1679-78252010000300001.

[54] E. Carrera, *Multilayered shell theories accounting for layerwise mixed description, Part 1: Governing equations*, AIAA J. 37(9) (1999), pp. 1107–1116. doi:10.2514/2.821.

[55] E. Carrera, Multilayered shell theories accounting for layerwise mixed description, Part 2: Numerical evaluations, AIAA J. 37(9) (1999), pp. 1117–1124. doi:10.2514/2.822.

[56] E. Carrera and S. Brischetto, *Analysis of thickness locking in classical, refined and mixed multilayered plate theories*, Composite Struct. 82 (2008), pp. 549–562. doi:10.1016/j.compstruct.2007.02.002.

[57] K.-J. Bathe and E. Dvorkin, *A formulation of general shell elements – the use of mixed interpolation of tensorial components*, Int. J. Numer. Methods Eng. 22 (1986), pp. 697–722. doi:10.1002/(ISSN)1097-0207.

[58] M.L. Bucalem and K.-J. Bathe, *Higher-order MITC general shell elements*, Int. J. Numer. Methods Eng. 36 (1993), pp. 3729–3754. doi:10.1002/(ISSN)1097-0207.

Appendix 1

Explicit form of stiffness fundamental nucleus

In order to write the fundamental nucleus $\boldsymbol{K}\ k\tau sij$ in compact form, the following integrals in the domain Ω_k are defined:

$$\left(W^k_{m1\,n1};\ W^k_{m1\,n2};\ W^k_{m2\,n1};\ W^k_{m2\,n2}\right) = \int_{\Omega_k} \left(N_{m1}N_{n1};\ N_{m1}N_{n2};\ N_{m2}N_{n1};\ N_{m2}N_{n2}\right)\,dx\,dy$$

$$\left(W^k_{m1\,n3};\ W^k_{m3\,n1};\ W^k_{m3\,n3};\ W^k_{m2\,n3};\ W^k_{m3\,n2}\right) = \int_{\Omega_k} \left(N_{m1}N_{n3};\ N_{m3}N_{n1};\ N_{m3}N_{n3};\ N_{m2}N_{n3};\ N_{m3}N_{n2}\right)\,dx\,dy$$

$$\left(W^k_{m1j};\ W^k_{m2j};\ W^k_{m3j}\right) = \int_{\Omega_k} \left(N_{m1}N_j;\ N_{m2}N_j;\ N_{m3}N_j\right)\,dx\,dy$$

$$\left(W^k_{in1};\ W^k_{in2};\ W^k_{in3};\ W^k_{ij}\right) = \int_{\Omega_k} \left(N_iN_{n1};\ N_iN_{n2};\ N_iN_{n3};\ N_iN_j\right)\,dx\,dy$$

$$\left(W^k_{m1j_{,x}};\ W^k_{m1j_{,y}};\ W^k_{m2j_{,x}};\ W^k_{m2j_{,y}}\right) = \int_{\Omega_k} \left(N_{m1}\frac{\partial N_j}{\partial x};\ N_{m1}\frac{\partial N_j}{\partial y};\ N_{m2}\frac{\partial N_j}{\partial x};\ N_{m2}\frac{\partial N_j}{\partial y}\right)\,dx\,dy$$

$$\left(W^k_{i_{,x}n1};\ W^k_{i_{,y}n1};\ W^k_{i_{,x}n2};\ W^k_{i_{,y}n2}\right) = \int_{\Omega_k} \left(\frac{\partial N_i}{\partial x}N_{n1};\ \frac{\partial N_i}{\partial y}N_{n1};\ \frac{\partial N_i}{\partial x}N_{n2};\ \frac{\partial N_i}{\partial y}N_{n2}\right)\,dx\,dy$$

Moreover, the integrals on the domain A_k, in the thickness direction, are written as:

$$\left(J^{k\tau s};J^{k\tau_z s};J^{k\tau s_z};J^{k\tau_z s_z}\right) = \int_{A_k} \left(F_\tau F_s;\ \frac{\partial F_\tau}{\partial z}F_s;\ F_\tau\frac{\partial F_s}{\partial z};\ \frac{\partial F_\tau}{\partial z}\frac{\partial F_s}{\partial z}\right)dz$$

The stiffness fundamental nucleus $\boldsymbol{K}_{\tau sij}$ is:

$$\boldsymbol{K}^{k\tau sij} = \begin{bmatrix} K_{11} & K_{12} & K_{13} & K_{14} \\ K_{21} & K_{22} & K_{23} & K_{24} \\ K_{31} & K_{32} & K_{33} & K_{34} \\ K_{41} & K_{42} & K_{43} & K_{44} \end{bmatrix}^{k\tau sij}$$

The elements of the nucleus are:

$$\begin{aligned} K^{k\tau s}_{uu_{11}} &= C^k_{55}N_i^{(m1)}N_j^{(n1)}\,W^k_{m1\,n1}J^{k\tau_z s_z} + C^k_{66}N_{i,y}^{(m3)}N_{j,y}^{(n3)}\,W^k_{m3\,n3}J^{k\tau s} \\ &+ C^k_{16}N_{i,x}^{(m1)}N_{j,y}^{(n3)}\,W^k_{m1\,n3}J^{k\tau s} + C^k_{16}N_{i,y}^{(m3)}N_{j,x}^{(n1)}\,W^k_{m3\,n1}J^{k\tau s} \\ &+ C^k_{11}N_{i,x}^{(m1)}N_{j,x}^{(n1)}\,W^k_{m1\,n1}J^{k\tau s} \end{aligned}$$

$$
\begin{aligned}
K^{k\tau s}_{uu_{12}} &= C^k_{45} N^{(m1)}_i N^{(n2)}_j W^k_{m1\,n2} J^{k\tau_z s_z} + C^k_{26} N^{(m3)}_{i,y} N^{(n2)}_{j,y} W^k_{m3\,n2} J^{k\tau s} \\
&+ C^k_{12} N^{(m1)}_{i,x} N^{(n2)}_{j,y} W^k_{m1\,n2} J^{k\tau s} + C^k_{66} N^{(m3)}_{i,y} N^{(n3)}_{j,x} W^k_{m3\,n3} J^{k\tau s} \\
&+ C^k_{16} N^{(m1)}_{i,x} N^{(n3)}_{j,x} W^k_{m1\,n3} J^{k\tau s}
\end{aligned}
$$

$$
\begin{aligned}
K^{k\tau s}_{uu_{13}} &= C^k_{45} N^{(m1)}_i N^{(n2)}_{j,y} W^k_{m1\,n2} J^{k\tau_z s} + C^k_{55} N^{(m1)}_i N^{(n1)}_{j,x} W^k_{m1\,n1} J^{k\tau_z s} + C^k_{36} N^{(m3)}_{i,y} W^k_{m3j} J^{k\tau s_z} \\
&+ C^k_{13} N^{(m1)}_{i,x} W^k_{m1j} J^{k\tau s_z}
\end{aligned}
$$

$$
\begin{aligned}
K^{k\tau s}_{uu_{21}} &= C^k_{45} N^{(m2)}_i N^{(n1)}_j W^k_{m2\,n1} J^{k\tau_z s_z} + C^k_{26} N^{(m2)}_{i,y} N^{(n3)}_{j,y} W^k_{m2\,n3} J^{k\tau s} \\
&+ C^k_{66} N^{(m3)}_{i,x} N^{(n3)}_{j,y} W^k_{m3\,n3} J^{k\tau s} + C^k_{12} N^{(m2)}_{i,y} N^{(n1)}_{j,x} W^k_{m2\,n1} J^{k\tau s} \\
&+ C^k_{16} N^{(m3)}_{i,x} N^{(n1)}_{j,x} W^k_{m3\,n1} J^{k\tau s}
\end{aligned}
$$

$$
\begin{aligned}
K^{k\tau s}_{uu_{22}} &= C^k_{44} N^{(m2)}_i N^{(n2)}_j W^k_{m2\,n2} J^{k\tau_z s_z} + C^k_{22} N^{(m2)}_{i,y} N^{(n2)}_{j,y} W^k_{m2\,n2} J^{k\tau s} \\
&+ C^k_{26} N^{(m3)}_{i,x} N^{(n2)}_{j,y} W^k_{m3\,n2} J^{k\tau s} + C^k_{26} N^{(m2)}_{i,y} N^{(n3)}_{j,x} W^k_{m2\,n3} J^{k\tau s} \\
&+ C^k_{66} N^{(m3)}_{i,x} N^{(n3)}_{j,x} W^k_{m3\,n3} J^{k\tau s}
\end{aligned}
$$

$$
\begin{aligned}
K^{k\tau s}_{uu_{23}} &= C^k_{44} N^{(m2)}_i N^{(n2)}_{j,y} W^k_{m2\,n2} J^{k\tau_z s} + C^k_{45} N^{(m2)}_i N^{(n1)}_{j,x} W^k_{m2\,n1} J^{k\tau_z s} + C^k_{23} N^{(m2)}_{i,y} W^k_{m2j} J^{k\tau s_z} \\
&+ C^k_{36} N^{(m3)}_{i,x} W^k_{m3j} J^{k\tau s_z}
\end{aligned}
$$

$$
\begin{aligned}
K^{k\tau s}_{uu_{31}} &= C^k_{36} N^{(n3)}_{j,y} W^k_{in3} J^{k\tau_z s} + C^k_{13} N^{(n1)}_{j,x} W^k_{in1} J^{k\tau_z s} + C^k_{45} N^{(m2)}_{i,y} N^{(n1)}_j W^k_{m2\,n1} J^{k\tau s_z} \\
&+ C^k_{55} N^{(m1)}_{i,x} N^{(n1)}_j W^k_{m1\,n1} J^{k\tau s_z}
\end{aligned}
$$

$$
\begin{aligned}
K^{k\tau s}_{uu_{32}} &= C^k_{23} N^{(n2)}_{j,y} W^k_{in2} J^{k\tau_z s} + C^k_{36} N^{(n3)}_{j,x} W^k_{in3} J^{k\tau_z s} + C^k_{44} N^{(m2)}_{i,y} N^{(n2)}_j W^k_{m2\,n2} J^{k\tau s_z} \\
&+ C^k_{45} N^{(m1)}_{i,x} N^{(n2)}_j W^k_{m1\,n2} J^{k\tau s_z}
\end{aligned}
$$

$$
\begin{aligned}
K^{k\tau s}_{uu_{33}} &= C^k_{33} W^k_{ij} J^{k\tau_z s_z} + C^k_{44} N^{(m2)}_{i,y} N^{(n2)}_{j,y} W^k_{m2\,n2} J^{k\tau s} + C^k_{45} N^{(m1)}_{i,x} N^{(n2)}_{j,y} W^k_{m1\,n2} J^{k\tau s} \\
&+ C^k_{45} N^{(m2)}_{i,y} N^{(n1)}_{j,x} W^k_{m2\,n1} J^{k\tau s} + C^k_{55} N^{(m1)}_{i,x} N^{(n1)}_{j,x} W^k_{m1\,n1} J^{k\tau s}
\end{aligned}
$$

$$
\begin{aligned}
K^{k\tau s}_{u\Phi_{14}} &= e^k_{25} N^{(m1)}_i W^k_{m1j_{,y}} J^{k\tau_z s} + e^k_{15} N^{(m1)}_i W^k_{m1j_{,x}} J^{k\tau_z s} + e^k_{36} N^{(m3)}_{i,y} W^k_{m3j} J^{k\tau s_z} \\
&+ e^k_{31} N^{(m1)}_{i,x} W^k_{m1j} J^{k\tau s_z}
\end{aligned}
$$

$$
\begin{aligned}
K^{k\tau s}_{u\Phi_{24}} &= e^k_{24} N^{(m2)}_i W^k_{m2j_{,y}} J^{k\tau_z s} + e^k_{14} N^{(m2)}_i W^k_{m2j_{,x}} J^{k\tau_z s} + e^k_{32} N^{(m2)}_{i,y} W^k_{m2j} J^{k\tau s_z} \\
&+ e^k_{36} N^{(m3)}_{i,x} W^k_{m3j} J^{k\tau s_z}
\end{aligned}
$$

$$
\begin{aligned}
K^{k\tau s}_{u\Phi_{34}} &= e^k_{33} W^k_{ij} J^{k\tau_z s_z} + e^k_{24} N^{(m2)}_{i,y} W^k_{m2j_{,y}} J^{k\tau s} + e^k_{25} N^{(m1)}_{i,x} W^k_{m1j_{,y}} J^{k\tau s} \\
&+ e^k_{14} N^{(m2)}_{i,y} W^k_{m2j_{,x}} J^{k\tau s} + e^k_{15} N^{(m1)}_{i,x} W^k_{m1j_{,x}} J^{k\tau s}
\end{aligned}
$$

$$K_{\Phi u_{41}}{}^{k\tau s} = e^k_{36} N^{(n3)}_{j,y} W^k_{in3} J^{k\tau_z s} + e^k_{31} N^{(n1)}_{j,x} W^k_{in1} J^{k\tau_z s} + e^k_{25} N^{(n1)}_{j} W^k_{i,y\,n1} J^{k\tau s_z}$$
$$+ e^k_{15} N^{(n1)}_{j} W^k_{i,x\,n1} J^{k\tau s_z}$$

$$K^{k\tau s}_{\Phi u_{42}} = e^k_{32} N^{(n2)}_{j,y} W^k_{in2} J^{k\tau_z s} + e^k_{36} N^{(n3)}_{j,x} W^k_{in3} J^{k\tau_z s} + e^k_{24} N^{(n2)}_{j} W^k_{i,y\,n2} J^{k\tau s_z}$$
$$+ e^k_{14} N^{(n2)}_{j} W^k_{i,x\,n2} J^{k\tau s_z}$$

$$K^{k\tau s}_{\Phi u_{43}} = e^k_{33} W^k_{ij} J^{k\tau_z s_z} + e^k_{24} N^{(n2)}_{j,y} W^k_{i,y\,n2} J^{k\tau s} + e^k_{14} N^{(n2)}_{j,y} W^k_{i,x\,n2} J^{k\tau s} + e^k_{25} N^{(n1)}_{j,x} W^k_{i,y\,n1} J^{k\tau s}$$
$$+ e^k_{15} N^{(n1)}_{j,x} W^k_{i,x\,n1} J^{k\tau s}$$

$$K^{k\tau s}_{\Phi\Phi_{44}} = -\ \varepsilon^k_{33} W^k_{ij} J^{k\tau_z s_z} - \varepsilon^k_{22} W^k_{i,y j,y} J^{k\tau s} - \varepsilon^k_{12} W^k_{i,x j,y} J^{k\tau s} - \varepsilon^k_{12} W^k_{i,y j,x} J^{k\tau s}$$
$$- \varepsilon^k_{11} W^k_{i,x j,x} J^{k\tau s}$$

Explicit form of mass fundamental nucleus

The mass fundamental nucleus $\boldsymbol{M}\tau sij$ is:

$$\boldsymbol{M}^{k\tau sij} = \begin{bmatrix} M_{11} & 0 & 0 & 0 \\ 0 & M_{22} & 0 & 0 \\ 0 & 0 & M_{33} & 0 \\ 0 & 0 & 0 & 0 \end{bmatrix}^{k\tau sij}$$

The elements of the nucleus are:

$$M_{11}{}^{k\tau sij} = M_{22}{}^{k\tau sij} = M_{33}{}^{k\tau sij} = \rho\, W^k_{ij} J^k_{\tau s}$$

2

Simultaneously improved actuated performance and mechanical strength of silicone elastomer by reduced graphene oxide encapsulated silicon dioxide

Nanying Ning[a,b], Minglu Wang, Jing Zhang, Liqun Zhang[a,b] and Ming Tian[a,b]*

[a]State Key Lab of Organic-Inorganic Composites, Beijing University of Chemical Technology, Beijing 100029, China; [b]Laboratory of Carbon Fiber and Functional Polymers, Ministry of Education, Beijing University of Chemical Technology, Beijing 100029, China

Herein, graphene oxide (GO)-encapsulated silica (SiO_2) hybrids (GO@SiO_2) were prepared via electrostatic self-assembly of the 3-aminopropyltriethoxysilane (APS)-modified SiO_2 and GO. The as-prepared GO@SiO_2 was introduced into polydimethylsiloxane (PDMS) elastomer to simultaneously increase the dielectric constant (k) and mechanical properties of PDMS. Then, the in situ thermal reduction of GO@SiO_2/PDMS composites was conducted at 180°C for 2 h to increase the interfacial polarizability of GO@SiO_2. As a result, the values of k at 1000 Hz are largely improved from 3.2 for PDMS to 13.3 for the reduced GO@SiO_2 (RGO@SiO_2)/PDMS elastomer. Meanwhile, the dielectric loss of the composites remains low (<0.2 at 1000 Hz). More importantly, the actuated strain at low electric field (5 kV/mm) obviously increases from 0.3% for pure PDMS to 2.59% for the composites with 60 phr of RGO@SiO_2, an eightfold increase in the actuated strain. In addition, both the tensile strength and elastic modulus are obviously improved by adding 60 phr of RGO@SiO_2, indicating a good reinforcing effect of RGO@SiO_2 on PDMS. Our goal is to develop a simple and effective way to improve the actuated performance and mechanical strength of the PDMS dielectric elastomer for its wider application.

Keywords: graphene oxide (GO)-encapsulated silica (SiO2) hybrids; electrostatic self-assembly; polydimethylsiloxane (PDMS) elastomer; dielectric constant (k) and mechanical properties; actuated strain at low electric field

1. Introduction

Dielectric elastomers (DEs), as an attractive branch of electro-active polymers (EAPs), can give rise to surprisingly large deformation when stimulated by an electric field, and work efficiently over a broad frequency range. Thus, they have received much attention in the past two decades [1,2]. A dielectric elastomer actuator (DEA) can shrink in the thickness direction and expand in the plane direction by applying an electric field across the film thickness [3]. By virtue of the excellent properties such as large strain, fast response, lightweight, reliability, high energy density, and high electromechanical coupling efficiency, DEAs find applications in artificial muscles, sensors, micro air vehicles, flat-panel speakers, micro-robotics, and responsive prosthetics [4–8]. A key limitation for the practical application of DEAs is the requirement of high electric field (>100 kV/mm) to drive them [9–11], which could be harmful to humans and can damage equipment,

*Corresponding author. Email: tianm@mail.buct.edu.cn (Ming Tian)

particularly in biological and medical fields [12]. Therefore, getting a large actuated strain at a low electric field is the biggest challenge for DEAs.

Based on the assumption of constant modulus and free boundary conditions for a DE film, the thickness strain Sz can be approximated by [13].

$$Sz = -P/Y = -\varepsilon_0\varepsilon_r E^2/Y \quad (1)$$

where P is Maxwell's stress on the material, ε_r is the relative permittivity, ε_0 is the permittivity of free space, E is the applied electric field, and Y is the elastic modulus of a DE film. According to the law of volume constancy [14], the planar strain Sp can be approximated by:

$$Sp = 1/(1 + Sz) - 1 \quad (2)$$

According to Equations (1) and (2), to obtain a DE with high actuated strain at a low electric field, a high electromechanical sensitivity (β) is required, which is defined as the ratio of the dielectric constant (k) to the elastic modulus (Y)($\beta = k/Y$) [1]. However, most DEs such as silicone elastomer and acrylic rubber have quite low values of k, which are 2.8 and 4.8, respectively [15]. Therefore, a key issue is to increase the k of DEs, while retaining other excellent properties such as low dielectric loss, low modulus (Y), and good flexibility. One common method to obtain a DE with high k is to introduce high-k ceramics such as TiO_2 into the elastomer matrix [10,15–17]. Usually, a high content (up to 50 vol%) of ceramics is required to obtain high k, leading to low flexibility and poor processability, and thus largely limiting the wide application of DEs [18]. Another method is to add conductive fillers such as carbon nanotubes (CNTs) into the matrix, or blending with chemical modification of silicone with organic dipoles is a promising approach to increase ε' (dielectric constant) [19]. Because of the curl and entanglement of CNTs with fibrous nanostructure, even quite a low content of CNTs can result in the formation of CNT network, thus leading to the large increase in k [19,20]. However, in this case, a high dielectric loss was also obtained owing to the high direct current (DC) conductance caused by the direct connection of CNTs [21].

Compared with CNTs, graphene sheets are usually synthesized from natural graphite, which is abundant, cheap, and easily available. In addition, graphene sheets have a larger aspect ratio, and thus have been considered as good fillers to improve the dielectric constant of elastomers, as has been reported in previous studies [19,20,22,23]. For instance, Romasanta et al. [24] reported that the value of k of polydimethylsiloxane (PDMS) increased from 2.7 to 23 at low frequency (10 Hz) by addition of 2.0 wt.% of thermally expanded graphene sheets. Obviously, the increase in the value of k of these polymers by addition of graphene oxide (GO) is far less than expected. The main reasons are the poor dispersion of graphene in matrix, the restacking of graphene sheets during preparation caused by π–π stacking and hydrophobic interactions, and/or weak interfacial adhesion between graphene sheets and the polymer matrix [25–27]. A good dispersion of GO in the polymer matrix and a strong interfacial adhesion between GO and the matrix can be obtained via the organic functionalization of GO [28,29]. A disadvantage of this approach is that it is complicated and time-consuming. In addition, the coating of GO on silica (SiO_2), metal oxide particles, or carbon nanosphere is an efficient method to prevent GO from self-agglomerating, as reported in previous studies [25,30–32].

In addition, a disadvantage of GO is that the graphite structure is disrupted by oxidation, resulting in a decrease in the dielectric constant [33,34]. Thus, reduction of

GO is required to obtain a DE with both high k and low dielectric loss. In most cases, GO was reduced via chemical reduction using hydrazine hydrate [35], Vitamin C [36], etc., leading to a high reduction degree, and thus a good electric conductivity of the composites. Thus, chemical reduction of GO can result in a high dielectric loss of the composites because of the large DC conductance [10]. In recent years, some studies reported that in situ thermal reduction of graphene oxide nanosheets (GONSs) at a moderate temperature (180–200°C) within a polymer is a simple and effective technique for the moderate reduction of graphitic structure [26,30].

PDMS, as one of the most important DEs, has a rather low modulus, a good thermal stability over a wide temperature range, a fast speed of response, high efficiency, and excellent biocompatibility for artificial muscle. However, its dielectric constant and mechanical strength are too low for practical application. Racles et al. [37] developed where cyanopropyl-groups were distributed along the backbone of PDMS chains. Good overall results were obtained for blends of the cyanopropyl-functional PDMS and PDMS.

Herein, GO-encapsulated silica (SiO_2) (GO@SiO_2) hybrids were prepared via electrostatic self-assembly in which the 3-aminopropyltriethoxysilane (APS) acts as a 'glue' molecule to connect GO and SiO_2.Therefore, in this study, GO@SiO_2 is introduced in the PDMS matrix to simultaneously increase the dielectric constant and mechanical properties of PDMS. Then, the reduced GO@SiO2 (RGO@SiO_2)/PDMS composites were obtained by in situ thermal reduction of GO@SiO_2/PDMS composites to increase the interfacial polarizability, and thus increase the k of the composites. Our goal is to develop a simple and effective way to improve the actuated performance and mechanical strength of PDMS DE for its wider application.

2. Experimental

2.1. *Materials*

Natural graphite flakes with a mean size of 18 μm were supplied by Huadong Graphite Factory (China). Submicron-sized silica (average diameter: 3 00 nm) was supplied by Guangdong Shengyi Technology Co., LTD (China). APS was supplied by Momentive Performance Materials (Shanghai). A commercial grade of siloxane (PDMS) (DC3481 184) and the curing agent (81-F) were purchased from Dow Corning Co., Ltd. (America). The viscosity and density of PDMS material used are 22.1 Pa s and 1.213 g/cm^3, respectively. Potassium permanganate ($KMnO_4$, 99.5%), sulfuric acid (H_2SO_4, 98%), sodium nitrate ($NaNO_3$, 99.0%), hydrogen peroxide (H_2O_2, 30%), hydrochloric acid (HCl, 37%), and methanol (CH_3OH, 99.5%), which were all used for the preparation of graphite oxide, were supplied by Beijing Chemical Reagents Co., Ltd. (China). Tetrahydrofuran (THF) was used as a solvent to dissolve PDMS and was purchased from Beijing Chemical Reagents Co., Ltd. (Beijing, China).

2.2. *Preparation of graphene oxide-encapsulated silica (GO@SiO_2) hybrids*

The graphene oxide-encapsulated silica (SiO_2–NH_2–GO) hybrids were prepared in two steps: the modification of silica particles with APS coupling agent and the surface assembly of GONSs with the modified silica microspheres.

Graphite oxide was prepared from natural graphite by Hummers method. The as-prepared graphite oxide was then dispersed in an aqueous solution (1 mg/mL) and exfoliated via sonication for 1 h using a GA 98-IIIultrasonic cell disruptor (Shangjia

Biological Technology, Shanghai, China) with a power of 1000 W to form a uniform colloidal suspension of GO. The nonexfoliated GO sheets were removed by centrifugation at 8000 rpm for 10 min.

The surface modification of SiO_2 with APS coupling agent was carried out in a mixture of ethyl alcohol and water. In a typical process, SiO_2 powder (1 g) was first dispersed well in 400 mL of ethyl alcohol by sonication for 1 h, followed by the addition of 40 mL water. Subsequently, 1 mL of APS coupling agent dissolved in a small amount of ethyl alcohol was added to the above solution, and the mixture was stirred at 60°C for 8 h until the grafting reaction was realized. The obtained particles (SiO_2–NH_2) were centrifuged from the mixture, washed with ethyl alcohol and deionized water thrice.

The GO@SiO_2 hybrids were simply fabricated by mixing the neutral aqueous suspension of SiO_2–NH_2 and GO solution. A total volume of 400 mL of SiO_2–NH_2 suspension (2.5 mg/mL) was added to 200 mL of aqueous GO solution (0.1 mg/mL) under mild stirring for 4 h. When stirring stopped, the GO precipitated with SiO_2–NH_2 at the bottom of the beaker, leaving a transparent aqueous solution. The sediment solid (GO@SiO_2) was collected and washed with deionized water several times to remove the unbound GO and then freeze-dried under vacuum.

2.3. Preparation of PDMS-based composites with solution compounding

The preparation of GO@SiO_2/PDMS composites is schematically illustrated in Figure 1. First, a certain weight of GO@SiO_2 hybrids was dispersed in THF by ultrasonic treatment and mild magnetic stirring. Meanwhile, 3 g of PDMS was dispersed in 20 ml THF by mild magnetic stirring for 1 h to form a stable solution. Then, the uniform GO@SiO_2/THF suspension was mixed with PDMS/THF solution under stirring for 2 h, followed by the addition of a curing agent. After mechanical stirring for 10 min, the GO@SiO_2/PDMS/THF solution was poured into a mold to remove the solvent and cure at room temperature in a fume cupboard with a thickness of 0.5 mm. The GO@SiO_2/PDMS film was obtained after the removal of solvent, and the RGO@SiO_2/PDMS composites were obtained by in situ thermal reduction of GO@SiO_2/PDMS composites at 180°C for 2 h at a pressure of

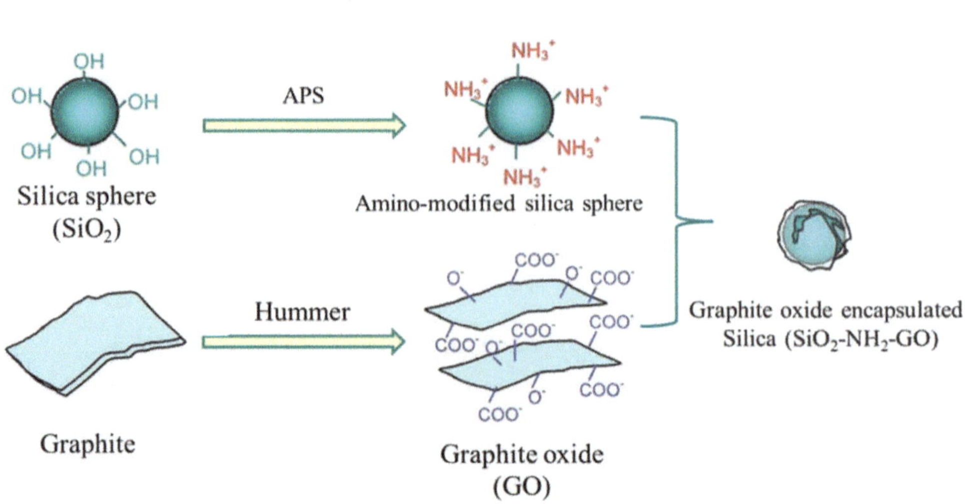

Figure 1. The schematic illustration of the fabrication of SiO_2–APS–GO hybrids.

4 MPa. The composites with different contents of GO@SiO_2 (30 wt%, 60 wt%) were achieved by changing the content of GO@SiO_2. For the purpose of comparison, a pure PDMS film was also prepared by the same method. Here, it should be noted that the GO content is only 0.6 and 1.2 wt% for the composites with 30 and 60 wt% of GO@SiO_2, respectively.

2.4. Characterization

The morphology and microstructure of SiO_2 and GO@SiO_2 were investigated by scanning electron microscopy (SEM, S-4800, Hitachi Co. Japan), transmission electron microscopy (TEM, Tecnai G^2 20, FEI Co. Hong Kong), and high-resolution transmission electron microscopy (HRTEM, JEM-3010, Hitachi, Japan). Samples for TEM and HRTEM observation were prepared by suspending GO@SiO_2 in water at a concentration of 0.005 mg/mL under ultrasonication for 30 min, dropping the suspension on micro grids, and then drying in a vacuum oven.

Zeta potential measurements were performed using a Zetasizer 90 (Malvern Instruments, Malvern, England). The aqueous suspensions of silica, SiO_2—NH_2, and GO were diluted to 0.05 mg/mL and sonicated for 30 min to form a homogeneous solution before conducting the measurements.

Raman spectra of GO and GO@SiO_2 dried powder samples were obtained by using a multichannel confocal micro spectrometer (Renishaw in Via, England) with a laser wavelength of 633 nm.

The chemical compositions of GO, SiO_2–NH_2, and GO@SiO_2 were obtained using an ESCALAB 250 X-ray photoelectron spectrometer (XPS) purchased from Thermo Fisher Scientific Company (American).

The volume resistivity (ρ_v) of GO@SiO_2/PDMS composites was measured using a high-resistance meter (EST121, Beijing Huajinghui Technology Co., Ltd, China). ρ_v is calculated as follows: $\rho_v = (R_V \cdot S)/L$, where R_V is the volume resistance of the specimen, S is the contact area between the composite and the electrode, which is 2122.64 mm^2 for this instrument, and L is the average thickness. The conductivity is the reciprocal of volume resistivity.

Dielectric properties were measured using an HP4294A impedance analyzer (Agilent, USA) in the frequency range of 102–106 Hz at room temperature. The volume resistivity of the composites was tested using a high-resistance meter (EST 121, Beijing HuaJingHui Scientific and Technical Co., Ltd., China) at room temperature. Mechanical properties were obtained using a tensile apparatus (CMT4104, Shenzhen SANS Testing Machine Co., Ltd., China); 5–10 samples were tested in the tensile test for each material at a strain rate of 50 mm/min, and the Y of pure PDMS and GO@SiO_2/PDMS composites was obtained by calculating the slopes of the stress–strain curves at the strain of 5%, which was obtained using a tensile apparatus.

The actuated strain was measured using a circular strain test and the diameter of the active area is 1 cm. The samples for actuated strain tests had a thickness of about 0.5 mm and a diameter of 5 cm, and compliant electrodes with a diameter of 11 mm were applied to each side of the DE. Before the test, two main surfaces of the films were sprayed with graphite suspension using an airbrush to fabricate compliant electrodes. The film was first fixed on two circular frames without a pre-strain, and electric field was applied on both sides of the film. The electric field needed was equipped with a high-voltage DC generator (DTZG-60, Wuhan Dotek Electric Co., Ltd.). The strain was calculated as the change of the pixel of the electrode area divided by the original pixel. A video camera was fixed at the same focal length to capture the actuator plain before and after voltage exertion. Three actuators were studied three times for each composition, and the average values are reported. The thickness of pure PDMS and GO@SiO_2/PDMS composites was 0.5 mm. The electric breakdown filed was obtained by testing the actuated stains of composites until electric breakdown occurred.

3. Results and discussion

3.1. Self-assembly of GO and APS-modified silica

The overall synthetic procedure of GO coated on silica is demonstrated in Figure 1. GONSs are negatively charged in aqueous solution because of the ionization of the carboxylic acid groups and phenolic hydroxy groups on their surface. When GO is mixed with other positively charged particles, the assembly can be realized by the electrostatic force. However, a silica (SiO_2) particle in an aqueous solution is also negatively charged because of the ionization of H_2SiO_3 on its surface. So the surface modification of SiO_2 is necessary realize the assembly of GO on silica. APS molecules with amino groups can ionize to form a positively charged surface and APS can also undergo condensation reaction with materials that have hydroxyl groups.

Here, SiO_2 was first modified by APS to form a positively charged surface. Therefore, it is easy to assume that electrostatic driving force can take place between SiO_2–NH_2 and GO. To clarify this process, zeta potential was tested to examine the surface charges of GO, SiO_2, and SiO_2–NH_2 aqueous suspensions, as shown in Table 1.

With a pH value of 7.0, GO had a highly negative surface charge with zeta value of −42.5 mV. Meanwhile, the surface charges of the SiO_2 switched from negative (zeta = − 26.8 mV) to positive (zeta = 39.1 mV) after the modification, suggesting the successful graft of APS on SiO_2. When the oppositely charged GONSs and SiO_2–NH_2 particles collided through simple solution mixing, the electrostatic assembly might have been triggered, forming the core–shell structure of GO@SiO_2 hybrid afterward.

The morphology and microstructure of the GO@SiO_2 hybrids and the pristine SiO_2 were observed by SEM, TEM, and HRTEM, as shown in Figure 2. Compared with the SiO_2 particles (specific area ~ 120 m^2/g) that have a microsphere morphology and a smooth surface (Figure 2a), GO@SiO_2 hybrids show crinkled and rough textures that are associated with the presence of flexible GO sheets. In most cases, the edges of individual as well as overlapping GO layers can be observed, particularly at the interface between aggregated particles, where the graphene layers appeared to link the neighboring spheres together (Figure 2(b)). The typical TEM and HRTEM images of GO@SiO_2 hybrid (Figure 2(c, d)) also confirm that the flexible GO sheets are successfully wrapped around the SiO_2–NH_2 microspheres, with the GO shell thicknesses of about 3 nm. The core–shell structure of the GO@SiO_2 hybrids can prevent the restacking of GO and the self-aggregation of SiO_2. Considering the large specific surface area gap between GO and silica (the specific surface area of an individual GO sheet is 2600 m^2/g), it is easy to

Table 1. The zeta value of SiO_2, SiO_2–NH_2, and GO.

Sample	SiO_2	SiO_2–NH_2	GO
Zeta value (mV)	−26.8	39.1	−42.5

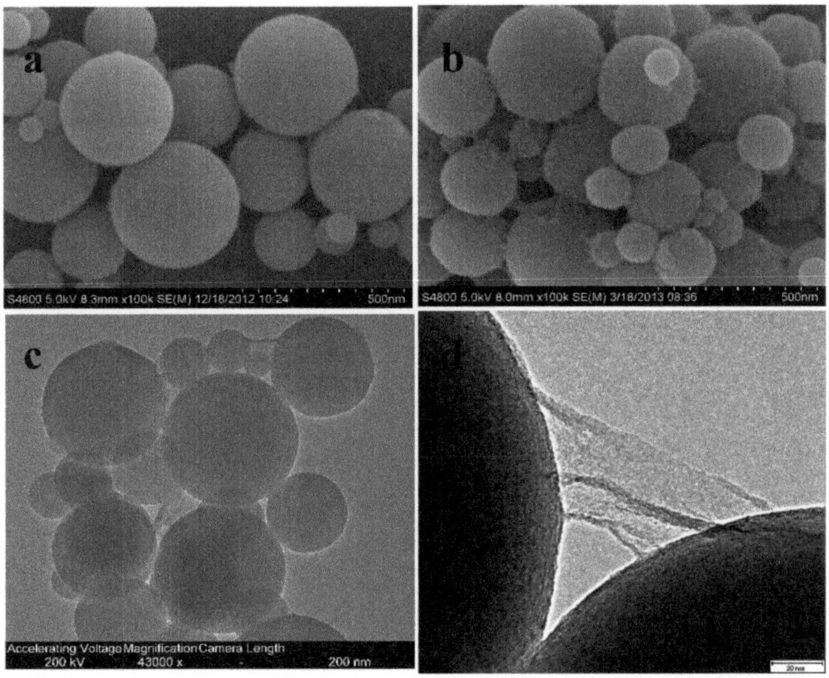

Figure 2. The SEM images of (a) pristine SiO_2 (b) GO@SiO_2 hybrid; (c)TEM; and (d) HRTEM of GO@SiO_2 hybrid.

Table 2. The element analysis of SiO_2, SiO_2–NH_2, GO@SiO_2, and GO.

	Relative atomic percentage (%)			
Samples	C	Si	O	N
GO	69.73	–	30.27	–
SiO_2	19.7	29.86	50.44	–
SiO_2–NH_2	25.12	25.98	45.51	3.39
GO@SiO_2	33.79	22.23	42.56	1.42

understand why only 2 wt% GO is needed to assemble on the SiO_2 when the equilibrium coverage is reached.

The encapsulation of GO on SiO_2 can be confirmed by XPS, as shown in Table 2. The nitrogen content of 3.39% in SiO_2–NH_2 suggests a successful introduction of amino groups on SiO_2. The increase in carbon content from 25.12% of SiO_2–NH_2 to 33.79% of GO@SiO_2 also indicates the immobilization of GO on SiO_2. Fourier transform infrared spectroscopy (FT-IR) was also used to confirm the covalently attached APS coupling agent and the GO wrapping on the SiO_2, as shown in Figure 3. The main peaks of SiO_2 spectra curve at 1621 and 3430 cm^{-1} are designed as stretching vibration of Si–O–Si, blending vibration of O–H, and stretching vibration of –OH, respectively. Several minor bands at around 2800–3000 cm^{-1} were detected in the spectra of SiO_2–NH_2, which were attributed to the C–H stretching vibration of the hydrocarbon chains of the grafting APS. A weak band at 1725 cm^{-1} appeared in the SiO_2@GO spectra curve, which is the

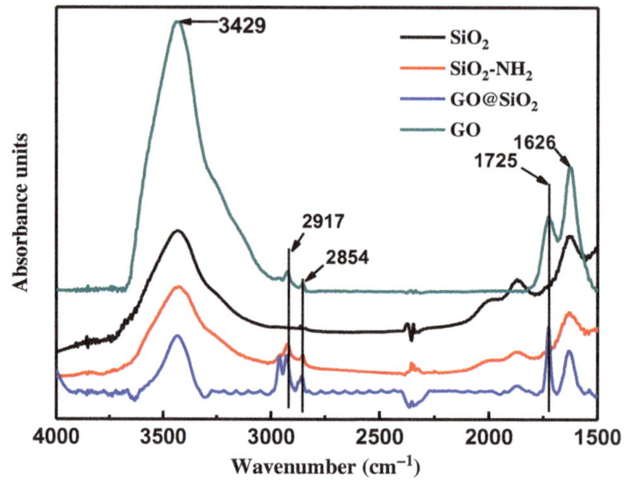

Figure 3. The FT-IR spectra curves of pristine SiO_2, SiO_2–NH_2, GO@SiO_2, and GO.

characteristic of the C=O stretching vibration band of GO. As GO has a large amount of –OH groups, the peak at 3429 cm^{-1} of SiO_2–GO is higher than that of SiO_2–NH_2.

Raman spectroscopy has been used as a powerful tool for the characterization of graphene and its derivatives. Therefore, it was also employed to further confirm the GO shells on SiO_2 hybrids. As shown in the Figure 4, there are two peaks at about 1337 and 1595 cm^{-1} in the Raman spectroscopy of graphite oxide, corresponding to the D and G bands. The existence of the two bands was also observed in the spectroscopy of GO@SiO_2, which demonstrates the successful assembly of GO on SiO_2. Meanwhile, it was reported that both G and 2D bands can be used to monitor the number of layers [38–40]. The G peak of the single-layer graphene shifts to lower wavenumbers, and the 2D band decreases in intensity and lower frequency peaks after stacking more GO layers. In this study, the peak positions of the G bands of the GO and GO@SiO_2 were centered at about 1595 and 1602 cm^{-1}, respectively. This indicates that GO@SiO_2

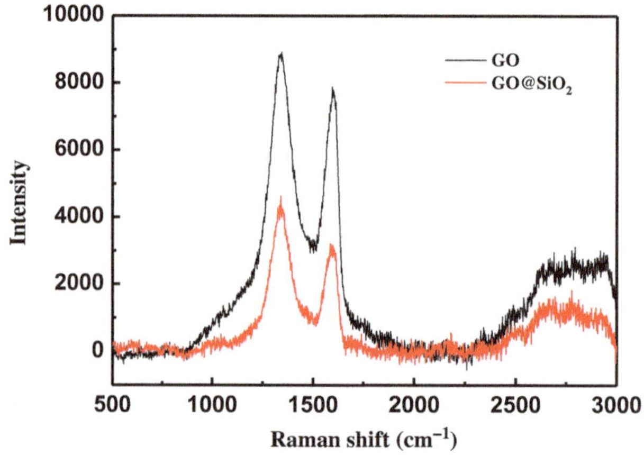

Figure 4. The Raman spectrum of GO and GO@SiO_2.

contains less layers compared to original GO. However, the 2D peaks of the two materials are not obvious in this study and cannot demonstrate the change in number of GO layers.

3.2. In situ thermal reduction of GO@SiO_2/PDMS

Many researchers have demonstrated that the increase in *k* is far less than that expected using GO as the dielectric filler because of the severe disruption of the graphite structure of GO [41]. Therefore, the reduction of GO@SiO_2/PDMS composites is required to increase the interfacial polarization ability of GO@SiO_2. So, the in situ thermal reduction of GO@SiO_2/PDMS composites was conducted at 180°C for 2 h because of the following advantages. First, the directional distribution of GO@SiO_2 around PDMS latex particles and network structure can be retained during in situ thermal reduction at 180°C. Second, the method is simple and efficient for the partial reduction of GO shells of GO@SiO_2

Thermal reduction of pure GO was conducted under the same conditions to evaluate the degree of reduction of GO shells. The degree of reduction was characterized using XPS and TGA, and the results are shown in Table 3 and Figure 5. The oxygen content obviously decreases from 28.88% for GO to 21.38% for RGO, as shown in Table 3, and the carbon oxygen ratio increases from 2.46% for GO to 3.73% for RGO, suggesting partial reduction of GO. Figure 5 shows the XPS C1s core-level spectra of pure GO. We can observe four characteristic peaks corresponding to carbon atoms in different functional groups: C–C (unoxidized graphite carbon skeleton) at 284.79 eV, C–O (hydroxyl

Table 3. The element analysis of GO and in situ thermal reduction of graphene.

	Relative atomic percentage (%)		
Samples	C	O	C/O
GO	71.12	28.88	2.46
180c 2h RGO	78.87	21.14	3.73

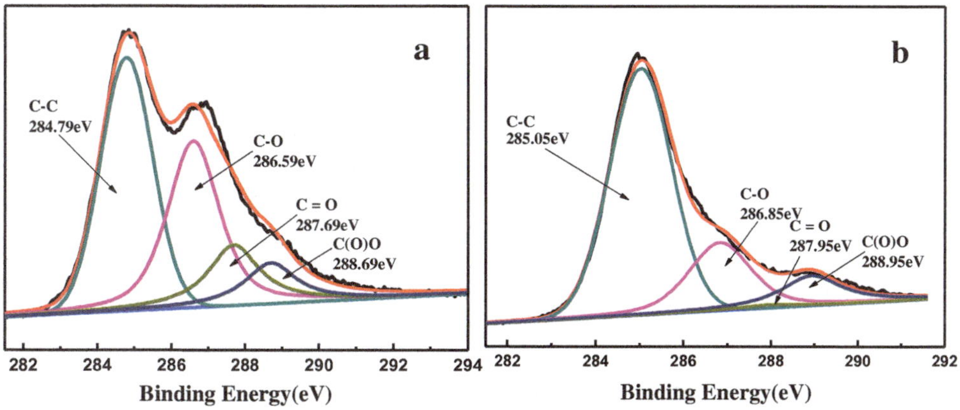

Figure 5. The C1s X-ray photoelectron spectroscopy of GO and RGO.

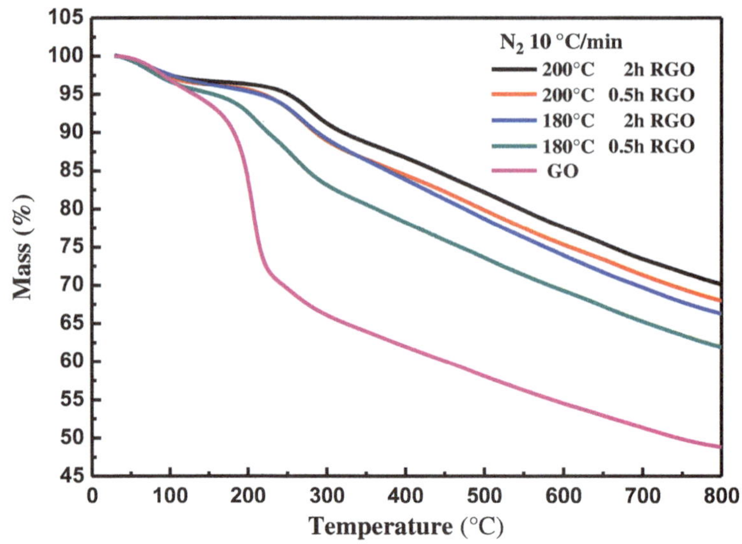

Figure 6. The TGA curves of GO and RGO.

and epoxy groups) at 286.59 eV, C=O (carbonyl group) at 287.69 eV and COOH (carboxyl and ester groups) at 288.69 eV. The presence of the same peak in Figure 5(b) suggests that there are still some oxygen-containing groups in RGO. But the peaks corresponding to the oxygen-containing groups in the spectrum of RGO are weakened, especially that of C=O (carbonyl), indicating that GOs have been partially reduced during thermal reduction at 180°C.

The mass loss of pure GO, RGO was measured at a heating rate of 10°C min^{-1} in nitrogen atmosphere, as shown in Figure 6. The mass loss below 100°C of GO is due to the removal of absorbed water. The maximum rate of mass loss, corresponding to the loss of most of the oxygen-containing groups, occurs at 150–200°C for both GO and RGO. The mass loss of RGO continues with further increase in temperature because of the remaining oxygen-containing groups caused by the partial reduction of GO during thermal reduction at 180°C

3.3. Microstructure of RGO@SiO$_2$/PDMS composites

The fracture surfaces of RGO@SiO_2/PDMS composites with different filler content were observed by SEM, and the results are shown in the Figure 7. A uniform distribution of RGO@SiO_2 in PDMS matrix is observed in both composites filled with 30 phr and 60 phr of RGO@SiO_2, as shown in Figure 7(a) and 7(b), respectively. To clearly observe the interspace between the fillers, the SEM images of the composites with larger magnification were obtained, as shown in Figure 7(c) and Figure 7(d), respectively. Obviously, although some agglomeration of RGO@SiO_2 in the matrix is observed with the increase in the content of RGO@SiO_2 to 60 phr, the interspace between the filler particles largely decreases. As a result, a strong filler network is observed in the composite with 60 phr of RGO@SiO_2, facilitating the increase in conductivity and dielectric constant.

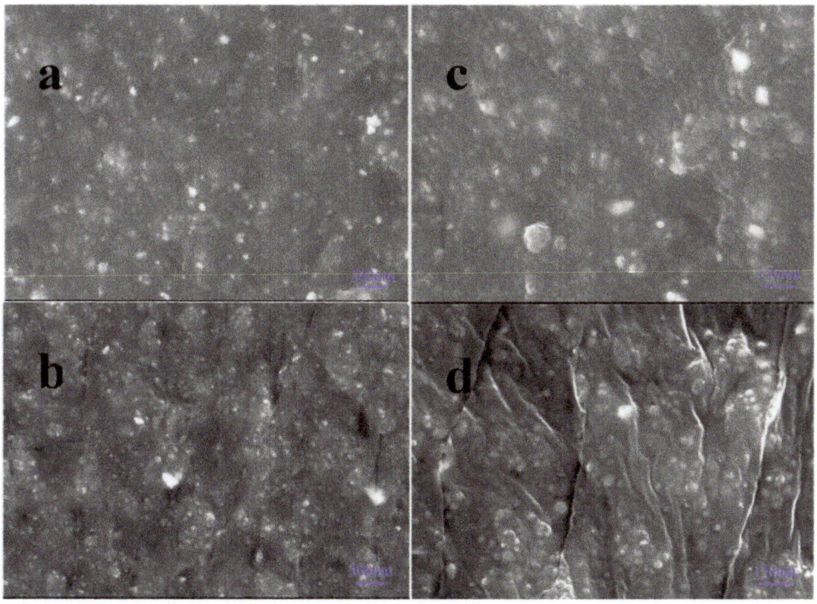

Figure 7. The SEM figures of fracture surfaces of RGO@SiO_2/PDMS composites with different filler content (a, b) 30 phr (c, d) 60 phr.

3.4. *Electromechanical properties of GO@SiO_2/PDMS composites*

The stress–strain curves of pure PDMS and the RGO@SiO_2/PDMS composites with 30 phr and 60 phr of RGO@SiO_2 are shown in Figure 8 and the corresponding mechanical properties are summarized in Table 5. We can observe that both the tensile strength and elastic modulus of PDMS obviously increase with the increase in the content of RGO@SiO_2, indicating the good reinforcing effect of RGO@SiO_2 on the PDMS matrix. For example, the tensile strength increases from 1.0 MPa for pure PDMS to 1.5 MPa for

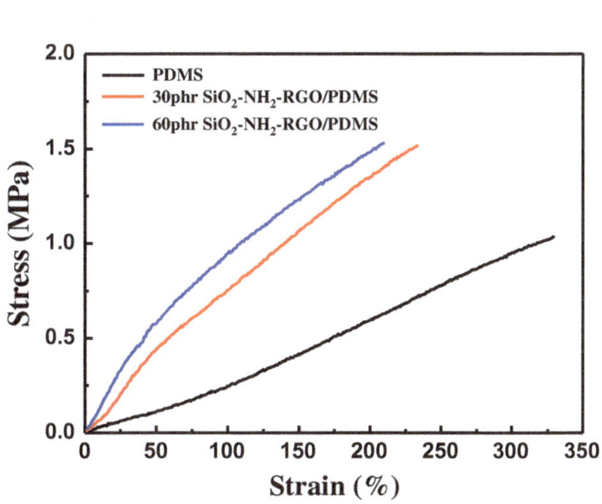

Figure 8. Strain–stress curve of PDMS and PDMS with 30 and 60 phr of RGO@SiO_2.

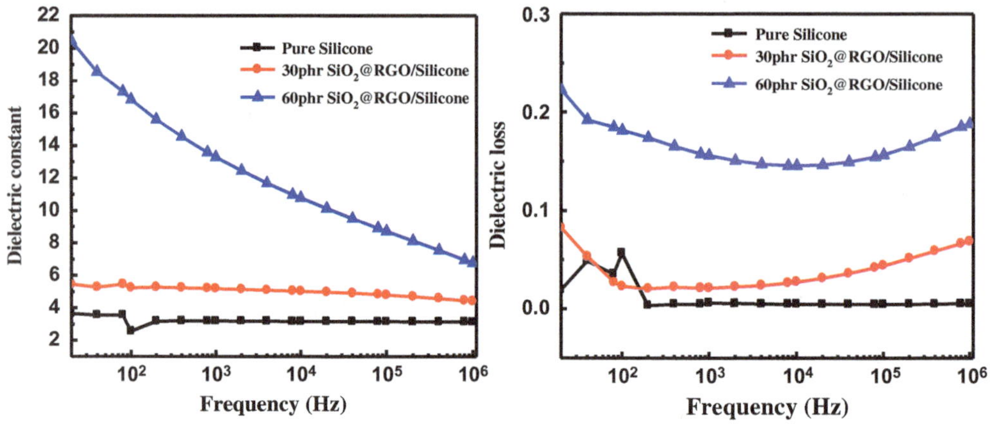

Figure 9. Frequency response of dielectric constant (a) and dielectric loss (b).

the composite with 60 phr of RGO@SiO_2, and the corresponding elastic modulus increases from 0.27 to 0.9 MPa. Although the elongation at break decreases with increasing the content of RGO@SiO_2, the elongation at break of the composites is larger than 200%, facilitating the practical application of DE. (The standard deviation is ±0.3.)

The dielectric properties of pure PDMS and RGO@SiO_2/PDMS composites as a function of frequency at room temperature are shown in Figure 9. As expected, the k of all the RGO@SiO_2/PDMS composites is largely decreased with an increase in frequency, suggesting a strong frequency dependence of k of all the composites. The strong dependence of k is ascribed to interfacial polarization effect (also named as the Maxwell–Wagner–Sillars (MWS) effect) of RGO@SiO_2 on PDMS molecules caused by the accumulation of many free charges at the internal interfaces between RGO@SiO_2 and PDMS [42]. Is has enough time to accumulate charge at the interfaces between RGO@SiO_2 and the rubber matrix at low frequency, resulting in the high k of the composites. However, the interfacial polarization cannot catch up with the change in electrical field frequency at high frequency, leading to the large decrease in k at high frequency [43].

Interestingly, a significant increase in k is observed with the increase in the content of RGO@SiO_2 at the same frequency (see Figure 9(a)) in the whole frequency range. The dielectric loss slightly increases with the increase of filler content, especially in the high frequency range of 10^3–10^6 Hz. For example, the k of the RGO@SiO_2/silicone rubber (SR) composite with 60 wt% of RGO@SiO_2 at 100 and 1000 Hz is 16.8 and 13, respectively, which is much higher than that of pure SR (3.6). The increase in k is ascribed to the large increase in the interfacial polarizability of RGO@SiO_2 caused by the in situ thermal reduction of GO. The dielectric loss of RGO@SiO_2/SR composite (see Figure 8(b)) at 10^3 Hz with 30 wt% and 60 wt% GO@SiO_2 remains low (<0.2), much lower than that of many other conductive filler/polymer dielectric composites. The relatively low loss is mainly attributed to the low DC conductance of RGO@SiO_2 caused by the partial reduction of GO shells by in situ thermal reduction at 180°C, as demonstrated by the relatively low electrical conductivity of all the composites, as shown in Table 4.

The elastic moduli of pure PDMS and RGO@SiO_2/PDMS composites are summarized in Table 5. We can observe that the elastic modulus increases from 0.27 MPa for pure PDMS to 0.6 and 0.9 MPa for the composites with 30 and 60 phr of RGO@SiO_2,

Table 4. The conductivity of $RGO@SiO_2/PDMS$ composites.

$RGO@SiO_2$ content (phr)	Conductivity (S/m)
0	8.5×10^{-12}
30	2.4×10^{-11}
60	1.2×10^{-10}

Table 5. The electromechanical properties of $RGO@SiO_2/PDMS$ composites.

$RGO@SiO_2$ content (phr)	Dielectric constant k at 10^3 Hz	Dielectric loss tangent at 10^3 Hz	Elastic modulus Y (MPa)	$\beta = k/Y$	Tensile strength (MPa)	Elongation at break (%)
0	3.2	0.006	0.27	5.3	1.03	329.4
30	5.2	0.021	0.60	8.6	1.52	234.1
60	13.3	0.156	0.90	14.8	1.53	210.3

respectively. There is a large increase in k and a mild increase in the elastic modulus of PDMS composites by addition of $RGO@SiO_2$.

Figure 10 shows the actuated strains of pure PDMS and $RGO@SiO_2/PDMS$ composites as a function of the electric field. The actuated strain of all the samples obviously increases with the increase in electric field, because it has a quadratic relationship with the applied electric field. Although the maximum actuated strain of the composites is lower than that of pure PDMS and the breakdown strength of PDMS obviously decreases with the increase in $RGO@SiO_2$ content, the actuated strain of the composites at low electric field is much higher than that of pure PDMS, as summarized in Table 6. For example, the actuated strain at 10 kV/mm obviously increases from 0.58% for pure PDMS to 3.48% for the composites with 60 phr of $RGO@SiO_2$. The k of the $RGO@SiO_2/SR$ composite with

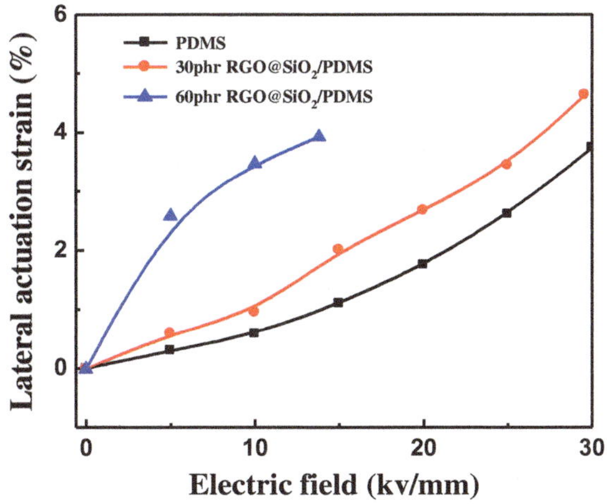

Figure 10. Lateral actuation strain of PDMS and $RGO@SiO_2/PDMS$ composites as a function of electric field.

Table 6. Comparison of actuated performances of advanced DE composites.

Composites	Maximum actuated strain (%)	Actuated strain at 10 kV/mm (%)
PMN/PDMS [13]	8.0 (53 kV/mm)	<0.5
PHT/PDMS [44]	8.0 (8 kV/mm)	0.4
PDVB@PANI/PDMS [43]	16.0 (43 kV/mm)	0.01
Dipoles–PDMS [45]	1.4 (16 kV/mm)	<1
TiO_2–PDMS (plasticized) [46]	18.0 (37 kV/mm)	<2.5
SR/BT [47]	57.0 (46 kV/mm)	<3
(60 phr)GO@SiO_2/PDMS in our study	3.9 (15 kV/mm)	3.3

Note: PMN, lead magnesium niobium; PHT, polymer (undoped poly(3-hexylthiophene)); PVDF, polydivinylbenzene; PANI, polyaniline; BT, barium titanate.

60 wt% of RGO@SiO_2 at 100 and 1000 Hz is 16.8 and 13, respectively, which is much higher than that of pure SR (3.6). According to the equation $Sz = -\varepsilon_r\varepsilon_0 E_2/Y$, the Y of composites with 60 wt% of RGO@SiO_2 is close to the Y of composites with 30 wt% of RGO@SiO_2. The β of 60 wt% RGO@SiO_2 has been increased due to the significantly improved dielectric constant of composites. So the actuator strain of 60 wt% RGO@SiO_2 composites has been enhanced, as the electric field strength increased, the composites with 60 wt% of RGO@SiO_2 possess high dielectric loss. So the electrical energy conversion to heat energy leads to the actuator strain non-linear increase. More interestingly, the actuated strain at 5 kV/mm obviously increases from 0.3% for pure PDMS to 2.59% for the composites with 60 phr of RGO@SiO_2, an eightfold increase in the actuated strain, much higher than that of other previously reported DEs [13,43–47], as summarized in Table 6. The improvement in the actuated strain at low electric fields is in favor of the application of DE in the biological and medical fields (such as synthesis of artificial skin), tactile displays, and braille displays, etc. [47].

3.5. Conclusions

We developed a simple and efficient method to simultaneously improve the actuated performance and mechanical strength of a silicone elastomer. GO@SiO_2 were first prepared via electrostatic self-assembly of APS-modified SiO_2 and GO, and then introduced into PDMS matrix to prepare GO@SiO_2/PDMS dielectric composites. RGO@SiO_2/PDMS composites are then prepared by in situ thermal reduction of GO@SiO_2/PDMS composites at 180°C for 2 h to increase the interfacial polarizability. The results show that the k of PDMS largely increases by adding 60 phr of RGO@SiO_2, whereas the dielectric loss of the composites remains low (<0.2 at 1000 Hz). The actuated strain at low electric field (5 kV/mm) obviously increases from 0.3% for pure PDMS to 2.59% for the composites with 60 phr of RGO@SiO_2, an eightfold increase in the actuated strain. On the other hand, both the tensile strength and elastic modulus are obviously improved by adding 60 phr of RGO@SiO_2 due to the good reinforcing effect of RGO@SiO_2 on PDMS.

Disclosure statement

No potential conflict of interest was reported by the authors.

Funding

We would like to express our sincere thanks to the National Natural Science Foundation of China [grant number 51173007], [grant number 51103090], [grant number 51221002] for financial support.

References

1. A. O'Halloran, F. O'Malley, and P. McHugh, *A review on dielectric elastomer actuators, technology, applications, and challenges*, J. Appl. Phys. 104 (7) (2008), pp. 071101. doi:10.1063/1.2981642
2. S. Bauer, S. Bauer-Gogonea, I. Graz, M. Kaltenbrunner, C. Keplinger, and R. Schwödiauer, *25th Anniversary Article: A Soft Future: From Robots and Sensor Skin to Energy Harvesters*, Adv. Mater. 26 (1) (2014), pp. 149–162. doi:10.1002/adma.201303349
3. K. Wongtimnoi, B. Guiffard, A. Bogner-Van De Moortèle, L. Seveyrat, C. Gauthier, and J.-Y. Cavaillé, *Improvement of electrostrictive properties of a polyether-based polyurethane elastomer filled with conductive carbon black*, Compos. Sci. Technol. 71 (6) (2011), pp. 885–892. doi:10.1016/j.compscitech.2011.02.003
4. L.Z. Chen, C.H. Liu, C.H. Hu, and S.S. Fan, *Electrothermal actuation based on carbon nanotube network in silicone elastomer*, Appl. Phys. Lett. 92 (26) (2008), pp. 263104. doi:10.1063/1.2955513
5. A. Tiwari, A.P. Mishra, S.R. Dhakate, R. Khan, and S.K. Shukla, *Synthesis of electrically active biopolymer–SiO2 nanocomposite aerogel*, Mater. Lett. 61 (23–24) (2007), pp. 4587–4590. doi:10.1016/j.matlet.2007.02.076
6. R. Palakodeti and M.R. Kessler, *Influence of frequency and prestrain on the mechanical efficiency of dielectric electroactive polymer actuators*, Mater. Lett. 60 (29–30) (2006), pp. 3437–3440. doi:10.1016/j.matlet.2006.03.053
7. S.M. Ha, W. Yuan, Q. Pei, R. Pelrine, and S. Stanford, *Interpenetrating Polymer Networks for High-Performance Electroelastomer Artificial Muscles*, Adv. Mater. 18 (7) (2006), pp. 887–891. doi:10.1002/(ISSN)1521-4095
8. R. Pelrine, *High-Speed Electrically Actuated Elastomers with Strain Greater Than 100%*, Science 287 (5454) (2000), pp. 836–839. doi:10.1126/science.287.5454.836
9. H. Stoyanov, M. Kollosche, S. Risse, D.N. McCarthy, and G. Kofod, *Elastic block copolymer nanocomposites with controlled interfacial interactions for artificial muscles with direct voltage control*, Soft Matter 7 (1) (2011), pp. 194–202. doi:10.1039/C0SM00715C
10. F. Carpi, P. Chiarelli, A. Mazzoldi, and D. De Rossi, *Electromechanical characterisation of dielectric elastomer planar actuators: Comparative evaluation of different electrode materials and different counterloads*, Sensors and Actuators A: Physical 107 (1) (2003), pp. 85–95. doi:10.1016/S0924-4247(03)00257-7
11. M. Molberg, D. Crespy, P. Rupper, F. Nüesch, J.-A.E. Månson, C. Löwe, and D.M. Opris, *High Breakdown Field Dielectric Elastomer Actuators Using Encapsulated Polyaniline as High Dielectric Constant Filler*, Adv. Funct. Mater. 20 (19) (2010), pp. 3280–3291. doi:10.1002/adfm.201000486
12. S. Liu, M. Tian, B. Yan, Y. Yao, L. Zhang, T. Nishi, and N. Ning, *High performance dielectric elastomers by partially reduced graphene oxide and disruption of hydrogen bonding of polyurethanes*, Polymer 56 (2015), pp. 375–384. doi:10.1016/j.polymer.2014.11.012
13. D.M. Opris, M. Molberg, C. Walder, Y.S. Ko, B. Fischer, and F.A. Nüesch, *New Silicone Composites for Dielectric Elastomer Actuator Applications In Competition with Acrylic Foil*, Adv. Funct. Mater. 21 (18) (2011), pp. 3531–3539. doi:10.1002/adfm.201101039
14. G. Kofod, H. Kollosche, M. Kollosche, S. Risse, H. Ragusch, D. Rychkov, M. Dansachmuller, and D.N. McCarthy, *Nano-scale materials science for soft dielectrics:Composites for dielectric elastomer actuators*. IEEE International Conference on Solid Dielectrics (ICSD 2010) 10th, 2010.
15. F. Carpi and D. De Rossi, *Dielectric elastomer cylindrical actuators: Electromechanical modelling and experimental evaluation*, Mater. Sci. Engineering: C 24 (4) (2004), pp. 555–562. doi:10.1016/j.msec.2004.02.005
16. P. Brochu and Q. Pei, *Advances in Dielectric Elastomers for Actuators and Artificial Muscles*, Macromol. Rapid Commun. 31 (1) (2010), pp. 10–36. doi:10.1002/marc.v31:1

17. S. Vudayagiri, S. Zakaria, L. Yu, S.S. Hassouneh, M. Benslimane, and A.L. Skov, *High breakdown-strength composites from liquid silicone rubbers*, Smart Mater. Struct. 23 (10) (2014), pp. 105017. doi:10.1088/0964-1726/23/10/105017
18. B. Li, H. Chen, and J. Zhou, *Electromechanical stability of dielectric elastomer composites with enhanced permittivity*, Composites Part A: Applied Science Manufacturing 52 (2013), pp. 55–61. doi:10.1016/j.compositesa.2012.11.013
19. K. Goswami, A.E. Daugaard, and A.L. Skov, *Dielectric properties of ultraviolet cured poly (dimethyl siloxane) sub-percolative composites containing percolative amounts of multi-walled carbon nanotubes*, RSC Adv 5 (17) (2015), pp. 12792–12799. doi:10.1039/C4RA14637A
20. H. Liu, Y. Shen, Y. Song, C.-W. Nan, Y. Lin, and X. Yang, *Carbon Nanotube Array/Polymer Core/Shell Structured Composites with High Dielectric Permittivity, Low Dielectric Loss, and Large Energy Density*, Adv. Mater. 23 (43) (2011), pp. 5104–5108. doi:10.1002/adma.201102079
21. N. Ning, X. Bai, D. Yang, L. Zhang, Y. Lu, T. Nishi, and M. Tian, *Dramatically improved dielectric properties of polymer composites by controlling the alignment of carbon nanotubes in matrix*, RSC Adv 4 (9) (2014), pp. 4543–4551. doi:10.1039/C3RA45769A
22. J.-K. Yuan, S.-H. Yao, A. Sylvestre, and J. Bai, *Biphasic Polymer Blends Containing Carbon Nanotubes: Heterogeneous Nanotube Distribution and Its Influence on the Dielectric Properties*, J. Phys. Chem. C 116 (2) (2012), pp. 2051–2058. doi:10.1021/jp210872w
23. I. Alig, D. Lellinger, S.M. Dudkin, and P. Pötschke, *Conductivity spectroscopy on melt processed polypropylene–multiwalled carbon nanotube composites: Recovery after shear and crystallization*, Polymer 48 (4) (2007), pp. 1020–1029. doi:10.1016/j.polymer.2006.12.035
24. L.J. Romasanta, M. Hernández, M.A. López-Manchado, and R. Verdejo, *Functionalised graphene sheets as effective high dielectric constant fillers*, Nanoscale Res. Lett. 6 (2011), pp. 508. doi:10.1186/1556-276X-6-508
25. S.K. Kumar, M. Castro, A. Saiter, L. Delbreilh, J.F. Feller, S. Thomas, and Y. Grohens, *Development of poly(isobutylene-co-isoprene)/reduced graphene oxide nanocomposites for barrier, dielectric and sensingapplications*, Mater. Lett. 96 (2013), pp. 109–112. doi:10.1016/j.matlet.2013.01.036
26. M. Tian, J. Zhang, L. Zhang, S. Liu, X. Zan, T. Nishi, and N. Ning, *Graphene encapsulated rubber latex composites with high dielectric constant, low dielectric loss and low percolation threshold*, J. Colloid Interface Sci. 430 (2014), pp. 249–256. doi:10.1016/j.jcis.2014.05.034
27. J. Oh, J.-H. Lee, J.C. Koo, H.R. Choi, Y. Lee, T. Kim, N.D. Luong, and J.-D. Nam, *Graphene oxide porous paper from amine-functionalized poly(glycidyl methacrylate)/graphene oxide core-shell microspheres*, J. Mater. Chem. 20 (41) (2010), pp. 9200–9204. doi:10.1039/c0jm00107d
28. D. Yang, M. Tian, D. Li, W. Wang, F. Ge, and L. Zhang, *Enhanced dielectric properties and actuated strain of elastomer composites with dopamine-induced surface functionalization*, J. Mater. Chem. A 1 (39) (2013), pp. 12276. doi:10.1039/c3ta12090b
29. F. B. Madsen, A. E. Daugaard, S. Hvilsted, and A. L. Skov, *Novel silicone compatible cross-linkers for controlled functionalization of PDMS networks*, Electroactive Polymer Actuators and Devices (EAPAD). 8687 (2013), pp. 86871H.
30. M. Tian, Q. Ma, X. Li, L. Zhang, T. Nishi, and N. Ning, *High performance dielectric composites by latex compounding of graphene oxide-encapsulated carbon nanosphere hybrids with XNBR*, J. Mater. Chem. A 2 (29) (2014), pp. 11144. doi:10.1039/c4ta01600a
31. L. Chu, Q. Xue, J. Sun, F. Xia, W. Xing, D. Xia, and M. Dong, *Porous graphene sandwich/poly(vinylidene fluoride) composites with high dielectric properties*, Compos. Sci. Technol. 86 (2013), pp. 70–75. doi:10.1016/j.compscitech.2013.07.001
32. L. Seveyrat, A. Chalkha, D. Guyomar, and L. Lebrun, *Preparation of graphene nanoflakes/polymer composites and their performances for actuation and energy harvesting applications*, J. Appl. Phys. 111 (10) (2012), pp. 104904. doi:10.1063/1.4718577
33. F.R. Carpi, *D.D., Improvement of electromechanical actuating performances of a silicone dielectric elastomer by dispersion of titanium dioxide powder. Dielectrics and Electrical Insulation*, IEEE Trans. 12 (2005), pp. 835–843.
34. L. Gu, T. Wang, W. Zhang, G. Liang, A. Gu, and L. Yuan, *Low-cost and facile fabrication of titanium dioxide coated oxidized titanium diboride–epoxy resin composites with high dielectric*

constant and extremely low dielectric loss, RSC Adv. 3 (19) (2013), pp. 7071. doi:10.1039/c3ra23239e

35. X. Li, A. Umar, Z. Chen, T. Tian, S. Wang, and Y. Wang, *Supramolecular fabrication of polyelectrolyte-modified reduced graphene oxide for NO2 sensing applications*, Ceram. Int. 41 (9) (2015), pp. 12130–12136. doi:10.1016/j.ceramint.2015.06.030
36. M. Tian, B. Yan, Y. Yao, L. Zhang, T. Nishi, and N. Ning, *Largely improved actuation strain at low electric field of dielectric elastomer by combining disrupting hydrogen bonds with ionic conductivity*, J. Mater. Chem. C 2 (39) (2014), pp. 8388–8397. doi:10.1039/C4TC01140F
37. C. Racles, M. Cazacu, B. Fischer, and D.M. Opris, *Synthesis and characterization of silicones containing cyanopropyl groups and their use in dielectric elastomer actuators*, Smart Mater. Struct. 22 (10) (2013), pp. 104004. doi:10.1088/0964-1726/22/10/104004
38. I. Calizo, A.A. Balandin, W. Bao, F. Miao, and C.N. Lau, *Temperature Dependence of the Raman Spectra of Graphene and Graphene Multilayers*, Nano Lett. 7 (9) (2007), pp. 2645–2649. doi:10.1021/nl071033g
39. S. Coh, L.Z. Tan, S.G. Louie, and M.L. Cohen, *Theory of the Raman spectrum of rotated double-layer graphene*, Phys. Rev. B. 88 (16) (2013). doi:10.1103/PhysRevB.88.165431
40. K.S. Kim, Y. Zhao, H. Jang, S.Y. Lee, J.M. Kim, K.S. Kim, J.-H. Ahn, P. Kim, J.-Y. Choi, and B.H. Hong, *Large-scale pattern growth of graphene films for stretchable transparent electrodes*, Nature 457 (7230) (2009), pp. 706–710. doi:10.1038/nature07719
41. C. Wu, X. Huang, G. Wang, X. Wu, K. Yang, S. Li, and P. Jiang, *Hyperbranched-polymer functionalization of graphene sheets for enhanced mechanical and dielectric properties of polyurethane composites*, J. Mater. Chem. 22 (14) (2012), pp. 7010–7019. doi:10.1039/c2jm16901k
42. X. Huang, P. Jiang, and L. Xie, *Ferroelectric polymer/silver nanocomposites with high dielectric constant and high thermal conductivity*, Appl. Phys. Lett. 95 (24) (2009), pp. 242901. doi:10.1063/1.3273368
43. Z.-M. Dang, L. Wang, Y. Yin, Q. Zhang, and -Q.-Q. Lei, *Giant Dielectric Permittivities in Functionalized Carbon-Nanotube/ Electroactive-Polymer Nanocomposites*, Adv. Mater. 19 (6) (2007), pp. 852–857. doi:10.1002/(ISSN)1521-4095
44. F. Carpi, G. Gallone, F. Galantini, and D. De Rossi, *Silicone–Poly(hexylthiophene) Blends as Elastomers with Enhanced Electromechanical Transduction Properties*, Adv. Funct. Mater. 18 (2) (2008), pp. 235–241. doi:10.1002/(ISSN)1616-3028
45. G. Gallone, F. Carpi, D. De Rossi, G. Levita, and A. Marchetti, *Dielectric constant enhancement in a silicone elastomer filled with lead magnesium niobate–lead titanate*, Mater. Sci. Engineering: C 27 (1) (2007), pp. 110–116. doi:10.1016/j.msec.2006.03.003
46. B. Kussmaul, S. Risse, G. Kofod, R. Waché, M. Wegener, D.N. McCarthy, H. Krüger, and R. Gerhard, *Enhancement Of Dielectric Permittivity And Electromechanical Response In Silicone Elastomers: Molecular Grafting Of Organic Dipoles To The Macromolecular Network*, Adv. Funct. Mater. 21 (23) (2011), pp. 4589–4594. doi:10.1002/adfm.v21.23
47. L. Jiang, A. Betts, D. Kennedy, and S. Jerrams, *The fabrication of dielectric elastomers from silicone rubber and barium titanate: Employing equi-biaxial pre-stretch to achieve large deformations*, J. Mater. Sci. 50 (24) (2015), pp. 7930–7938. doi:10.1007/s10853-015-9357-6

3

Influence of in-flight particle state diagnostics on properties of plasma sprayed YSZ-CeO_2 nanocomposite coatings

S. Mantry[a]*, B.B. Jha[a], A. Mandal[b], D.K. Mishra[c,d], B.K. Mishra[a] and M. Chakraborty[b]

[a]CSIR-Institute of Minerals & Materials Technology, Bhubaneswar 751013, Odisha, India; [b]School of Minerals, Metallurgical and Materials Engineering, Indian Institute of Technology Bhubaneswar, Bhubaneswar 751007, Odisha, India; [c]Department of Physics, Institute of Technical Education and Research, Siksha 'O'Anusandhan University, Khandagiri Square, Bhubaneswar 751030, Odisha, India; [d]Department of Physics, College of Science, Engineering and Technology, University of South Africa, Johannesburg, South Africa

This article describes the influence of controlling in-flight hot particle characteristics on properties of plasma sprayed nanostructured yttria stabilized zirconia (YSZ) coatings. This article depicts dependence of adhesion strength of as-sprayed nanostructured YSZ coatings on particle temperature, velocity and size of the splat prior to impact on the metallic substrate. Particle temperature measurement is based on two-color pyrometry and particle velocities are measured from the length of the particle traces during known exposure times. The microstructure and adhesion strength of as-sprayed nano-YSZ coatings were studied. Field emission scanning electron microscopy results revealed that morphology of coating exhibits bimodal microstructure consisting of nano-zones reinforced in the matrix of fully melted particles. The coating adhesion strength is noticed to be greatly affected by the melting state of agglomerates. Maximum adhesion strength of 42.39 MPa has been experimentally found out by selecting optimum levels of particle temperature and velocity. The enhanced bond strength of nano-YSZ coating may be attributed to higher interfacial toughness due to cracks being interrupted by adherent nano-zones.

Keywords: CeYSZ; nano-zones; spray diagnostics; adhesion; plasma spraying

1. Introduction

Increased operating temperatures and hence, improved performance of gas turbines or diesel engines can be realized by using thermal barrier coatings (TBCs) [1–6]. Plasma sprayed TBCs based on yttrium stabilized zirconia (YSZ) have been applied to hot section industrial components. Zirconium based ceramics are considered to be best suitable for TBC applications due to its low density, high hardness and low thermal conductivity [7–11]. The above properties are dependent on the microstructure of the ceramic system. Hence, it is imperative to have a control over the particle state in order to achieve reproducible microstructure and properties in coatings. It is evident from literature that both performance and characteristics of coatings are influenced by microstructure, which in turn, are affected by the particle state along with substrate and spray conditions [12]. Hence, controlling each parameter and evaluating its role in performance of coatings

*Corresponding author. Email: mantrysisir@gmail.com

becomes difficult. Substrate and deposition conditions are relatively easier to control in comparison to the complex spray stream. Some uncontrollable factors like nozzle wear and gas flow stability make the process more difficult for assessment.

The important control parameters in plasma spraying are the flow rates of plasma and powder carrier gas, torch input power, dwell time and powder feed rate. There have been a few reports regarding the studies on structural, micro-structural, thermal and mechanical properties by controlling particle temperature and velocity [13–15]. *However, optimum spray conditions which would yield required physical and mechanical properties have not been identified.* Hence, a practical way of obtaining *in-situ* information from the individual in-flight coating particles provides the input for desired coating properties [16–18]. Amongst the parameters of utmost interest are the particle temperature and velocity distributions, size distribution of particles prior to the impact on the substrate. These parameters play an important role in enhancement of interfacial bond strength of nanostructured coatings as compared to conventional coatings. The increase in bond strength for the nanostructured coatings was approximately 1.8 times higher for atmospheric plasma spraying (APS) nano Al_2O_3–13 wt.% TiO_2 [19] and 2.4 times higher for high velocity oxy-fuel (HVOF)-sprayed nano TiO_2 [20]. The higher bond strength achieved may be attributed to higher interfacial toughness due to cracks being interrupted by strong adherent nano-zones [21]. The presence of dense nano-zones (bimodal distribution of microstructure reflected in our earlier work [22]) will have an effect on their mechanical properties and the microstructure which in turn depends on particle temperature and velocity. In this article, an attempt has been made to address the effect of controlling in-flight hot particle characteristics (monitored by CCD camera) on adhesion strength of as-sprayed CeO_2 doped YSZ coatings on Inconel 718 substrates.

2. Experimental details

2.1. Synthesis of nanostructured powders by sol-gel route

The nanostructured YSZ powders employed in this study were synthesized through sol-gel technique [23]. A water based solution of zirconium oxychloride and yttrium oxide was prepared in order to obtain 8 mol% YSZ. Precursor materials were taken according to the proper stoichiometric ratio and an excess amount of nitric acid was added to the solution. Final solution was homogenized by constant stirring at 100–120°C. After 5–6 h of constant stirring and heating, the translucent solution (with little amount of citric acid) was heated on a hot plate at about 200–250°C until it turned into a black viscous gel, which on continued heating burnt due to a vigorous exothermic reaction. Black ashes obtained after combustion were treated at 350°C in air for 1 h to eliminate the low temperature volatile residues and calcined at 600°C for 2 h to free the matrix from carbon, resulting in YSZ powder.

The transmission electron micrograph of as-synthesized YSZ powders by sol-gel technique is shown in Figure 1. The average grain size of the nanoparticles is on the order of 20–30 nm. The calcined YSZ and commercial CeO_2 powders were taken as the raw ingredients for the plasma spraying process and were mixed properly and heat treated at 1400°C.

2.2. Reconstitution of nano powders

Nano-sized powder of 10CeYSZ [90 wt.% YSZ (5.4 wt.% Y_2O_3–ZrO_2) + 10 wt.% CeO_2] was plasma sprayed on Inconel 718 substrates. The finely dispersed nanoparticles were agglomerated using spray-drying technique [24] to a size of ~6–10 μm as required for

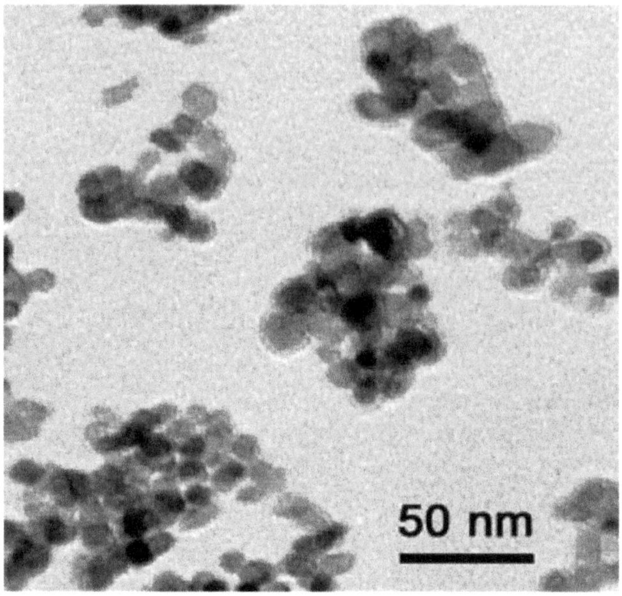

Figure 1. Transmission electron micrograph of as-synthesized YSZ particles.

plasma spraying. The agglomeration process was carried out in a Buchi B-290 (M/s BUCHI, Flawil, Switzerland) research model spray dryer. Spray-drying suspensions were prepared by mixing nano CeO_2 and nano-YSZ powders with 2 wt.% polyethylene glycol (PEG) binder and 1 wt.% ammonium citrate. Then the suspension was ball-milled for more than 24 h. The nano-CeYSZ slurry was attained via vigorous magnetic stirring for 30 min and heated up to 90°C. The preheating of solution assists in the slurry formation and in lowering the enthalpy needed during the drying process for moisture removal. The polymeric binder, PEG, in solution uses van der Waals forces to bind the nanoparticles together and forms spherical droplets during atomization. The suspension was spray-dried, and the granules were collected and sieved to a size of ~10 μm spherical agglomerates containing nanograins of size 20–30 nm as shown in Figure 2(a). Surface

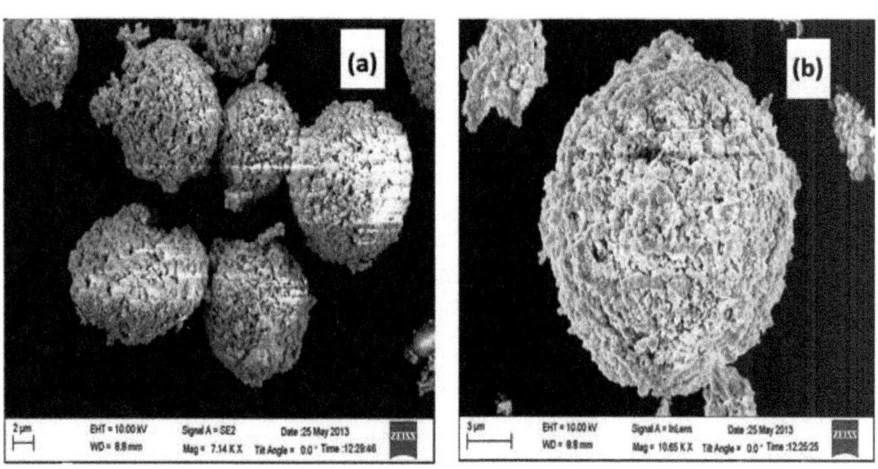

Figure 2. Field emission scanning electron micrographs of spray-dried CeYSZ particles.

morphology of one agglomerated CeYSZ particle is shown in Figure 2(b). Nanocrystalline particles on the surface of CeYSZ agglomerate are clearly visible.

2.3. Development of coatings

The surface of the substrate was subjected to grit blasting to make the surface rough. In this grit blasting method, highly compressed air carrying alumina particles were bombarded on the surface to remove some material which made the surface rough. A uniform roughness of 6–8 μm was maintained in order to provide better adhesion at the interface. The coating process was carried out at CSIR-IMMT, Bhubaneswar, using an 80 kW plasma spray system supplied by M/s Metallization, UK. In this study, high purity argon and helium were used as primary and secondary plasmagen gas, respectively, at an outlet pressure of 4 kg/cm^2. A roughened Inconel 718 substrate of dimension 120 × 60 × 5 mm^3 was fixed on the turn table and CeYSZ nanocomposite powders was sprayed by varying process parameters. The process parameters are listed in Table 1. Prior to top coat, NiCrCoAlY bond coats of thickness ~100 μm were applied on Inconel 718 substrates by HVOF (Hipojet 2700, M/s MEC, Jodhpur, India) by using set of process parameters given in Table 2. The number of passes was kept constant for each sample in order to make the thickness of all coatings (~250 μm including bond coat) within similar range.

Table 1. Selected operating parameters for plasma spray coating process (top coat).

Parameter	Operating range
Operating power	35–45 kW
Current	700–900 amps
Primary plasmagen gas (argon) flow rate	40–60 lpm
Secondary plasmagen gas (hydrogen) flow rate	0.5–2 lpm
Carrier gas flow rate	5 lpm
Nozzle to substrate distance (stand-off distance)	80 mm
Powder feed rate	25 g/min

Table 2. Selected operating parameters for HVOF process (bond coat).

Parameter	Operating range
Oxygen pressure	10 bar
Oxygen flow rate	265 lpm
LPG pressure	7 bar
LPG flow rate	70 lpm
Air pressure	5 bar
Air flow rate	700 lpm
Carrier gas (nitrogen) pressure	6 bar
Carrier gas flow rate	6.5 lpm
Air cap	9.3 mm

2.4. Adhesion test

To evaluate the coating adhesion strength, universal testing machine (make: INSTRON 8801; M/s Instron India Pvt Ltd., Chennai, India) was used. The test was conducted by the pull-out method as per ASTM C633 standard using two cylindrical specimens. The face of one of the cylinders is coated by plasma spraying with the material under investigation. This coated face is glued with a resin HTK Ultra Bond 100 (M/s HTK, Hamburg, Germany) to the face of the other uncoated cylindrical specimen and kept in furnace at 150°C for approximately 1.5 hours for the setting of the glue. This uncoated face was sand blasted prior to gluing. The assembly of two cylinders is then subjected to gradual tensile load. The cross head speed is kept constant at 1 mm/min. The tensile strength, i.e. the coating adhesion strength is calculated by dividing the maximum load applied at the rupture (i.e. failure occurs only at the coating–substrate interface) by the cross-sectional area of the cylindrical specimen considered.

2.5. Online particle diagnostics

The coating properties are intimately linked to the properties of as-sprayed lamellae, which in turn depend on in-flight particle properties as well as substrate temperature during spraying. It is known that both performance and properties of coatings are influenced largely by the microstructure, which in turn is influenced by the particle state along with substrate and deposition conditions. Online diagnostics using SprayWatch 2i equipment were carried out at different spray conditions to measure the particle temperature and velocity. Primary gas flow (Ar) strongly influences the average particle velocity whereas secondary gas flow (H_2) and current strongly influence the average particle temperature [12]. These three parameters are varied to achieve adequate temperature and velocity of melted agglomerates. Distance of the camera from the spray gun was kept between 150 and 200 mm. Particle temperature determination is based on the two-color pyrometry and in-flight particle velocities are measured from the length of particle traces during known exposure times using a single high-speed CCD camera [25]. The average particle temperature is 2650 ± 50°C – which is below the melting point of YSZ [17] – and velocity prior to impact is 200 ± 50 m/sec.

3. Results and discussion

3.1. X-ray diffraction (XRD)

The XRD patterns of agglomerated CeYSZ composite powder, bulk YSZ, Nanocrystalline YSZ and the CeYSZ nanocomposite coatings are shown in Figure 3(a) and (b). The commercial bulk polycrystalline YSZ coating is taken as a standard to compare the change in XRD patterns of the nanocrystalline YSZ and CeYSZ nanocomposite coatings. The nanostructured coatings obtained from reconstituted YSZ synthesized by chemical technique show cubic phases of zirconia. Impurity phases of Y_2O_3 are hardly observed in the XRD pattern. However, the high intensity peak (111) shifts towards the lower 2θ due to the development of strain in the nano-YSZ coating. The nanocrystalline CeYSZ composite powder shows high intensity peaks of tetragonal YSZ phases and cubic CeO_2. When the nanocrystalline composite powders were taken for coating, the XRD pattern clearly reveals the presence of tetragonal zirconia phases. The pattern is well matched with the JCPDS File no. 30–1468 which corresponds to tetragonal YSZ and no peaks corresponding to Y_2O_3 are detected. It is also evident that no monoclinic phase of

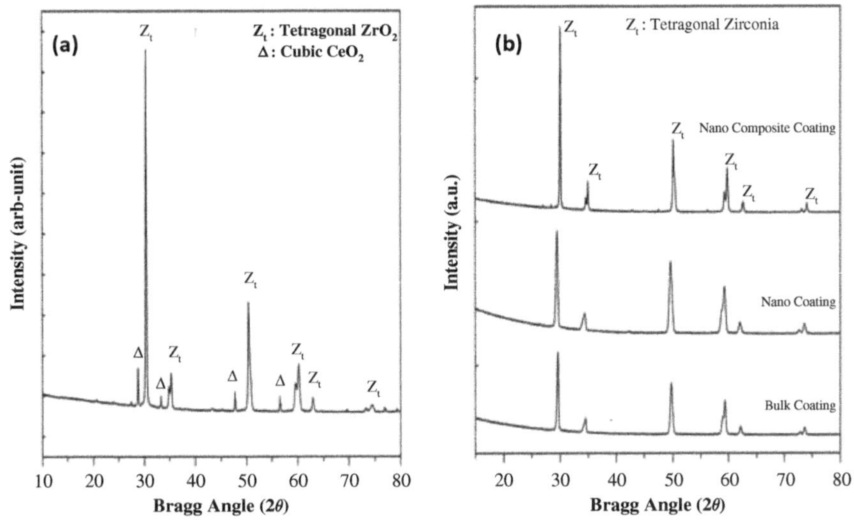

Figure 3. XRD patterns of (a) Nanocrystalline CeYSZ powder (b) Bulk YSZ coating, nanostructured YSZ coating and 10CeYSZ nanocomposite coating.

zirconia exists in the nanocomposite coating. It indicates that no CeO_2 peaks appear in the pattern, which confirms that CeO_2 is in solid solution with ZrO_2 and does not form any cluster in the CeYSZ coating.

3.2. Adhesion test results

Adhesion strength for as-sprayed coatings with the variation of particle temperature and velocity is shown in Figures 4 and 5, keeping other process parameters constant. Based on

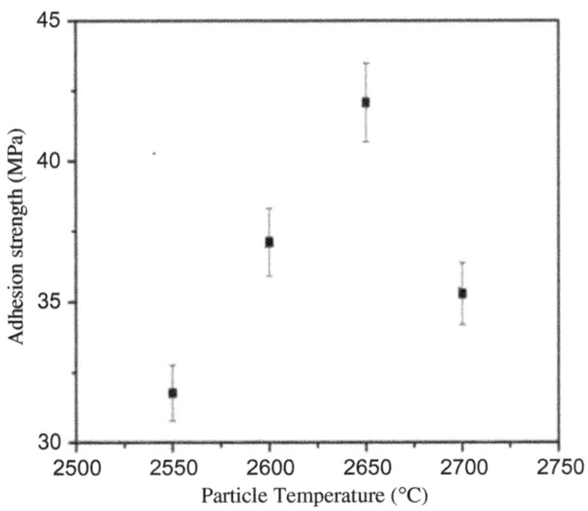

Figure 4. Variation in adhesion strength with temperature of agglomerates prior to impact on the substrate.

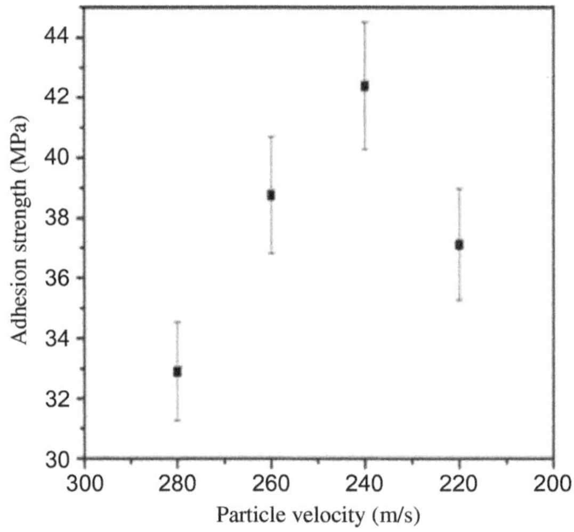

Figure 5. Variation in adhesion strength with velocity of melted agglomerates prior to impact on the substrate.

the present understanding, it can be achieved by fixing one of the three torch parameters. Of the three significant torch parameters, primary gas flow (Ar) strongly influences the average particle velocity whereas secondary gas flow (H_2) and current strongly influence the average particle temperature. It is found that bond strength increases from 2550 ± 50°C to 2650 ± 50°C and then decreases after 2700°C (~ melting point of feed stock). Similarly, it is maximum at particle velocity of 240 m/sec and seems to attain minimum value at 220 m/sec (may be fully melted) and 280 m/sec (may be un-melted). A series of coating are developed in the broad range of temperature and velocity. The test results for the adhesion of the APS nano-YSZ-coated substrate using Taguchi's L_{16} orthogonal design along with the corresponding S/N ratios are described elsewhere, where maximum adhesion strength is found to be 40.56 MPa [26]. The improvement of adhesion strength compared to nano-YSZ coatings is attributed to the compactness and uniform microstructure of as-sprayed coatings and formation of Ce-diffusion layers within the splats, which enhances inter splat adhesive forces. *The semi-melted particles (nano-zones) and the substrate were adherent to each other for which no micro cracks or gaps have been observed.* Therefore, those cracks observed in conventional coating may be *arrest or* interrupted *in* embedded dense nano-zones having high toughness. The maximum adhesion strength of nano-CeYSZ nanocomposite coating is found to be 42.39 MPa at 40 kW torch input power corresponding to particle temperature of 2686°C and velocity of 241.6 m/s.

3.3. Surface morphology analysis

The field emission scanning electron micrographs of YSZ-CeO_2 nanocomposite coatings are shown in Figure 6, which represents bimodal microstructure consisting of dense and smooth zones, indicating good molten state of particles and the rough and porous zones, indicating unmolten or semi-molten state of particles (Figure 6(a) and (c)). The percentage of nano-zones are less in case of coatings shown in Figure 6(a) than that in Figure 6(c), which may be attributed to their corresponding particle temperature and velocity. At temperature ~2550°C and velocity 280 m/sec, the particle residence time is less and not

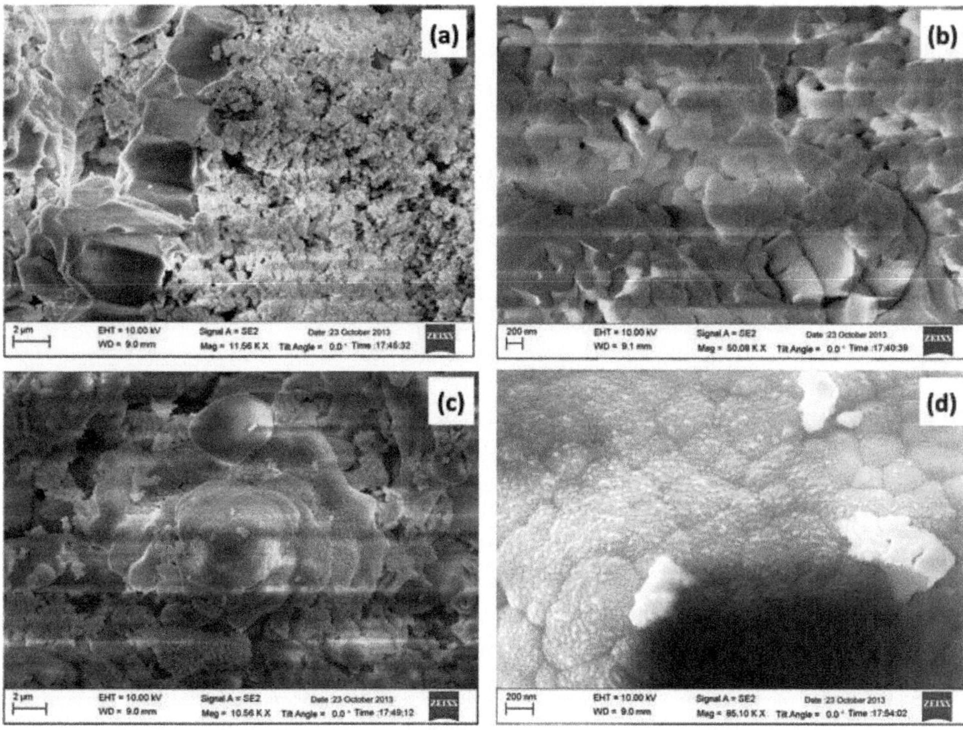

Figure 6. (a, b) Field emission scanning electron microscopy micrographs of YSZ-CeO_2 nanocomposite coatings at particle temperature 2550 ± 50°C and particle velocity 280 m/sec, (c, d) at particle temperature 2650 ± 50°C and particle velocity 240 m/sec.

molten and hence does not exhibit appreciable amount of adhesion, and the same holds good for high particle temperature and low velocity, i.e. particles become fully molten and nanostructure is lost. At higher magnification (Figure 6(b) and (d)), the micrographs show densely packed nanograins having average grain size of ~90–120 nm. The grains are very closely packed with development of distinct grain boundaries without having any voids and porous structure (Figure 6(d)). In Figure 6(b), the presence of grains and nano-zones on the surface is evident, but micro-pores are present, indicating the presence of some amount of unmolten particles.

4. Conclusions

This study indicates that successful deposition of nanocomposite CeYSZ powder by plasma spraying route is possible on Inconel 718 substrates. Particle temperature and velocity seem to be most significant parameter affecting adhesion in case of CeYSZ nanocomposite coatings. Optimum adhesion strength of 42.39 MPa has been experimentally found out at 40 kW torch input power corresponding to particle temperature of 2686°C and velocity of 241.6 m/sec. XRD pattern of nanocomposite coating clearly reveals the cubic phase transformation of tetragonal and monoclinic YSZ with a small percentage of tetragonal phases. Ce is in solid solution with ZrO_2 and helps in stabilizing ZrO_2 and does not cluster in as-sprayed nanocomposite coating. Micrograph of surface reveals the average grain size to be in the range of ~90–120 nm. The grains are very closely packed with distinct grain boundaries. The grains are interconnected with tetragonal grain boundary junctions. In summary, the enhancement of adhesion strength may be attributed to compact structure and high toughness of embedded nano-zones.

References

[1] M. Gell, *Application opportunities for nanostructure materials and coatings*, Mater. Sci. Eng. 204(1–2) (1995), pp. 246–251. doi:10.1016/0921-5093(95)09969-7.

[2] D. Zhu and R.A. Miller, *Thermal Conductivity and Elastic Modulus Evolution of Thermal Barrier Coatings under High Heat Flux Conditions, NASA TM-209069* 1999, NASA Glenn Research Center, Cleveland, OH, 1999.

[3] N.P. Padture, M. Gell, and E.H. Jordan, *Thermal barrier coatings for gas-turbine engine applications*, Science 296 (2002), pp. 280–284. doi:10.1126/science.1068609.

[4] J.A. Haynes, M.K. Ferber, and W.D. Porter, *Thermal cycling behavior of plasma-sprayed TBCs with various MCrAlX bond coats*, J. Therm. Spray. Technol. 9(1) (2000), pp. 38–48. doi:10.1361/105996300770350041.

[5] W.J. Brindley, *Properties of plasma sprayed bond coats*, J. Therm. Spray. Technol. 6(1) (1997), pp. 85–90. doi:10.1007/BF02646316

[6] V. Tolpygo and D. Clarke, *The effect of oxidation pre-treatment on the cyclic life of EB-PVD thermal barrier coatings with platinum–aluminide bond coats*, Surf. Coat. Technol. 200 (2005), pp. 1276–1281. doi:10.1016/j.surfcoat.2005.07.088.

[7] L.Y. Ni, C. Liu, H. Huang, and C.G. Zhou, *Thermal cycling behavior of thermal barrier coatings with HVOF NiCrALY bond coat*, J. Therm. Spray. Technol. 20(5) (2011), pp. 1133–1138. doi:10.1007/s11666-011-9647-8.

[8] D. Clarke, *Materials selection guidelines for low thermal conductivity thermal barrier coatings*, Surf. Coat. Technol. 163–164 (2003), pp. 67–74. doi:10.1016/S0257-8972(02)00593-5.

[9] S. Sodeoka, M. Suzuki, K. Ueno, H. Sakuramoto, T. Shibata, and M. Ando, *Thermal and mechanical properties of ZrO_2-CeO_2 plasma-sprayed coatings*, J. Thermal Spray Technol. 6(3) (1997), pp. 361–367. doi:10.1007/s11666-997-0071-z.

[10] A.L. Vasiliev and N.P. Padture, *Coatings of metastable ceramics deposited by solution-precursor plasma spray: II. Ternary ZrO_2-Y_2O_3-Al_2O_3 system*, Acta Mater. 54(18) (2006), pp. 4921–4928. doi:10.1016/j.actamat.2006.06.026.

[11] M.R. Winter and D.R. Clarke, *Oxide materials with low thermal conductivity*, J. Am. Ceram. Soc. 90(2) (2007), pp. 533–540. doi:10.1111/j.1551-2916.2006.01410.x.

[12] V. Srinivasan, A. Vaidya, T. Streibl, M. Friis, and S. Sampath, *On the reproducibility of air plasma spray process and control of particle state*, J. Therm. Spray. Technol. 15(4) (2006), pp. 739–743. doi:10.1361/105996306X147199.

[13] M. Prystay, P. Gougeon, and C. Moreau, *Structure of plasma-sprayed zirconia coatings tailored by controlling the temperature and velocity of the sprayed particles*, J. Thermal Spray Tech. 10(1) (2001), pp. 67–75. doi:10.1361/105996301770349529.

[14] Y. Su, T. Bernecki, and K. Faber, *In situ characterization of small-particle plasma sprayed powders*, J. Thermal Spray Tech. 11(1) (2002), pp. 52–61. doi:10.1361/105996302770348970.

[15] C. Moreau, J.-F. Bisson, R.S. Lima, and B.R. Marple, *Diagnostics for advanced materials processing by plasma spraying*, Pure Appl. Chem. 77(2) (2005), pp. 443–462. doi:10.1351/pac200577020443.

[16] C. Moreau, *Towards a better control of thermal spray processes,* in *Thermal Spray: Meeting the Challenges of the 21st Century*, C. Coddet, ed., ASM International, Materials Park, OH, 1998, pp. 1681–1693.

[17] P. Fauchais, J.F. Coudert, and M. Vardelle, *Plasma Diagnostics*, Academic Press, New York, NY, 1989, pp. 349–445.

[18] J.R. Fincke, D.C. Haggard, and W.D. Swank, *Particle temperature measurement in the thermal spray process*, J. Therm. Spray. Technol. 10(2) (2001), pp. 255–266. doi:10.1361/105996301770349358.

[19] E.H. Jordan, M. Gell, Y.H. Sohn, D. Goberman, L. Shaw, S. Jiang, M. Wang, T.D. Xiao, Y. Wang, and P. Strutt, *Fabrication and evaluation of plasma sprayed nanostructured alumina–titania coatings with superior properties*, Mater. Sci. Eng. A. 301 (2001), pp. 80–89. doi:10.1016/S0921-5093(00)01382-4.

[20] R.S. Lima and B.R. Marple, *Enhanced ductility in thermally sprayed titania coating synthesized using a nanostructured feedstock*, Mater. Sci. Eng. A. 395 (2005), pp. 269–280. doi:10.1016/j.msea.2004.12.039.

[21] R.S. Lima and B.R. Marple, *Thermal spray coatings engineered from nanostructured ceramic agglomerated powders for structural, thermal barrier and biomedical applications: A review*, J. Therm. Spray. Technol. 16(1) (2007), pp. 40–63. doi:10.1007/s11666-006-9010-7.

[22] S. Mantry, A. Mandal, D.K. Mishra, B.B. Jha, B.K. Mishra, and M. Chakraborty, *Microstructure and thermal characterization of plasma sprayed nanostructured $La_2Ce_2O_7$ doped YSZ coatings*, J. Therm. Spray. Technol. (2014) (accepted) doi:10.1007/s11666-014-0070-9.

[23] R.E. Juárez, D.G. Lamas, G.E. Lascalea, and N.E. Walsöe De Reca, *Synthesis of nanocrystalline zirconia powders for TZP ceramics by a nitrate-citrate combustion route*, J Eur. Ceram. Soc. 20 (2000), pp. 133–138. doi:10.1016/S0955-2219(99)00146-6.

[24] V. Viswanathan, E. Keith, A.V. Rea, and S. Seal, *Role of spray drying of nanoagglomerates in morphology evolution in nanostructured APS coatings*, J. Am. Ceram. Soc. 91(2) (2008), pp. 379–386. doi:10.1111/j.1551-2916.2007.02090.x.

[25] J. Vattulainen, E. Hämäläinen, R. Hernberg, P. Vuoristo, and T. Mäntylä, *Novel method for in-flight particle temperature and velocity measurements in plasma spraying using a single CCD camera*, J. Therm. Spray Technol. 10(1) (2001), pp. 94–104. doi:10.1361/105996301770349556.

[26] S. Mantry, B.K. Mishra, and M. Chakraborty, *Parametric appraisal of process parameters for adhesion of plasma sprayed nanostructured YSZ coatings using Taguchi experimental design*, Sci. World J. Article No. 527491, 2013 (2013), pp. 1–8. doi:10.1155/2013/527491.

4

Microstructural designs of plate-type elastic metamaterial and their potential applications

R. Zhu[a]*, X.N. Liu[b], G.K. Hu[b], F.G. Yuan[a] and G.L. Huang[c]*

[a]Department of Mechanical and Aerospace Engineering, North Carolina State University, Raleigh, NC 27695, USA; [b]School of Aerospace Engineering, Beijing Institute of Technology, Beijing 100081, People's Republic of China; [c]Department of Mechanical and Aerospace Engineering, University of Missouri, Columbia, MO 65211, USA

Elastic metamaterials are of growing interest due to their unique effective properties and wave manipulation abilities. Unlike phononic crystals based on the Bragg scattering mechanism, elastic metamaterials (EMMs) are based on the locally resonant (LR) mechanism and can fully control elastic waves at a subwavelength scale. Microstructural designs of EMMs in plate-like structures have attracted a great deal of attention. In this paper, the recent advances in the microstructural designs of LR-based EMM plates are reviewed. Their potential applications in the fields of low frequency guided wave attenuation, wave manipulation and energy trapping at a subwavelength scale, and structural health monitoring are discussed.

Keywords: metamaterial; composite; structural dynamics; nondestructive evaluation; structural health monitoring

1. Introduction

The research on metamaterials has grasped a lot of attentions recently. With artificially designed microstructures, metamaterial has the ability to achieve bizarre but very useful dynamic material properties. Initially, the strangeness of metamaterial was discovered when researchers investigated its effective electromagnetic (EM) properties at certain frequency ranges where the EM metamaterial behaves as if it has negative permittivity [1] and/or negative permeability [2]. Later on, similar unnatural phenomena were found in acoustic metamaterials which have negative mass density (NMD) [3] and/or negative bulk modulus (NBM) [4]. Anisotropic mass density was also achieved in non-resonant acoustic metamaterials containing fluid components [5,6,7]. Unlike the essential similarity between the governing equations in EM metamaterials and those in acoustic metamaterials, applying the concept of metamaterial in solids is more complex [8]. This is mainly due to the inherent coupling between longitudinal and shear wave modes and wave mode conversion. However, driven by the curiosity as well as the promising applications in mechanical and aerospace engineering, researchers have become more and more interested in the abnormal effective material properties of elastic metamaterials (EMMs). Many EMM-based devices have been proposed for promising applications with elastic wave controls and manipulations such as: elastic wave flat-lens focusing [9], EMM waveguides [10], EMM cloaking [11], and sub-wavelength imaging [12].

*Corresponding authors. Email: rzhu2@ncsu.edu (R. Zhu); HuangG@missouri.edu (G.L. Huang)

One of the biggest advantages of EMM over other wave-control structures, such as phononic crystals (PCs), is that the size of the metamaterial can be very small even in the low frequency range. Like photonic crystals, PCs have periodic structures with spatially modulated mechanical properties and therefore, can block or redirect acoustic/elastic waves based on the Bragg scattering mechanism. However, a reliance on Bragg scattering makes PC unsuited for either wave manipulation or vibration isolation at low frequency range due to the long wavelength necessitating impractically huge structures. By introducing the specially designed locally resonant (LR) microstructures into the building block of a PC, bandgap can be created at an extremely low-frequency range [13] where the wavelength becomes much larger than the size of the microstructure inside the EMM. The advantage of the subwavelength-scale LR microstructure is twofold: LR EMM based devices can be made small enough to be easily implanted into the engineering structures without sacrificing their lightness and compactness, and with an inhomogeneity scale that is much smaller than the wavelength of interest, EMM can be treated as a homogeneous medium with effective dynamic material properties.

With specially designed LR microstructures, unique negative effective material properties can be achieved in solid EMMs such as NMD, NBM and negative shear modulus. The frequency ranges where EMM demonstrates negative material property are found to be overlapped with the bandgaps that could, therefore, be easily tuned by changing the effective material properties through proper subwavelength-scale microstructural design. To understand the relationships between the negative effective material properties and the bandgaps in EMM, Milton and Willis [14] employed a simple one-dimensional (1D) mass-spring model to produce NMD. Yao *et al.* [15] experimentally examined the model in the low frequency regime, NMD was confirmed at the bandgap region of a finite periodic system composed of mass-spring units. For continuum media, Wang *et al.* [16] investigated the propagation of longitudinal and transverse elastic waves in a 1D infinite periodically layered system and the difference between the Bragg scattering mechanism in PC and the LR mechanism in EMM are discussed. Zhu *et al.* [17] proposed a 1D multilayered EMM design to achieve multiple bandgaps based on NMD retrieved from the theoretical reflection and transmission analysis of the longitudinal elastic waves. Using a rectangular solid waveguide with clamped boundary conditions, Yao *et al.* [18] demonstrated that extreme low frequency bandgap of the clamped waveguide can be attributed to NMD that follows the Drude-form below a cutoff frequency. For EMM with complex microstructures, Liu *et al.* [19] developed a finite element (FE)-based effective medium model to obtain the effective dynamic material properties. More recently, Yang *et al.* [20] designed a new class of active EMM with shunted piezoelectric patches in which bandgaps can be actively controlled and tuned by varying effective stiffness constant of the linear spring through appropriately selecting the value of negative capacitance of the piezoelectric patch.

In two-dimensional (2D) and three-dimensional (3D) EMMs, the LR-based bandgaps may not occur in the same frequency range when the waves propagate along different directions, which implies that the resonant frequencies of the microstructure are different along principle directions and therefore, the effective mass density is anisotropic in the EMM. Based on the relations between the effective mass density and the LR microstructures, the anisotropic effective mass density of an EMM can be achieved by changing position or distribution array of the microstructure in a unit cell. A conceptual microstructural design of the EMM with anisotropic mass density is a 2D mass-in-mass lattice system consisting of two different elastic springs in the two principle directions. The first appearance of this microstructural design was made explicitly by Milton and Willis [14].

Then, Huang *et al.* [21] modeled the lattice system as an equivalent 2D elastic solid with its effective mass density characterized as a second-order tensor to describe the dynamic behavior of the original lattice system in all directions. A multi-displacement microstructure continuum model was proposed [22] and the anisotropic effective dynamic mass density tensor of a continuous EMM made of lead cylinders coated with elliptical rubbers in an epoxy matrix was calculated based on the microstructure continuum model [23]. In addition, Milton [24] has shown that all mechanical materials can in principle be synthesized by combing pentamodes that are artificial special solid with almost infinitely large ratio of the bulk modulus to the shear modulus. By building 3D pentamode metamaterials, mass density and bulk modulus can be tailed independently over a large range [25,26]. With 3D dilational EMM, an ultimate Poisson' ratio of −1 has been experimentally demonstrated [27].

Most designs of the EMM devices are based on the bulk waves that propagate in infinite or semi-infinite media with no boundary effects considered. However, in the real world, the wave medium is always finite, such as plate-like structures where the propagations of elastic waves are guided by two parallel free surfaces. On one hand, reflections on the two free surface boundaries will generate different Lamb wave modes and therefore complicate the problem. On the other hand, studies on plate-like structures consisting of bandgap materials are very important due to their more promising applications in the real-world engineering fields. Compared with an EMM plate, elastic wave propagation in the PC plate has been well studied by many researchers. The band structure of a 2D solid/solid PC plate has been analyzed by using the FE method [28]. Hsiao *et al.* [29] studied the band structure of steel/epoxy PC plates experimentally. Good agreement with the numerical simulation was found. Due to the wave attenuation mechanism in PC plates being attributed to the Bragg scattering bandgaps, the wave isolation ability of PC plate is limited in the mid-high frequency ranges. Different LR-based microstructural designs of EMM plates were proposed to block low frequency guided waves as well as control wave propagations in thin plates. Such designs can be cataloged into three groups: perforated microstructural design [30,31,32], stubbed microstructural design [33,34,35], and binary/ternary insert microstructural design [36,37].

In this paper, we present a comprehensive review of the microstructural designs of LR-based EMM plates and their potential applications in low frequency guided wave attenuation, subwavelength-scale wave manipulation, and energy trapping. The paper is arranged in five sections including this introduction: in Section 2, the relation between the LR mechanism and the abnormal effective material properties of EMMs are explained through simple mass-spring lattices; in Section 3, three groups of LR-based microstructural design for EMM plates are reviewed; in Section 4, the potential applications of the EMM plates with the reviewed microstructures are discussed. Finally, conclusions are presented in Section 5.

2. Effective material properties of locally resonant EMMs

In order to explain the origin of the abnormal effective material properties of LR EMMs, such as NMD and anisotropic mass density and their effects on the wave propagations, simple mass-spring models have been studied [21,38]. Negative effective mass can be demonstrated in a 1D mass-in-mass model while the anisotropic effective mass can be explained through the analysis on a 2D mass-spring lattice.

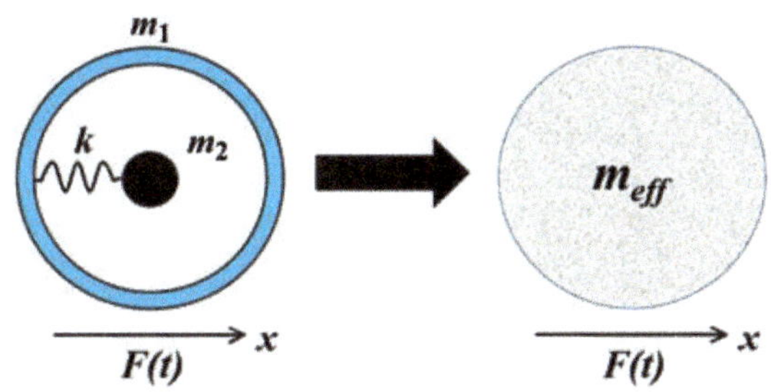

Figure 1. Single spring-mass system and its effective single mass model. Taken from Huang *et al.* [38].

2.1. One-dimensional mass-in-mass model and negative effective mass

Consider a single mass-in-mass unit as shown in Figure 1. Both masses m_1 and m_2 are rigid and the spring with linear spring constant k is massless. $F(t)$ is an external harmonic force with frequency ω and the mass-in-mass unit moves only along x direction. For the steady-state harmonic movement, the governing equation can be written as:

$$\left[\bar{F} + \left(m_1 + \frac{k}{\omega_0^2 - \omega^2}\right)\omega^2\bar{u}\right]e^{-i\omega t} = 0, \tag{1}$$

where $\bar{F}$ and $\bar{u}$ are the amplitudes of the harmonic external force and the harmonic displacement along x-direction, respectively. $\omega_0 = \sqrt{k/m_2}$ is the LR frequency of the inner mass m_2. When the mass-in-mass unit is represented by an equivalent model with single effective mass m_{eff} that is defined to produce the same displacement of m_1 in the lattice system, the corresponding governing equation for the equivalent model can be represented as:

$$\left[\bar{F} + m_{\text{eff}}\omega^2\bar{u}\right]e^{-i\omega t} = 0 \tag{2}$$

Therefore, the effective mass can be obtained by comparing Equations (1) and (2):

$$m_{\text{eff}} = m_1 + \frac{m_2\omega_0^2}{\omega_0^2 - \omega^2}. \tag{3}$$

It can be found that the effective mass becomes negative when the frequency ω approaches the LR frequency ω_0. According to Newton's second law of motion, the acceleration would be in the opposite direction to the external force when the negative mass occurs therefore the propagating wave is prevented.

2.2. Two dimensional mass-spring lattice and anisotropic effective mass

In a 2D mass-spring lattice as shown in Figure 2, two different massless springs with spring constants k_1 and k_2 connect the inner mass m_2 to the outer rigid mass m_1 along the horizontal (x) and the vertical (y) directions, respectively. If an equivalent anisotropic lattice with different effective masses along x- and y-directions is defined, the effective masses along the two principal directions can be obtained as [21]:

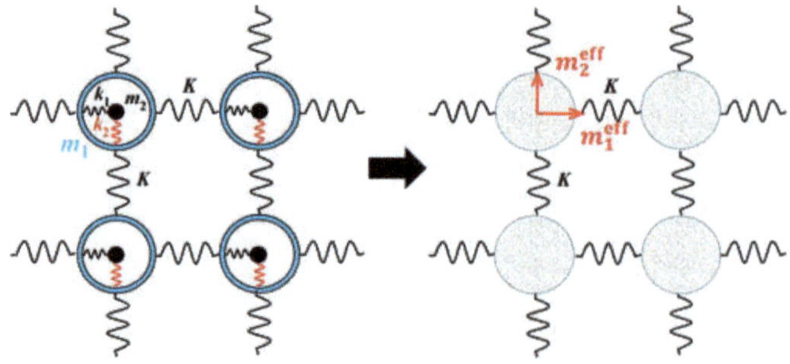

Figure 2. 2D mass-in-mass lattice system and effective medium with anisotropic effective mass.

$$m_1^{\text{eff}} = m_1 + \frac{m_2 k_1}{k_1 - m_2\omega^2} \tag{4a}$$

$$m_2^{\text{eff}} = m_1 + \frac{m_2 k_2}{k_2 - m_2\omega^2}, \tag{4b}$$

respectively. Comparing Equation (4a) and (4b), it is noticed that m_1^{eff} and m_2^{eff} are generally different. In a quasi-static case, the effective mass is almost isotropic. When the frequency approaches either LR frequency $\omega_{0,1} = \sqrt{k_1/m_2}$ or $\omega_{0,2} = \sqrt{k_2/m_2}$, the anisotropy of the effective mass becomes much stronger. In a certain frequency range, the effective masses along the two principal directions may have opposite signs. Therefore, the anisotropy between the effective mass densities along different principle directions of the lattice system is mainly caused by the difference between the LR frequencies that can be tuned through the design of the internal springs in the 2D lattice system.

3. Microstructural designs of locally resonant EMM plates

The mass-spring models perfectly complete their job that is illustrating the critical relation between negative effective material properties and the LR mechanism. However, applying these discrete designs into realistic engineering structure that requires high mechanical strength is unreliable and may introduce safety hazards. Therefore, designing an EMM with continuum medium having enough elastic strength is very critical for real-world engineering applications. In this section, various microstructural designs of solid EMMs in plate-like structures are reviewed and cataloged into three groups: perforated design, stubbed design, and binary/ternary insert design.

3.1. Perforated microstructural design

By periodically drilling holes through a metal plate, a PC bandgap can be created based on the Bragg scattering mechanism [39]. When the guided plate wave at the Bragg bandgap region hits the periodic holes, the reflected waves interfere with each other destructively and therefore, no propagating wave can be found in the PC plate. In order to generate the destructive interference of the reflected guided waves, the size of the through hole needs to be comparable with the wavelength of the guided wave that makes the PC plate impractically large and heavy at low frequency range. In order to produce a low-frequency bandgap with a compact unit cell size, LR microstructures can be introduced into the perforated unit cell of the PC plate that can exhibit spectral gap at frequencies two orders of magnitude lower than those by Bragg scattering.

Zhu *et al.* [30] introduced a cantilever-mass LR microstructure inside the perforated cell and, therefore, both in-plane and out-of-plane guided waves are forbidden in the low frequency range. Inspired from the well-studied three-phase EMM unit cell consisting of rubber-coated lead core embedded in an epoxy matrix, they proposed a single-phase EMM unit cell consisting of only one medium – stainless steel and used cantilever-mass microstructure to introduce the LR motion. In order to study the working mechanism of the proposed single-phased EMM plate and to provide an optimized design for the low-frequency bandgap application, a FE-based dispersion calculation method was applied to five unit cells with different microstructure configurations, as shown in the second row in Figure 3.

In Figure 3 (a), the unit cell is a plate with an interior square cutout. Based on the unit cell in Figure 3 (a), an interior cantilever is subsequently added in the unit cell as shown in Figure 3 (b). In Figure 3 (c)–(e), the unit cell is a plate with an interior cantilever-mass system. As shown in Figure 3 (c)–(e), the masses at the tips of the cantilevers are m_1, $m_2 = 2m_1$, and $m_3 = 3m_1$, respectively. The thin plate was made of stainless steel (ρ = 8000 kg/m^3, E = 193 GPa, and ν = 0.31) with the thickness h = 1 mm. Thirty unit cells in length direction and two unit cells in width direction were used for the simulations. The spacing of the adjacent unit cells is a = 5.6 mm.

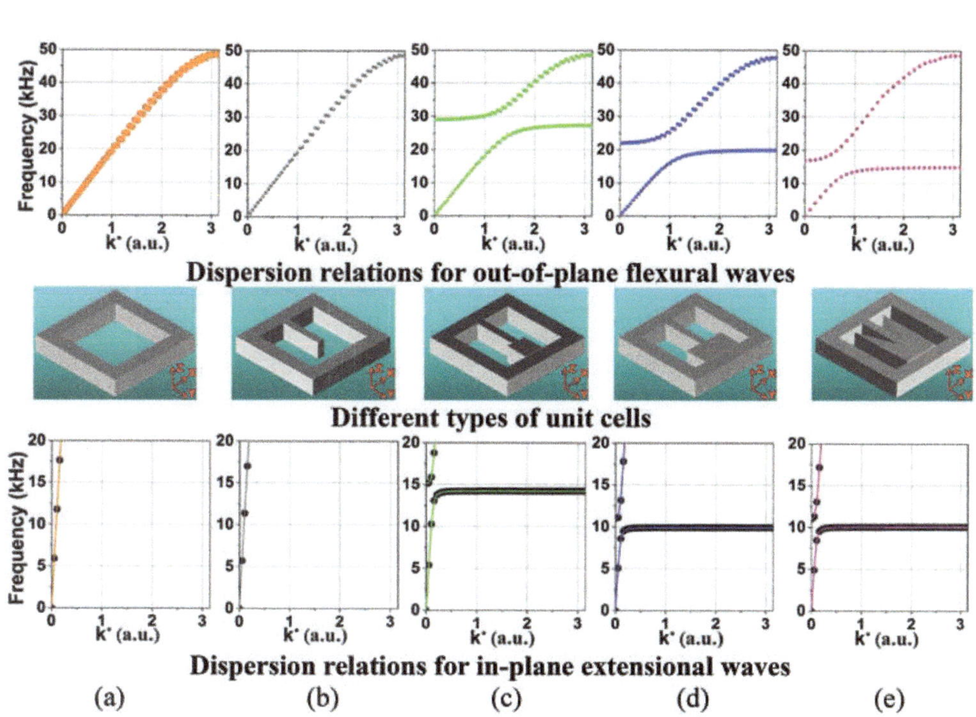

Figure 3. Dispersion relations of a thin plate with different types of unit cells: (a) only an interior square cutout, (b) added interior cantilever, (c) added interior cantilever with a tip mass m_1, (d) added interior cantilever with a tip mass m_2 = $2m_1$, (e) added interior cantilever with a tip mass m_3 = $3m_1$. Taken from Zhu *et al.* [30].

Commercial FE software, ANSYS v11.0 was used for the simulation. In the FE simulation, the 3D solid element SOLID45 was chosen to discretize the unit cell of the EMM plate and five elements were used in the thickness direction. Manual mesh strategy was used to better control mesh density and element shape for the complex interior cantilever-mass geometry. Guided waves propagate along the x-direction as shown in Figure 3. The dispersion relations of the metamaterial plates with different types of unit cells for out-of-plane flexural waves and in-plane extensional waves are plotted in the first row and the third row in Figure 3, respectively. The phase velocities of the out-of-plane and in-plane guided waves in Figure 3 (a) are 670 m/s and 3900 m/s, respectively. Comparing with the phase velocities measured on the pure stainless steel plate, the interior square cutout reduces the effective elastic modulus of the unit cell hence decreases the phase velocities of the guided waves propagating along the plate. Attention is focused on the frequency range near the bandgap. The normalized wave number k^* is defined as $k^* = k \cdot a$, where k is the wave number. It can be noticed that no low-frequency bandgap can be found for the plates with the first two types of the unit cells. The low-frequency bandgap only appears in the plates that contain a mass at the tip of the cantilever in the unit cell. It is also noted that, as the tip mass increases, the bandgap frequencies are lowered. This can be easily explained by the resonance mechanism. For the cantilever-mass system, the LR frequency can be approximately estimated by $\omega_0 = \sqrt{\bar{K}_{\text{eff}}/M}$ where M is the tip mass, $\bar{K}_{\text{eff}} = 3EI/l^3$, E, I and l are elastic modulus, moment of inertia and length of the cantilever in Figure 3 (e), respectively. It should be mentioned that the moment of inertia I for in-plane wave is different from that of the flexural wave.

LR mechanisms cannot only form low frequency bandgaps but also introduce slow propagating modes inside the complete PC bandgaps with specially designed resonant microstructures. By replacing the periodical through holes in the PC plate with blind holes, Sun *et al.* [31] proposed an LR PC plate with a membrane type resonant microstructure in the unit cell to propagate resonant slow modes of concentrated acoustic energy. Figure 4 (a) demonstrates the schematic of the LR PC plate's unit cell. The unit cell contains a cylindrical hole and a membrane at one end of the hole. A steel specimen consisting of 16 × 10 unit cells were made by mechanical machining, as shown in Figure 4 (b). The lattice constant of the unit cell is a = 9.52 mm. The radius of the blind hole is r = 4.475 mm. The thickness of the plate and the membrane is H = 5.05 mm

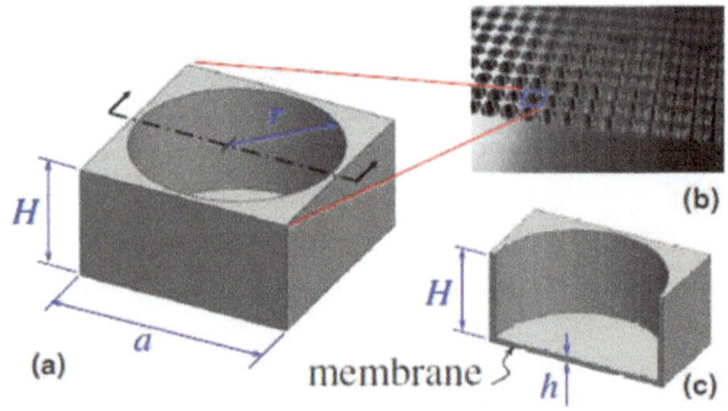

Figure 4. (a) Schematic of the unit cell of the square-lattice PC plate with periodic membranes. (b) Picture of the specimen. (c) Full section view of the unit cell. Taken from Sun *et al.* [31].

and h = 0.29 mm, respectively. The small thickness ratio between the plate and the membrane h/H = 1/17.4 makes the rigidity of the membrane very weak comparing with that of the rest of the plate.

The band structures of the PC plate with through holes and the LR PC plate with blind holes are shown in Figure 5 (a) as solid curves and dot curves, respectively. Commercial FE software, COMSOL MULTIPHYSICS (COMSOL Inc., Burlington, MA, USA) was used to obtain the band structures. In the calculation, the mass density, Young's modulus, and Poisson's ratio of steel were set to be 7750 kg/m^3, 226 GPa, and 0.29, respectively. Bloch periodic boundary conditions [40] were applied to the unit cell shown in Figure 4 (a). In Figure 5 (a), some near-flat bands can be found in the band structure of the LR PC plate, which are labeled by *A, B*1, *B*2, *C*1, *C*2, *D, E*1, *E*2, and *F.* These bands are associated with the vibration modes of the membranes that interact with the propagation modes of the PC plate and, therefore, result in additional bands in the band structure of the LR PC plate. When one compares the band structure of the PC plate and that of the proposed LR PC plate, eigenmodes of the LR PC plate, the modes *F*, can be found in the complete PC bandgap between 173 and 212 kHz. It should be noticed that these eigenmodes are not standing waves of energy localized inside the membranes but are propagating resonant modes with very slow group velocities. Figure 5 (b) and (c) shows the mode shapes of the resonant slow modes *F* labeled on the band structure in Figure 4 (a). The same periodic boundary conditions were used in the simulations of the mode shapes. The patterns of the mode shapes coincide with the eigenmodes of a clamped circular membrane that indicates that their eigen

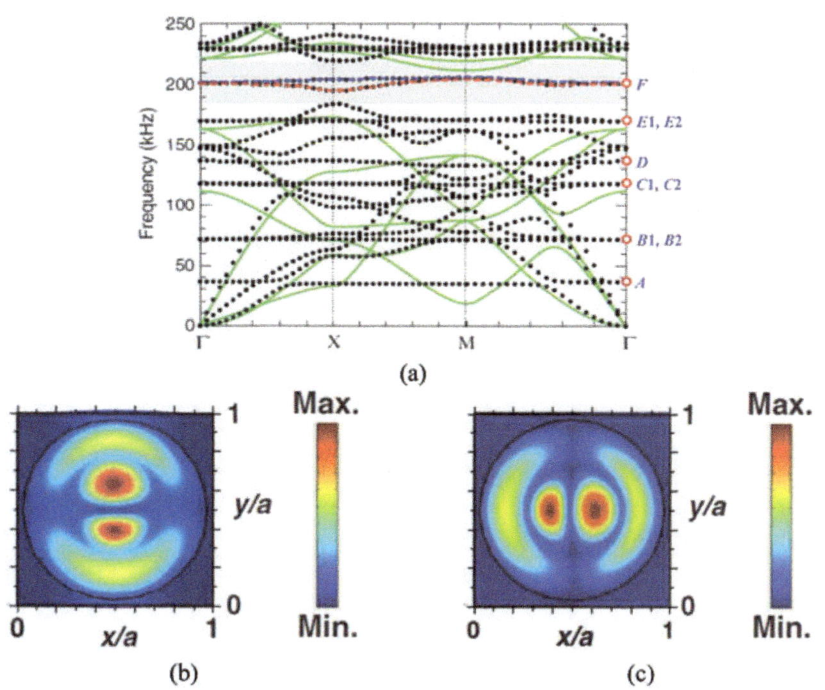

Figure 5. (a) Band structures of the PC plate with through holes (solid curves) and the LR PC plate with blind holes (dot curves). (b) Mode shapes of the lower resonant slow mode (red band). (c) Mode shapes of the higher resonant slow mode (blue band). Taken from Sun *et al.* [31].

frequencies of resonance are relevant [36]. According to the symmetries of the mode shapes, the resonant membrane modes can be classified into monopole, dipole, quadrupole mode, and so on. The red and blue bands of the modes F are found to be the lower and higher second-order dipole modes, respectively.

Most recently, Zhu *et al.* [32] introduced a subwavelength microstructure in a perforated hexagonal unit cell to produce simultaneous translational and rotational resonances and, therefore, create a double-negative EMM plate. The schematic top view of the unit cell of the proposed EMM plate is shown in Figure 6 (a). The unit cell is designed by properly slot-cutting in a hexagonal area and forming three chirally distributed ribs to connect the center piece (functioned as a mass) and the frame. The lattice constant is a, and widths of the slot, ribs, and frames are s, r, and t_f, respectively. In this design, the three ribs are very critical since they will induce not only the translational resonance but also the rotational resonance to the center piece, as shown in Figure 6 (b) and (c), respectively. Therefore, by carefully selecting the dimension of the chirally distributed ribs, it is possible to achieve and enhance simultaneous translational and rotational resonances at a desired low frequency range.

In order to calculate the effective medium parameters of the EMM plate with such a complicated microstructure, the use of analytical methods is not practicable. Instead, a numerical-based effective medium method based on the micromechanics approach is adopted, which is under the long-wavelength assumption [19]. The effective medium theory links the microscopic resonant structures to the EMM's macroscopic parameters. In the unit cell analysis, the global (macroscopic) stress, strain, resultant force, and acceleration of the unit cell can be numerically calculated by averaging local quantities on the external boundary of the hexagonal cell. Then, based on the constitutive relationship, the effective bulk modulus, shear modulus, and mass density of the proposed chiral EMM can be determined from the calculated macro-quantities. Based on the numerical-based effective model, the effective bulk modulus and mass density of the chiral EMM were calculated and are shown in

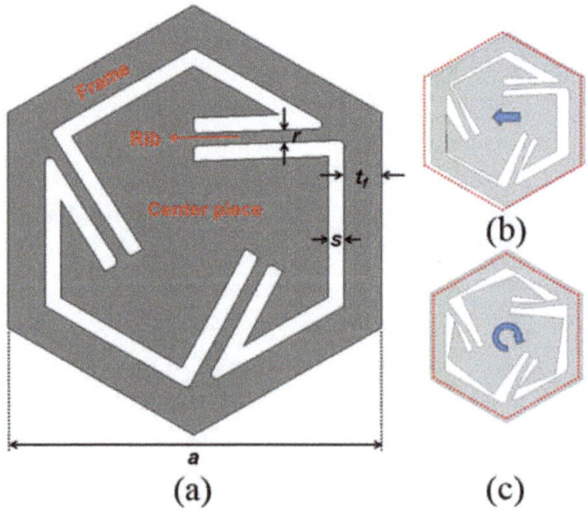

Figure 6. (a) Schematic top view of the EMM's unit cell. (b) The mode shape of the translational resonance. (c) The mode shape of the rotational resonance. Taken from Zhu *et al.* [32].

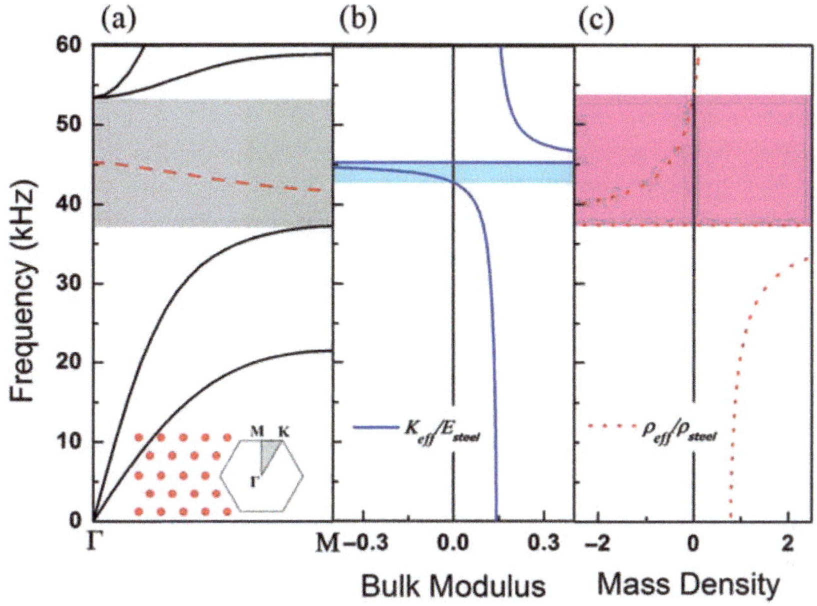

Figure 7. (a) Dispersion curves of the EMM plate along ΓM direction. (b) Effective bulk modulus of the chiral EMM. (c) Effective mass density of the chiral EMM. Taken from Zhu *et al.* [32].

Figure 7 (b) and (c), respectively. The effective mass density ρ_{eff} and bulk modulus K_{eff} were normalized with the density and Young's modulus of the single-phase medium (stainless steel), respectively. The dispersion curves of the system along the ΓM direction were also numerically obtained based on the Floquet theory, as shown in Figure 7 (a). The lattice array of the chiral EMM and its irreducible Brillouin zone (IBZ) were also inserted in Figure 7 (a).

As evidenced in Figure 7, the frequency range of the bandgap predicted from the dispersion calculation (gray area in 37.2–53.6 kHz) is almost the same as the frequency range with negative ρ_{eff} (pink area in 37.4–54.1 kHz) as shown in Figure 7 (c), which can be explained by the out-of-phase translational motion of the center piece against the frame. As shown in Figure 7 (b), K_{eff} becomes negative in the frequency range of 42.9–45.0 kHz (highlighted with blue color). The effective shear modulus μ_{eff} was calculated as 16.23 GPa in the considered frequency range with a small variation. Eventually, both effective longitudinal modulus $E_{\mathrm{eff}} = K_{\mathrm{eff}} + \mu_{\mathrm{eff}}$ and effective mass density ρ_{eff} become negative in the regime from 43.6 to 45.0 kHz, which implies a negative band for the longitudinal wave in this frequency region. It is also noticed that the double-negative frequency region (43.6–45.0 kHz) is not the same as the total negative passband frequency region (the red dashed line, 41.8–45.0 kHz) from the dispersion curves, as shown in Figure 7 (a). The difference can be attributed to the strong chirality and anisotropy of the current microstructural design, which cannot be exactly captured by using the effective isotropic medium [41]. For the chiral topology, the number of the chirally distributed ribs, *N*r, is an important parameter for the global property of the chiral EMM. The EMM becomes more isotropic when *N*r is a larger value, while displaying stronger anisotropy when *N*r is a smaller value.

3.2. *Stubbed microstructural design*

Although the perforated designs have the advantages of using single type of material and are easy to be fabricated, removing part of the pristine structure through perforation can significantly reduce the mechanical strength of the structure and, therefore, is strictly prohibited in many engineering applications such as structural health monitoring (SHM). In order to fabricate a practical plate-type EMM without sacrificing the strength of the original plate, local resonance is introduced into the PC plate with surface-bonded or stubbed microstructures.

Wu *et al.* [33] numerically and experimentally demonstrated the existence of complete bandgaps and resonances in a PC plate with a periodically stubbed surface. In their design, stubbed Al cylinders were periodically arranged in a square lattice on one side of a thin Al plate. Band structures of the unit cells with stubs of different heights were calculated along the ΓX direction to understand the influence of the stubs' height on the dispersion, as shown in Figure 8. Upon comparison with the dispersion relation of a homogeneous thin plate, as shown in Figure 8 (a), formation of the special coupling modes can be identified. For example, in Figure 8 (b), when the stub height $h_2 = 0.25\ h_1$, the purple line marked with S0+A0 indicates the mode is transformed from the S0 mode and A0 modes shown in Figure 8 (a). Similarly, the yellow line marked with A0+S0+A1 in Figure 8 (b) denotes the mode is transformed from the A0, S0, and A1 modes. The evolution of the coupling modes as well as formation of level repulsions [42] can be identified by comparing Figure 8 (a–c).

In the numerical study, the height of the stubs was increased from one quarter of the plate thickness gradually up to ten times of the thickness. As the height of stubs increased, some resonances were formed, which resulted in slower wave velocity and flatter modes

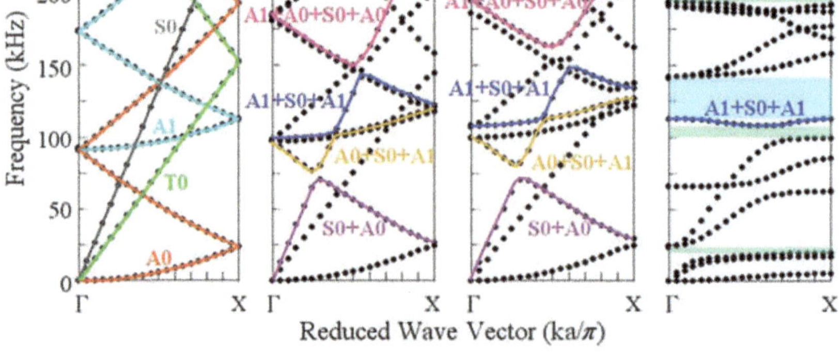

Figure 8. Dispersion relations of the thin PC plate with stubs of different heights along ΓX direction: (a) flat thin plate, (b) stub height $h_2 = 0.25\ h_1$, (c) stub height $h_2 = 0.5\ h_1$, and (d) stub height $h_2 = 10\ h_1$. Lattice constant $a = 10$ mm, plate thickness $h_1 = 1$ mm, and diameter of the cylindrical stub is 7 mm. Taken from Wu *et al.* [33].

[43] near the edge of the Brillouin zone. It can be found that as the stub height approaches about triple the plate thickness, a narrow complete bandgap ranging from 167.5 to 171 kHz forms. Moreover, when the stub height is equal to nine times of the plate thickness, the complete bandgap is the largest, with a range from 119 to 157.5 kHz and a relative bandgap width equal to 0.278. When the stub height is ten times the thickness, a secondary large complete bandgap appeared, ranging from 119 to 143 kHz (blue region), and three partial bandgaps (green region) also appeared, as shown in Figures 8 (d) and 9.

By periodically depositing two-layer composite stubs on the surface of a thin plate, Oudich *et al.* [34] opened a complete bandgap at extremely low frequency range by the LR mechanism. The schematic view of the LR PC plate with two-layer stubs is shown in Figure 10. The proposed LR PC is constructed by depositing the cylinder resonant stubs squarely onto the surface of a thin epoxy plate. The stub is a composite cylinder with a uniform radius of r and a height of $h = h_1 + h_2$, where h_1 and h_2 are the thicknesses of the silicone rubber and Pb layers, respectively. The lattice constant of the unit cell is a, and the plate has a thickness of $e = 0.05a$. Because the stiffness of the capped Pb is far greater than that of rubber, the effective stiffness of the composite stub is determined mainly by

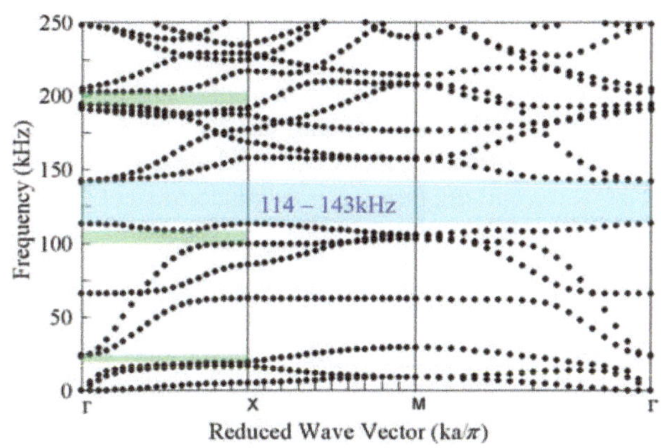

Figure 9. Dispersion relations of the thin plate with a periodic stubbed surface in a square array made of Al 6061 with $a = 10$ mm, $h_1 = 1$ mm, $h_2 = 10\ h_1$, and $d = 7$ mm. Complete bandgap and partial bandgap along the ΓX direction are shown as the blue region and green region, respectively. Taken from Wu *et al.* [33].

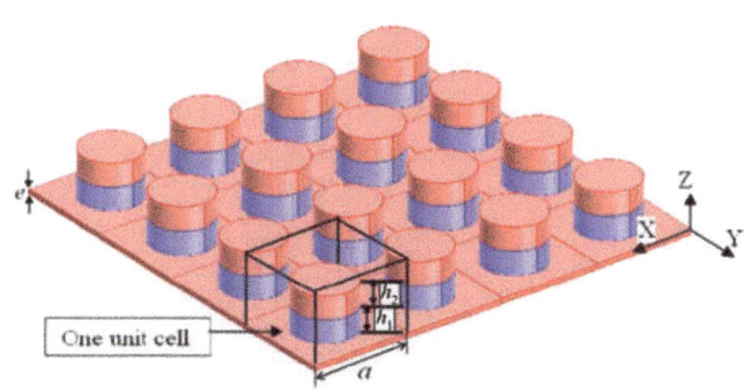

Figure 10. Schematic view of the LR PC plate with two-layer composite stubs. Taken from Oudich *et al.* [34].

the rubber layer. Comparing with the low frequency bandgap obtained in the single-layer stub LR PC plate like the one in Figure 8 (d), the bandgap in the two-layer composite stub LR PC plate can be pressed into even lower frequency region by either increasing the thickness of the Pb layer which is equivalent to increasing the mass of the stub or by increasing the thickness of the rubber layer which is equivalent to decreasing the effective elastic constant of the stub.

In Figure 11, the band structures for the composite stubs are given for fixed r and h_1 ($r = 0.3a$ and $h_1 = 0.1a$), but various values of h_2 ($h_2 = 0.1a$, $0.2a$, $0.3a$ and $0.4a$). It can be clearly found that the bandgap being pushed into lower frequency range as h_2 increases.

By changing the shape and the geometrical parameters of the stub, the location of the low frequency bandgap can also be shifted into the desired frequency range. Hsu *et al.* [35] numerically investigated a 2D LR PC plate composed of periodically arranged stepped resonant stubs. Because of the simultaneous mechanisms of the local resonances and Bragg scattering, the proposed LR PC plates exhibit low frequency forbidden bands and Bragg bandgaps.

Figure 12 illustrates the schematics of a unit cell of the LR PC plate. The unit cell is composed of stepped resonant stubs on a unit slab of thickness h. The resonant stub consists of two cylinder segments with different lengths and radii, termed as the neck and the cavity, respectively. l and r are the length and radius of the neck, respectively. L and R are those of the cavity, respectively. The lattice spacing between two nearby square lattices is a. Figure 13 shows the band structure of the unit cell described in Figure 12. The wave vectors are restricted within the x–y plane and point to the boundary ΓXMΓ of the IBZ [44]. Both the stubs and the thin plate are made of steel. Two complete bandgaps

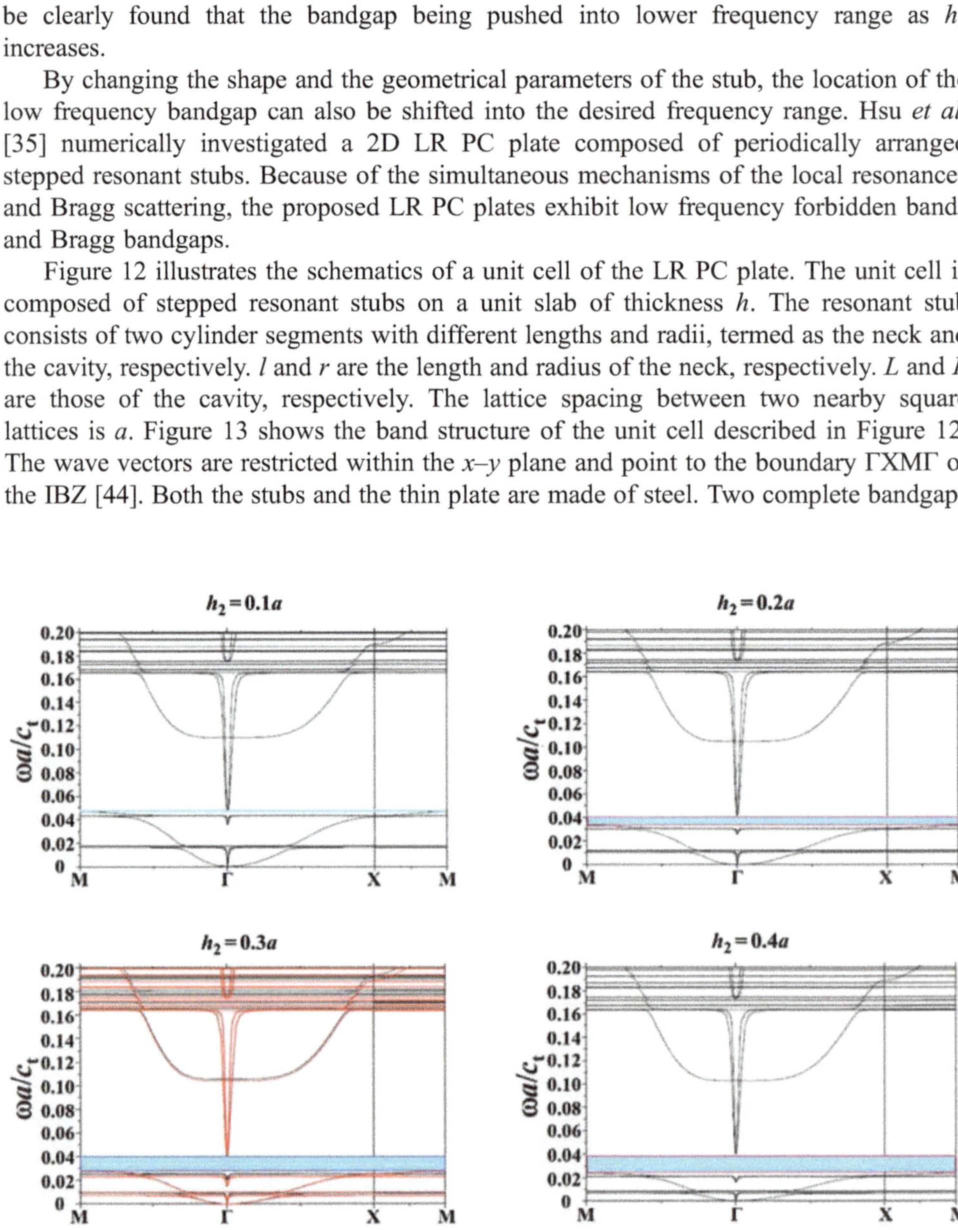

Figure 11. Band structures for LR PC plate with two-layer composite stubs. Taken from Oudich *et al.* [34].

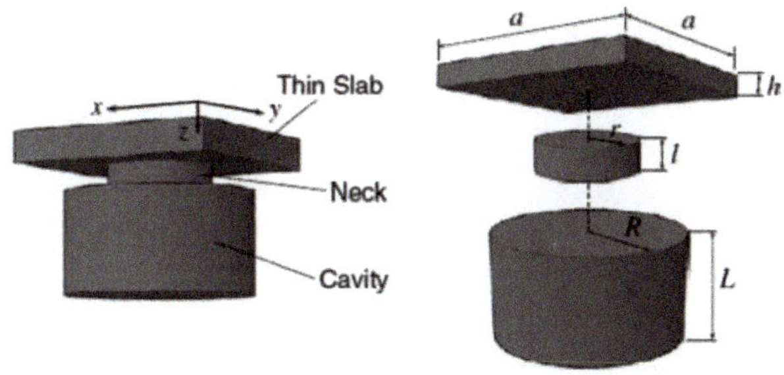

Figure 12. Schematics of the unit cell of the LR PC plate with stepped resonant stubs. Taken from Hsu *et al.* [35].

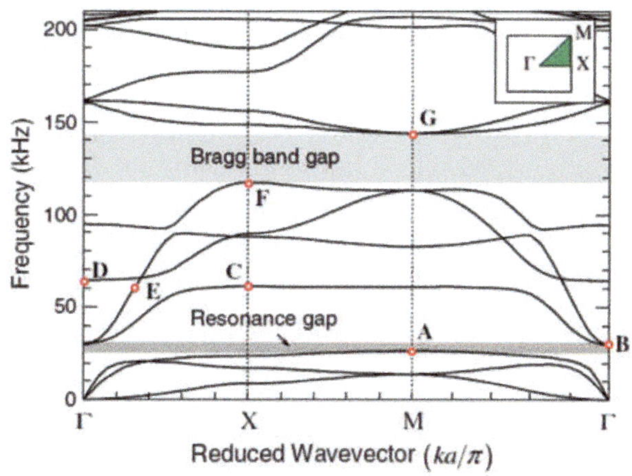

Figure 13. Band structure with two kinds of complete band gaps (LR and Bragg) of the PC plate with stepped resonant stubs. Taken from Hsu *et al.* [35].

can be found in the range from 26.4 to 30.2 kHz and range from 117.5 to 143.9 kHz. The lower gap is the local resonance gap while the higher gap is the Bragg gap. The effect of the geometrical factors on the variation of the bandgap width and frequency was also studied and both the widths and frequencies of the lower and higher bandgaps were found to be dependent on these geometrical factors [35].

3.3. *Binary/ternary insert microstructural design*

In 2000, Liu *et al.* [13] first demonstrated the extremely low frequency bandgap using an EMM with a ternary LR microstructure – soft-rubber-coated lead core embedded in epoxy matrix. The low frequency bandgap can also be made by a PC with binary LR microstructure. Wang *et al.* [16] investigated a binary 1D PC with periodic microstructure consisting of Pb and epoxy layers. A bandgap is observed at an extremely low frequency range due to the LR mechanism introduced by the large mismatch between the elastic modulus of the two media. Inspired by the 1D microstructural design, Hsu *et al.* [36] investigated the propagation of Lamb waves in 2D LR PC plates composed of periodic

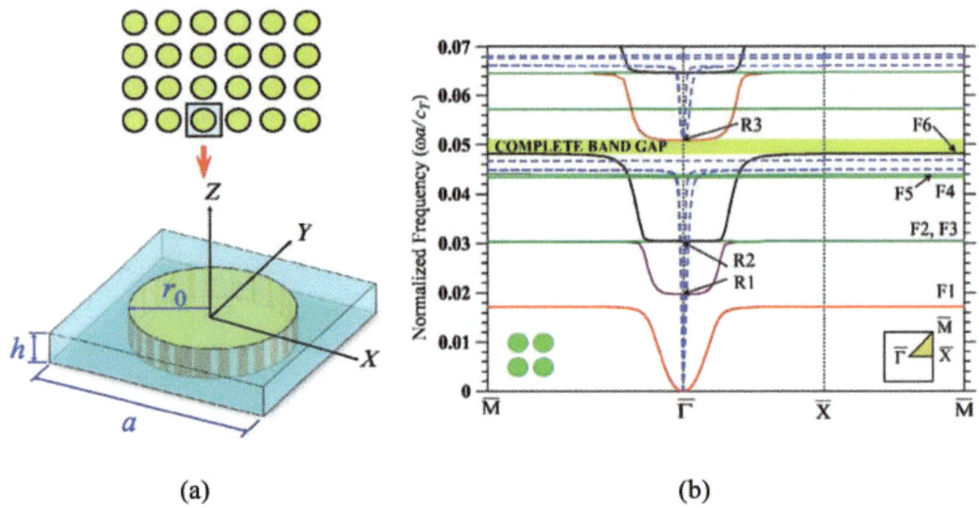

Figure 14. (a) Schematic of a unit cell of binary LR PC plate. (b) Band structure of Lamb waves in the LR PC plate with $r_0/a = 0.4$ and $h/a = 0.115$. Taken from Hsu *et al.* [36].

soft rubber fillers in epoxy host based on the plane wave expansion formulation utilizing Mindlin's theory [45].

The schematic of a unit cell of the proposed binary LR PC plate is illustrated in Figure 14 (a). The lattice constant and the radius of the circular fillers are denoted by a and r_0, respectively. The plate thickness is h. In calculations of the band structures, the density ρ of soft rubber is assumed as 1300 kg/m^3, and that of epoxy is 1180 kg/m^3. The phase velocities of longitudinal and transverse waves in soft rubber are $c_L^r = 33$ m/s and $c_T^r = 5$ m/s, respectively [46]. In epoxy, $c_L^e = 2534$ m/s and $c_T^e = 1157$ m/s, respectively. Large mismatch between the wave velocities in the two material components leads to local resonance phenomenon inside the metamaterial unit cell [47]. With a properly selected ratio between the plate thickness and the lattice constant, such as $h/a = 0.115$, a complete bandgap can be obtained in the low frequency range from 0.04812 to 0.05095 which is two orders of magnitude lower than the that resulting from Bragg scattering, as shown in Figure 14 (b). The frequencies in the figure are given in the normalized unit $\omega a/c_T$, where $c_T = 1157$ m/s. There are many flat bands in the band structure and they indicate the resonant frequencies of the localized plate mode. The flexure-dominated branches that have a larger out-of-plane component of displacement are labeled by F1–F6 and they exhibit sensitivity to the variation of the plate thickness. R1–R3 denote the flexure-dominated modes with a zero group velocity near the $\bar{\Gamma}$ point, which presents a localized pattern.

Based on the modifications of the well-known ternary EMM proposed by Liu *et al.* [13], Zhu *et al.* [37] designed EMM plates with anisotropic dynamic mass density. Figure 15 shows the unit cell of the ternary EMM plate with an elliptical coating layer. The in-plane x_1 and x_2 axes are parallel to the elliptical semimajor axis and the semiminor axis, respectively. This design introduces local stiffness anisotropy in the proposed unit cell that creates different LR frequencies $\omega_{0,1}$ and $\omega_{0,2}$ along the two in-plane principal directions of the EMM plate and, therefore, generates the anisotropic effective mass

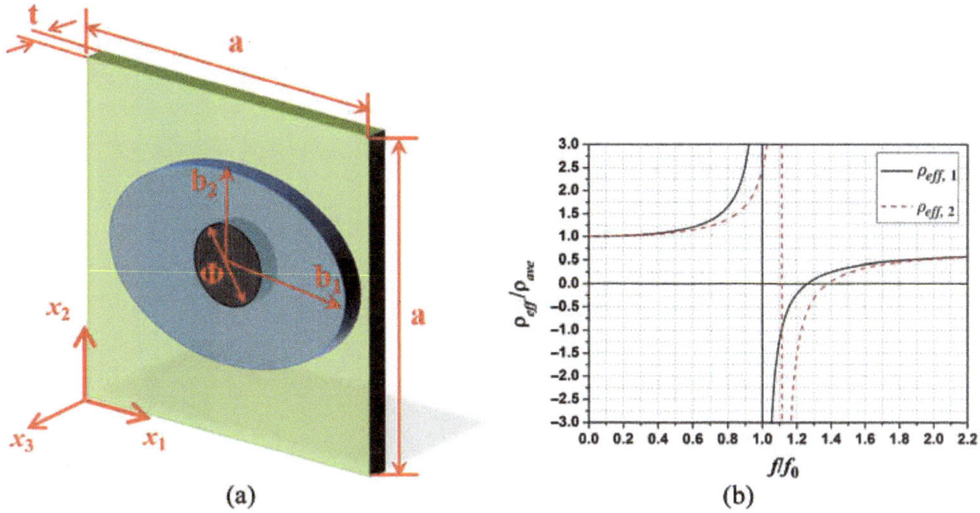

Figure 15. (a) Unit cell of the EMM plate with elliptical coating layer. (b) Normalized effective mass densities as functions of the normalized frequency along different principal directions. Taken from Zhu *et al.* [37].

density. By using a numerically based effective medium model [48], the normalized effective mass densities along x_1 and x_2 directions as functions of the normalized frequency are calculated, as shown in Figure 15 (b). In the figure, ρ_{ave} is the average static mass density for the composite and f_0 is the LR frequency of the microstructure along x_1 direction. It is noticed that $\rho_{eff,1}$ and $\rho_{eff,2}$ have different values in the normalized frequency range from $f/f_0 = 0.4$ to $f/f_0 = 1.8$. However, the anisotropic design is quite limited to the ratio of the semi-major axis to the semi-minor axis of the coating ellipse. To achieve a more strongly anisotropic effective mass density, a more modified microstructure design is needed.

Figure 16 (a) shows a new microstructure design in the elliptical coating layer with four symmetric micro-pores for greater anisotropy in mass density. The same geometrical and material parameters in Table 1 are used. The diameter of the micro-pore is $D = 2.15$ mm and the centers of the four symmetric micro-pores are located at $r = 3.13$ mm with the orientation angles being 25°, 155°, 205°, and 335°, respectively. Figure 16 (b) shows the in-plane normalized effective mass density of the EMM plate with the micro-pores as a function of the normalized frequency. Comparing with the results in Figure 15 (b), it can be observed that the lower band of the frequency range of $\rho_{eff,1}$ and $\rho_{eff,2}$ being different values keeps the same; however, the upper band of the frequency range is dramatically increased from $f/f_0 = 1.8$ to $f/f_0 = 3.0$, which shows that stronger anisotropy of the effective mass density can be achieved through the current microstructure design. The same lower band of the frequency range is expected because of the frequency normalization with respect to the average static mass density. It is also interesting to note that the effective mass density becomes isotropic when the frequency is close to the static case or much larger than the resonant frequencies. It should be mentioned that the anisotropy can be further tuned through the change of the positions, shapes, and sizes of the micro-pores in the coating layer and the inner mass.

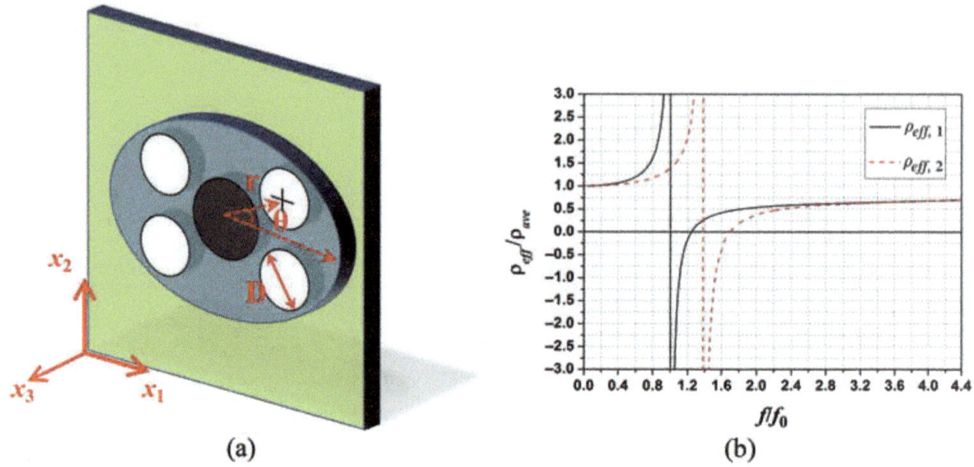

Figure 16. (a) Unit cell of the EMM plate with micro-pores in the coating layer. (b) Normalized effective mass densities as functions of the normalized frequency along different principal directions. Taken from Zhu *et al.* [37].

Table 1. Microstructure geometrical and material parameters.

Geometrical parameters		Material parameters			
a	11 mm		Matrix: aluminum	Coating: epoxy	Core: lead
b_1	4.95 mm	Mass density	2700 kg/m^3	11310 kg/m^3	11310kg/m^3
b_2	3.5 mm	Young's modulus	71GPa	0.595GPa	13GPa
Φ	3.175 mm	Poisson's ratio	0.32	0.38	0.435
t	3.175 mm				

4. Applications of locally resonant EMM plates

Designing EMM devices to control guided elastic wave propagation in finite medium with boundaries is very important for real-world engineering applications. In this section, we will review various engineering applications of LR EMM plates, such as low frequency guided wave attenuation, waveguide of Lamb waves, flat-lens focusing, subwavelength-scale negative refraction, and energy trapping.

4.1. Low frequency guide wave attenuation

By using the perforated LR microstructures in a single-phase thin steel plate as shown in Figure 3 (e), Zhu *et al.* [30] demonstrated the low frequency wave attenuation ability of the EMM plate for both in-plane and out-of-plane guided waves. Numerical investigation was conducted by considering free harmonic in-plane wave propagation. Since the wavelength of the harmonic wave generated at the excitation frequency is much larger than the size of the microstructure of the plate, the EMM plate can be regarded as an effectively homogeneous continuous medium. For a harmonic longitudinal wave propagating in the x-direction, let the global in-plane displacement at the $(j)^{\text{th}}$ unit cell of the EMM plate be expressed as:

$$u_x^{(j)} = Be^{-i\omega t}, \tag{5}$$

where ω is angular frequency and B denotes wave amplitude. Then, the wave motion of the $(j + n)^{\text{th}}$ unit cell can be written as:

$$u_x^{(j+n)} = Be^{i(nka-\omega t)}, \tag{6}$$

where n is an integer representing the periodic number of unit cells, k is the wave number, and a is the lattice constant of the unit cell. If the excitation frequency is in the bandgap, then the wave number would become complex, that is, $ka = \alpha + i\beta$, and we have

$$\left|\frac{u_x^{(j+n)}}{u_x^{(j)}}\right| = e^{-n\beta}, \tag{7}$$

From Equation (7), it can be noticed that the displacement decays spatially as an exponential function when wave frequency is within the bandgap. In order to confirm the attenuation mechanism, a numerical analysis was performed by extracting displacements of the nodes along the wave propagation direction. In the numerical simulation, the metamaterial plate consisted of 30 × 2 unit cells with cantilever-mass microstructures. The plate was free at one end and the sinusoidal displacement excitation was applied at the other end of the plate. The spatial attenuation of the EMM plate under different exciting frequencies for the x-directional in-plane extensional wave is shown in Figure 17 (a). In the figure, the normalized displacement field is defined as $U^* = |u_x|/|u_{x0}|$ with u_x being displacement responses at different locations in the plate and u_{x0} the excitation displacement. To gain a direct insight of the spatial attenuation, the entire deformation fields in the EMM plate for different excitation frequencies are schematically displayed in

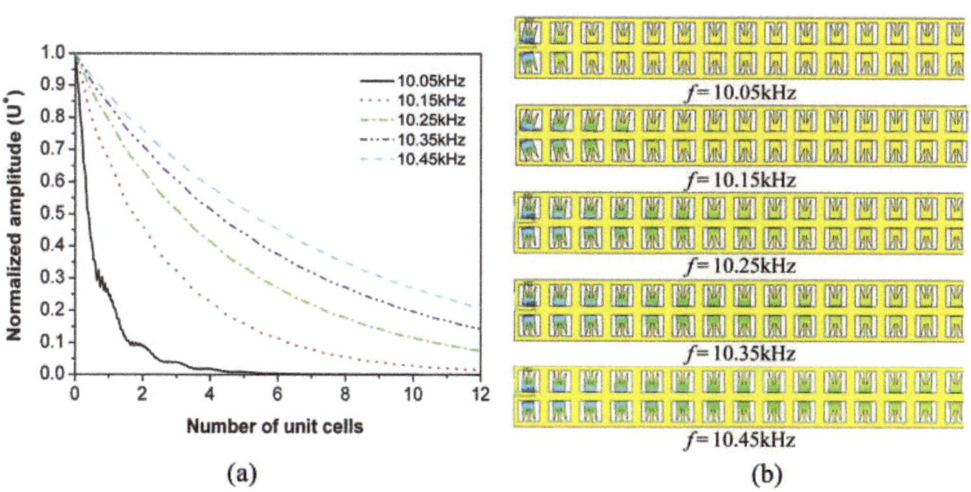

Figure 17. (a) Normalized x-directional in-plane guided wave displacements at different frequencies; (b) x-directional displacement fields in the plate at different frequencies. Taken from Zhu *et al.* [30].

Figure 17 (b). It is obvious that the displacement response exhibits significant attenuation when the frequency is close to the LR frequency (10.05 kHz for the in-plane extensional wave from numerical simulation in Figure 3 (e)). The wave amplitude is seen to decay significantly within two or three unit cells when the excitation frequency is almost equal to the resonance frequency. Similar phenomenon can be also observed for the out-of-plane guided wave propagation. From the figure, it can be observed that most of the energy is localized in the cantilever inclusion near where the load is applied. Very little energy is stored in the host plate. Since the system is purely elastic with the absence of material damping, the attenuation process within the bandgap provides an effective damping. This phenomenon has been explained by Huang *et al.* [38] using a discrete mass-spring metamaterial model.

4.2. Waveguide of Lamb waves

By using a LR stubbed microstructural design similar to those discussed in Section 3.2, Oudich *et al.* [49] constructed an LR PC plate to present an analysis of the Lamb waves in both straight and bending waveguides. First, a straight waveguide was created by removing three rows of the LR stubs in the Lamb wave propagation direction (ΓX direction). Due to the removal of the three rows of stubs, a confined mode was introduced in the complete bandgap of the original PC. In order to clearly understand what is happening in the waveguide, the total displacement fields of the LR PC in different reduced wave numbers, inside and outside of the complete bandgap, were demonstrated in Figure 18 (a) and (b), respectively. Inside the bandgap, the displacement field shows a very confined mode in the waveguide even though there is a very small penetration of waves in the first neighboring stubs of the waveguide, as shown in Figure 18 (a). When the frequency moves away from the bandgap, the confined mode in the waveguide becomes less confined and propagates throughout the LR PC plate, as shown in Figure 18 (b). It is noticed that waveguiding in a LR PC plate can be achieved for a larger waveguide width than for the traditional PC waveguide based on Bragg scattering. This result can be supported by the fact that in the LR mechanism, one encounters larger wavelengths compared to the PC lattice period.

The bent waveguide through four 90° corners was also investigated. Figure 19 (a) and (b) shows the schematic of the bent waveguide and the displacement field for the guiding mode, respectively. The width of the waveguide is still three rows. From the figure, perfect guiding of Lamb waves can be achieved by using the LR PC bent waveguide.

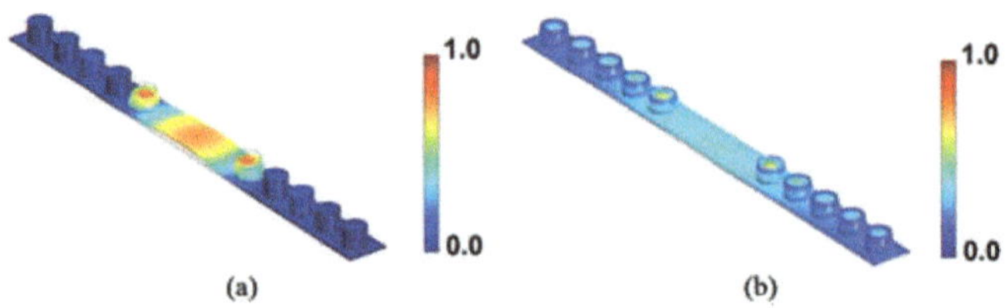

Figure 18. Absolute value of the total displacement field of the defect mode (a) inside and (b) outside the bandgap. Taken from Oudich *et al.* [49].

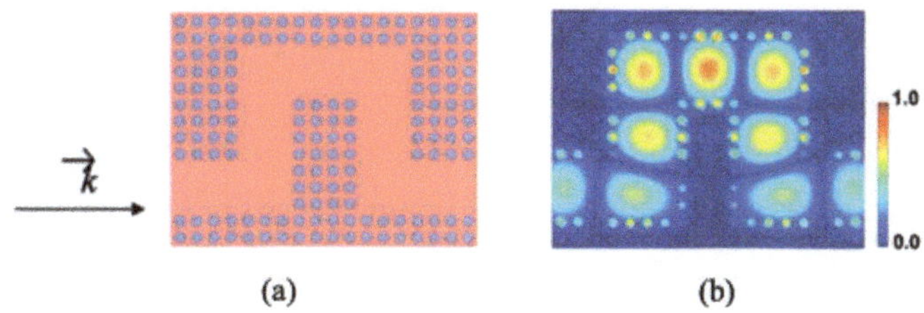

Figure 19. Schematic of the bent waveguide made in LR PC plate. (b) Absolute value of the displacement field for the guiding mode. Taken from Oudich *et al.* [49].

4.3. Flat-lens focusing

Flat-lens focusing of Lamb waves has great potential applications in the field of SHM by increasing the SNR for detecting plate damage in high resolution. Focusing of the lowest flexural (or antisymmetric) Lamb waves in the gradient-index (GRIN) PC plate has been successfully demonstrated [50,51,52]. However, the inherent Bragg scattering mechanism of PC significantly increases the size of the lens at low frequency range and, therefore, makes it unsuitable for embedded monitoring system. In addition, many proposed GRIN PC lenses are based on the perforated designs that degrade the strength of the pristine structure. In order to overcome the problems, Yan *et al.* [53] proposed a LR EMM flat lens by bonding a 2D planar array of the two-layer composite stubs (Pb disc deposited on the silicon rubber layer) for the A0 Lamb wave focusing. Different from the hyperbolic refractive index used in PC flat lenses, the proposed EMM lens for Lamb waves focusing is to vary effective mass densities that can be tailored from different EMM cells. The effective out-of-plane mass density profile is in the form of a hyperbolic secant as:

$$\rho_{\mathrm{eff}} = \rho_{\mathrm{eff},0}\mathrm{sech}^2(\alpha y), \tag{8}$$

where $\rho_{\mathrm{eff},0}$ is the effective mass density of the reference cells in an array located at $y = 0$ and α is the gradient coefficient characterizing the degree of focusing. Figure 20 (a) shows the hyperbolic secant effective mass density profile. Figure 20 (b) shows the schematic view of the proposed EMM flat lens.

A FE method was adopted to simulate the 30 kHz A0 wave mode through the EMM plate. Displacement contours in out-of-plane *z*-direction at different time steps were used to demonstrate the flat-lens focusing phenomenon, as shown in Figure 21. It can be

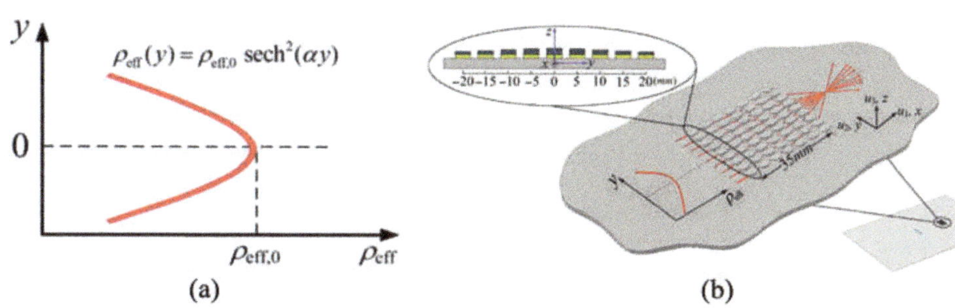

Figure 20. (a) Hyperbolic secant effective mass density profile. (b) Schematic view of the proposed EMM flat lens. Taken from Yan *et al.* [53].

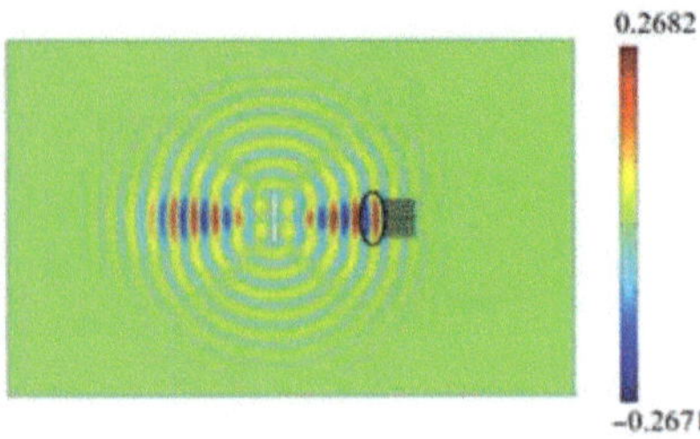

(a) Displacement contour in z direction at $t = 140$ μs

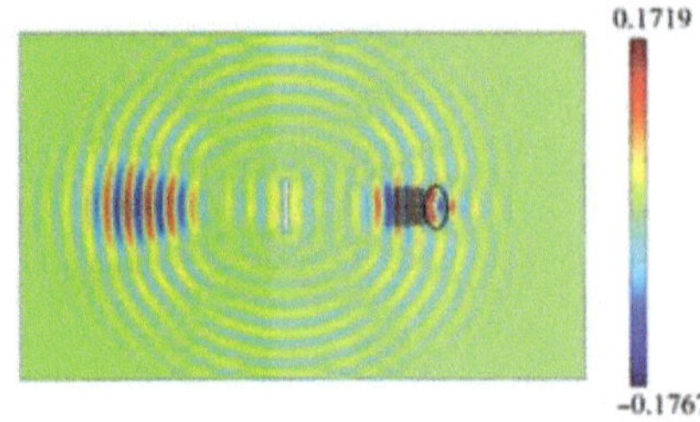

(b) Displacement contour in z direction at $t = 197$ μs

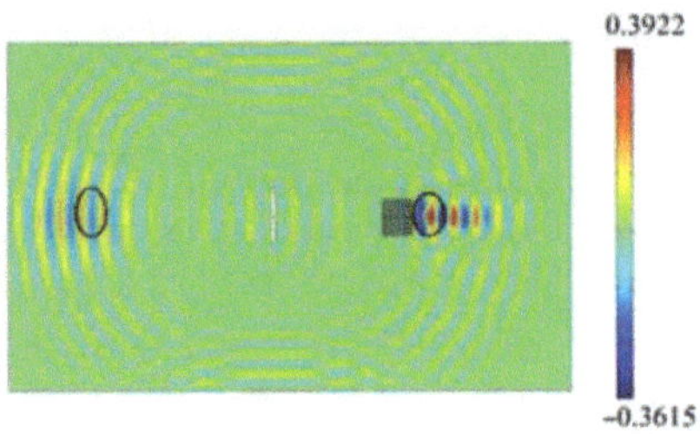

(c) Displacement contour in z direction at $t = 259$ μs

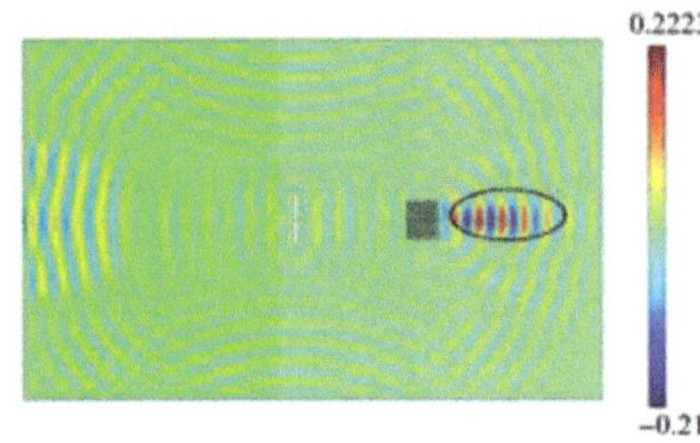

(d) Displacement contour in z direction at $t = 290$ μs

Figure 21. Snapshot of out-of-plane displacement contour (μm) for Lamb waves focusing. Taken from Yan *et al.* [53].

observed in Figure 21 (b) that Lamb waves are bent gradually toward the center axis. Figure 21 (c) shows that the waves are eventually focused to a focal spot (area) with the largest out-of-plane displacement of 0.3922 μm that is more than two times than the amplitude of the backward propagating waves without focusing.

4.4. *Subwavelength negative refraction*

The discovery of negative refractions of EM waves and mechanical waves opens exciting avenues to various applications such as super-resolution imaging to overcome diffraction limit [54,55,56,57]. Although a large amount of research work has been reported on negative refraction of EM waves and acoustic waves in fluid-like media, relatively fewer experimental demonstrations exist on negative refraction of elastic waves in solid media where the coupled longitudinal and transverse waves bring challenging but richer wave phenomena. Zhu *et al.* [32] experimentally achieved negative refraction of the longitudinal elastic waves at the deep-subwavelength-scale in the LR EMM fabricated in a stainless steel plate. The microstructural design of the proposed EMM is one of the perforated designs discussed in Section 3.1 and is shown in Figure 6 (a). With the double-negative effective material properties produced by the simultaneous translational and rotational resonances at a desired low frequency range, the negative refraction was obtained at a deep subwavelength scale, as shown in Figure 22 (a). The longitudinal elastic wave was incident from the left side of the EMM prism and a 50-cycle narrow band tone burst excitation signal with the central frequency inside the double-negative region (red dashed band in Figure 7 (a)) was used. Figure 22 (b) shows the positive refraction when the central frequency of the excitation signal was outside the double-negative region.

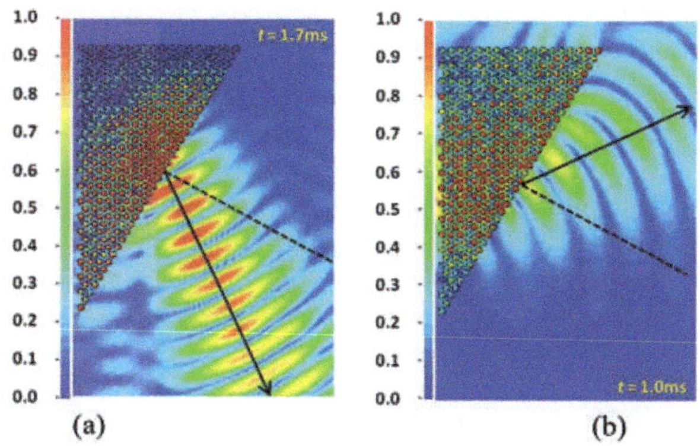

Figure 22. (a) Negative refraction of the longitudinal elastic wave at an excitation frequency inside the double-negative region. (b) Positive refraction of the longitudinal elastic wave at an excitation frequency outside the double-negative region. Taken from Zhu *et al.* [32].

Experimental validations were conducted on a prism-shaped EMM plate fabricated in a thin stainless steel (Grade 304) plate. The specimen was manufactured by using a precision laser cutting system, as shown in Figure 23 (a). Zoomed-in views of the specimen at different scale bars are demonstrated in Figure 23 (b–c). A 3D laser scanning vibrometer (Polytec PSV-400-3D) was used for the experimental measurement of the velocity field on the fabricated EMM plate. The experimentally measured velocity fields show the same negative refraction phenomena and therefore validated numerical simulation and the microstructural design.

4.5. Subwavelength energy trapping

By removing one scatterer, a defect is created inside the PC and the energy within the frequency bandgap is confined inside the defect region in the bandgap. This energy trapping phenomena through the defect mode has been recently studied in a

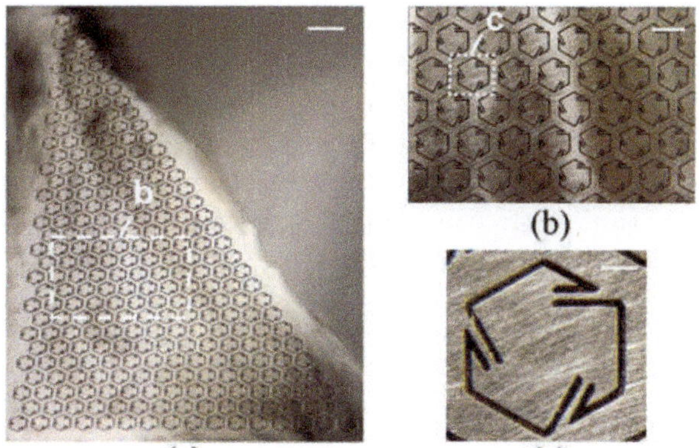

Figure 23. (a) The fabricated triangular array of the EMM. Scale bar, 20 mm. (b) Zoomed unit cells. Scale bar, 10 mm. (c) Zoomed single unit cell. Scale bar, 2 mm. Taken from Zhu *et al.* [32].

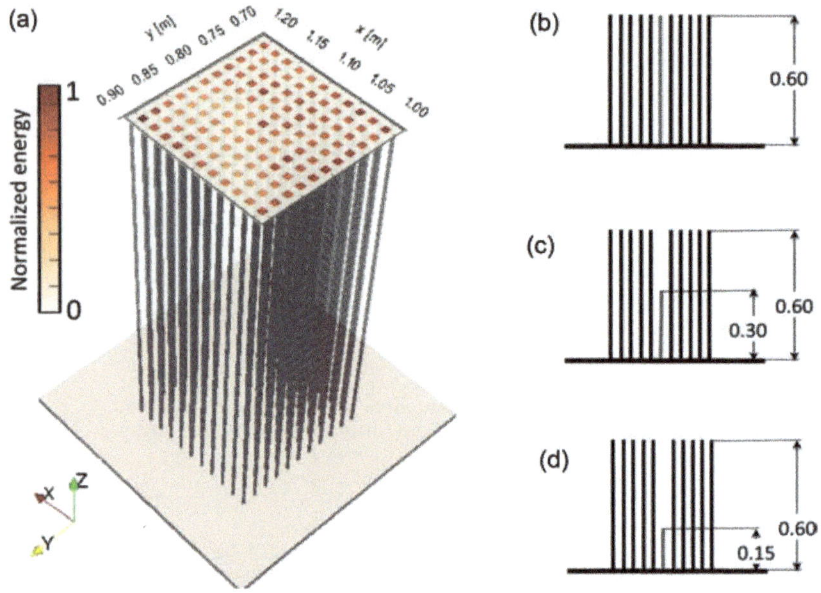

Figure 24. (a) EMM plate with 11 × 11 subwavelength resonator attached to the surface of the Al plate. The value of the energy beam is superimposed on the EMM region. (b) Slice view of the reference EMM (no defect). (c) Slice view of the defect EMM with short central beam. (d) Slice view of the defect EMM with shorter central beam. Taken from Colombi *et al.* [58].

subwavelength EMM plate by Colombi *et al.* [58]. The EMM was realized by a stubbed microstructural design by attaching a 11 × 11 vertical narrow beams (L = 60 cm long, 4 × 4 mm square section) perpendicular to a 6-mm-thick Al plate, as shown in Figure 24 (a). In order to demonstrate the energy trapping inside the EMM region, the local energy E in each resonant beam was calculated as:

$$E(x,y) = \sum_{i=1}^{3}\sum_{z_i=0}^{L}\int_{t_0}^{T} u_i^2(x,y,z_i,t)dt, \tag{9}$$

where L is the total length of the beam, (x, y) which are selected pairs on the x–y plane, z_i is the vertical component that is regularly sampled between 0 and L, u_i is the ith component of the displacement and $[t_0, T]$ is the time window of the integration on the transient response. Two configurations of the defect were analyzed by applying different lengths for the beam in the defect unit cell, as shown in Figure 24 (c–d).

The anticipated energy trapping can be found in the plot of $E(x, y)$ in Figure 25. Figure 25 (a) shows the energy is concentrated in the short central beam resonator that has the height shown in Figure 24 (c). Figure 25 (b) shows the energy concentration in the configuration shown in Figure 24 (d). In order to distinguish the energy trapping in EMM plate from that in PC plate that relies on the periodicity due to the Bragg scattering, the configuration of Figure 24 (c) was modified with a spatially random arrangement and the result energy plot is shown in Figure 25 (c). It can be found that the energy boost is similar to the ordered case, and hence this confirms that the local resonances are at the origin of the phenomenon.

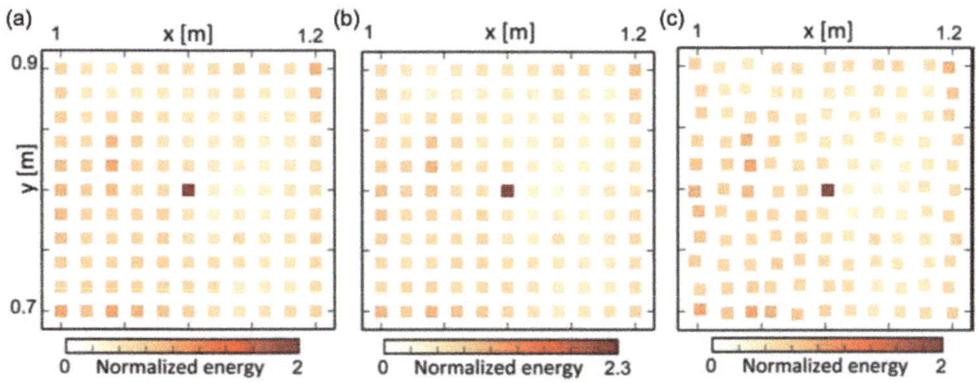

Figure 25. (a) The value of the energy $E\ (x, y)$ for the configuration in Figure 24 (c). (b) Same as (a), but for the configuration in Figure 24 (d). (c) Same as (a), but using a spatially random arrangement. Taken from Colombi *et al.* [58].

5. Conclusions

Microstructure designs of plate-type EMMs open an exciting avenue for applying metamaterial concept into real-world engineering applications. In this paper, different microstructural designs of LR-based EMM plates are reviewed. Some abnormal wave phenomenon caused by the metamaterial plates are demonstrated such as attenuation of low-frequency guided wave, subwavelength negative refraction, and energy trapping, and their potential applications are also summarized in vibration/noise isolation and SHM.

Disclosure statement

No potential conflict of interest was reported by the author(s).

Funding

This work was supported by the Air Force Office of Scientific Research [grant no. AF 9550-15-1-0061] with Program Manager Dr. Byung-Lip (Les) Lee. XN and GK are grateful to the National Natural Science Foundation of China [grants no. 11221202, 11290153, 11372035].

References

[1] J.B. Pendry, A.J. Holden, D.J. Robbins, and W.J. Stewart, *Magnetism from conductors and enhanced nonlinear phenomena*, IEEE Trans. Microwave Theory Tech. 47 (1999), pp. 2075–2084. doi:10.1109/22.798002.

[2] D.R. Smith, J.B. Pendry, and M.C.K. Wiltshire, *Metamaterials and negative refractive index*, Science 305 (2004), pp. 788–792. doi:10.1126/science.1096796.

[3] S.H. Lee, C.M. Park, Y.M. Seo, Z.G. Wang, and C.K. Kim, *Acoustic metamaterial with negative density*, Phys. Lett. A. 373 (2009), pp. 4464–4469. doi:10.1016/j.physleta.2009.10.013.

[4] N. Fang, D. Xi, J. Xu, M. Ambati, W. Srituravanich, C. Sun, and X. Zhang, *Ultrasonic metamaterials with negative modulus*, Nature Mater. 5 (2006), pp. 452–456. doi:10.1038/nmat1644.

[5] D. Torrent and J. Sánchez-Dehesa, *Anisotropic mass density by two-dimensional acoustic metamaterials*, New J. Phys. 10 (2008), pp. 023004. doi:10.1088/1367-2630/10/2/023004.

[6] L. Zigoneanu, B.I. Popa, and S.A. Cummer, *Design and measurements of a broadband two-dimensional acoustic metamaterial with anisotropic effective mass density*, J. Appl. Phys. 109 (2011), pp. 054906.
[7] C. Shen, J. Xu, N. Fang, and Y. Jing, *Anisotropic complementary acoustic metamaterial for cancelling out aberrating layers*, Phys. Rev. X. 4 (2014), pp. 041033.
[8] M. Kadic, T. Bückmann, R. Schittny, and M. Wegener, *Metamaterials beyond electromagnetism*, Rep. Prog. Phys. 76 (2013), pp. 126501. doi:10.1088/0034-4885/76/12/126501.
[9] D. Bigoni, S. Guenneau, A.B. Movchan, and M. Brun, *Elastic metamaterials with inertial locally resonant structures: Application to lensing and localization*, Phys. Rev. B. 87 (2013), pp. 174303. doi:10.1103/PhysRevB.87.174303.
[10] F. Bongard, H. Lissek, and J.R. Mosig, *Acoustic transmission line metamaterial with negative/zero/positive refractive index*, Phys. Rev. B. 82 (2010), pp. 094306. doi:10.1103/PhysRevB.82.094306.
[11] M. Farhat, S. Guenneau, and S. Enoch, *Ultrabroadband elastic cloaking in thin plates*, Phys. Rev. Lett. 103 (2009), pp. 024301. doi:10.1103/PhysRevLett.103.024301.
[12] J.H. Oh, H.M. Seung, and Y.Y. Kim, *A truly hyperbolic elastic metamaterial lens*, Appl. Phys. Lett. 104 (2014), pp. 073503. doi:10.1063/1.4865907.
[13] Z.Y. Liu, X.X. Zhang, Y. Mao, Y.Y. Zhu, Z. Yang, C.T. Chan, and P. Sheng, *Locally resonant sonic materials*, Science 289 (2000), pp. 1734–1736. doi:10.1126/science.289.5485.1734.
[14] G.W. Milton and J.R. Willis, *On modifications of Newton's second law and linear continuum elastodynamics*, Proc. Roy. Soc. A. 463 (2007), pp. 855–880. doi:10.1098/rspa.2006.1795.
[15] S.S. Yao, X.M. Zhou, and G.K. Hu, *Experimental study on negative effective mass in a 1D mass–spring system*, New J. Phys. 10 (2008), pp. 043020. doi:10.1088/1367-2630/10/4/043020.
[16] G. Wang, D. Yu, J. Wen, Y. Liu, and X. Wen, *One-dimensional phononic crystals with locally resonant structures*, Phys. Lett. A. 327 (2004), pp. 512–521. doi:10.1016/j.physleta.2004.05.047.
[17] R. Zhu, G.L. Huang, and G.K. Hu, *Effective dynamic properties and multi-resonant design of acoustic metamaterials*, ASME J. Vibr. Acoust. 134 (2012), pp. 031006. doi:10.1115/1.4005825.
[18] S.S. Yao, X.M. Zhou, and G.K. Hu, *Investigation of the negative-mass behaviors occurring below a cut-off frequency*, New J. Phys. 12 (2010), pp. 103025. doi:10.1088/1367-2630/12/10/103025.
[19] X.N. Liu, G.K. Hu, C.T. Sun, and G.L. Huang, *Wave propagation characterization and design of two-dimensional elastic chiral metacomposite*, J. Sound Vib. 330 (2011), pp. 2536–2553. doi:10.1016/j.jsv.2010.12.014.
[20] Y.Y. Chen, G.L. Huang, and C.T. Sun, *Band gap control in an active elastic metamaterial with negative capacitance piezoelectric shunting*, ASME J. Vib. Acoust. 136 (2014), pp. 061008. doi:10.1115/1.4028378.
[21] H.H. Huang and C.T. Sun, *Locally resonant acoustic metamaterials with 2D anisotropic effective mass density*, Philos. Mag. 91 (2011), pp. 981–996. doi:10.1080/14786435.2010.536174.
[22] R. Zhu, H.H. Huang, G.L. Huang, and C.T. Sun, *Microstructure continuum modeling of an elastic metamaterial*, Int. J. Eng. Sci. 49 (2011), pp. 1477–1485. doi:10.1016/j.ijengsci.2011.04.005.
[23] A.P. Liu, R. Zhu, X.N. Liu, G.K. Hu, and G.L. Huang, *Multi-displacement microstructure continuum modeling of anisotropic elastic metamaterials*, Wave Motion 49 (2012), pp. 411–426. doi:10.1016/j.wavemoti.2011.12.006.
[24] G. Milton, *The Theory of Composites*, Cambridge University Press, Cambridge, 2002.
[25] M. Kadic, T. Bückmann, N. Stenger, M. Thiel, and M. Wegener, *On the practicability of pentamode mechanical metamaterials*, Appl. Phys. Lett. 100 (2012), pp. 191901. doi:10.1063/1.4709436.
[26] M. Kadic, T. Bückmann, R. Schittny, P. Gumbsch, and M. Wegener, *Pentamode metamaterials with independently tailored bulk modulus and mass density*, Phys. Rev. Appl. 2 (2014), pp. 054007.
[27] T. Bückmann, R. Schittny, M. Thiel, M. Kadic, G.W. Milton, and M. Wegener, *On three-dimensional dilational elastic metamaterials*, New J. Phys. 16 (2014), pp. 033032. doi:10.1088/1367-2630/16/3/033032.

[28] J.O. Vasseur, P.A. Deymier, B. Djafari-Rouhani, Y. Pennec, and A.-C. Hladky-Hennion, *Absolute forbidden bands and waveguiding in two-dimensional phononic crystal plates*, Phys. Rev. B. 77 (2008), pp. 085415. doi:10.1103/PhysRevB.77.085415.
[29] F. Hsiao, A. Khelif, H. Moubchir, A. Choujaa, C. Chen, and V. Laude, *Waveguiding inside the complete band gap of a phononic crystal slab*, Phys. Rev. E. 76 (2007), pp. 056601. doi:10.1103/PhysRevE.76.056601.
[30] R. Zhu, G.L. Huang, H.H. Huang, and C.T. Sun, *Experimental and numerical study of guided wave propagation in a thin metamaterial plate*, Phys. Lett. A. 375 (2011), pp. 2863–2867. doi:10.1016/j.physleta.2011.06.006.
[31] C.-Y. Sun, J.-C. Hsu, and T.-T. Wu, *Resonant slow modes in phononic crystal plates with periodic membranes*, Appl. Phys. Lett. 97 (2010), pp. 031902. doi:10.1063/1.3464955.
[32] R. Zhu, X.N. Liu, G.K. Hu, C.T. Sun, and G.L. Huang, *Negative refraction of elastic waves at the deep subwavelength scale in a single-phase metamaterial*, Nature Commun. 5 (2014), pp. 5510. doi:10.1038/ncomms6510.
[33] T.-T. Wu, Z.-G. Huang, T.-C. Tsai, and T.-C. Wu, *Evidence of complete band gap and resonances in a plate with periodic stubbed surface*, Appl. Phys. Lett. 93 (2008), pp. 111902. doi:10.1063/1.2970992.
[34] M. Oudich, Y. Li, B.M. Assouar, and Z.L. Hou, *A sonic band gap based on the locally resonant phononic plates with stubs*, New J. Phys. 12 (2010), pp. 083049. doi:10.1088/1367-2630/12/8/083049.
[35] J.-C. Hsu, *Local resonances-induced low-frequency band gaps in two-dimensional phononic crystal slabs with periodic stepped resonators*, J. Phys. D: Appl. Phys. 44 (2011), pp. 055401. doi:10.1088/0022-3727/44/5/055401.
[36] J.-C. Hsu and T.-T. Wu, *Lamb waves in binary locally resonant phononic plates with two-dimensional lattices*, Appl. Phys. Lett. 90 (2007), pp. 201904. doi:10.1063/1.2739369.
[37] R. Zhu, X.N. Liu, G.L. Huang, H.H. Huang, and C.T. Sun, *Microstructural design and experimental validation of elastic metamaterial plates with anisotropic mass density*, Phys. Rev. B. 86 (2012), pp. 144307. doi:10.1103/PhysRevB.86.144307.
[38] H.H. Huang, C.T. Sun, and G.L. Huang, *On the negative effective mass density in acoustic metamaterials*, Int. J. Eng. Sci. 47 (2009), pp. 610–617. doi:10.1016/j.ijengsci.2008.12.007.
[39] R. Marchal, O. Boyko, B. Bonello, J. Zhao, L. Belliard, M. Oudich, Y. Pennec, and B. Djafari-Rouhani, *Dynamics of confined cavity modes in a phononic crystal slab investigated by in situ time-resolved experiments*, Phys. Rev. B. 86 (2012), pp. 224302. doi:10.1103/PhysRevB.86.224302.
[40] A. Khelif, B. Aoubiza, S. Mohammadi, A. Adibi, and V. Laude, *Complete band gaps in two-dimensional phononic crystal slabs*, Phys. Rev. E. 74 (2006), pp. 046610. doi:10.1103/PhysRevE.74.046610.
[41] X.N. Liu, G.L. Huang, and G.K. Hu, *Chiral effect in plane isotropic micropolar elasticity and its application to chiral lattices*, J. Mech. Phys. Solids. 60 (2012), pp. 1907–1921. doi:10.1016/j.jmps.2012.06.008.
[42] T.T. Wu and Z.G. Huang, *Level repulsions of bulk acoustic waves in composite materials*, Phys. Rev. B. 70 (2004), pp. 214304.
[43] G. Wang, X. Wen, J. Wen, L. Shao, and Y. Liu, *Two-dimensional locally resonant phononic crystals with binary structures*, Phys. Rev. Lett. 93 (2004), pp. 154302. doi:10.1103/PhysRevLett.93.154302.
[44] M.S. Kushwaha, P. Halevi, L. Dobrzynski, and B. Djafari-Rouhani, *Acoustic band structure of periodic elastic composites*, Phys. Rev. Lett. 71 (1993), pp. 2022–2025. doi:10.1103/PhysRevLett.71.2022.
[45] J.-C. Hsu and T.-T. Wu, *Efficient formulation for band-structure calculations of two-dimensional phononic-crystal plates*, Phys. Rev. B. 74 (2006), pp. 144303. doi:10.1103/PhysRevB.74.144303.
[46] C. Goffaux and J. Sánchez-Dehesa, *Two-dimensional phononic crystals studied using a variational method: Application to lattices of locally resonant materials*, Phys. Rev. B. 67 (2003), pp. 144301. doi:10.1103/PhysRevB.67.144301.
[47] L. Fok, M. Ambati, and X. Zhang, *Acoustic metamaterials*, MRS Bull. 33 (2008), pp. 931–934. doi:10.1557/mrs2008.202.

[48] X.N. Liu, G.K. Hu, G.L. Huang, and C.T. Sun, *An elastic metamaterial with simultaneously negative mass density and bulk modulus*, Appl. Phys. Lett. 98 (2011), pp. 251907. doi:10.1063/1.3597651.

[49] M. Oudich, M.B. Assouar, and Z.L. Hou, *Propagation of acoustic waves and waveguiding in a two-dimensional locally resonant phononic crystal plate*, Appl. Phys. Lett. 97 (2010), pp. 193503. doi:10.1063/1.3513218.

[50] T.-T. Wu, Y.-T. Chen, J.-H. Sun, S.C.S. Lin, and T.J. Huang, *Focusing of the lowest antisymmetric Lamb wave in a gradient-index phononic crystal plate*, Appl. Phys. Lett. 98 (2011), pp. 171911. doi:10.1063/1.3583660.

[51] J.F. Zhao, R. Marchal, B. Bonello, and O. Boyko, *Efficient focalization of antisymmetric Lamb waves in gradient-index phononic crystal plates*, Appl. Phys. Lett. 101 (2012), pp. 261905. doi:10.1063/1.4773369.

[52] S.-C. Lin, T.J. Huang, J.-H. Sun, and T.-T. Wu, *Gradient-index phononic crystals*, Phys. Rev. B. 79 (2009), pp. 094302. doi:10.1103/PhysRevB.79.094302.

[53] X. Yan, R. Zhu, G.L. Huang, and F.-G. Yuan, *Focusing guided waves using surface bonded elastic metamaterials*, Appl. Phys. Lett. 103 (2013), pp. 121901. doi:10.1063/1.4821258.

[54] J.B. Pendry, *Negative refraction makes a perfect lens*, Phys. Rev. Lett. 85 (2000), pp. 3966–3969. doi:10.1103/PhysRevLett.85.3966.

[55] R.A. Shelby, D.R. Smith, and S. Schultz, *Experimental verification of a negative index of refraction*, Science 292 (2001), pp. 77–79. doi:10.1126/science.1058847.

[56] J.-F. Robillard, J. Bucay, P.A. Deymier, A. Shelke, K. Muralidharan, B. Merheb, J.O. Vasseur, A. Sukhovich, and J.H. Page, *Resolution limit of a phononic crystal superlens*, Phys. Rev. B. 83 (2011), pp. 224301. doi:10.1103/PhysRevB.83.224301.

[57] S. Zhang, L.L. Yin, and N. Fang, *Focusing ultrasound with an acoustic metamaterial network*, Phys. Rev. Lett. 102 (2009), pp. 194301. doi:10.1103/PhysRevLett.102.194301.

[58] A. Colombi, P. Roux, and M. Rupin, *Sub-wavelength energy trapping of elastic waves in a metamaterial*, J. Acoust. Soc. Am. 136 (2014), pp. EL192–EL198. doi:10.1121/1.4890942.

5

Recent progress on the magnetorheological plastomers

Shouhu Xuan, Yangguang Xu, Taixiang Liu and Xinglong Gong*

CAS Key Laboratory of Mechanical Behavior and Design of Materials, Department of Modern Mechanics, University of Science and Technology of China, Hefei 230026, P. R. China

Different from the traditional magnetorheological (MR) fluids and elastomers, the magnetic particles in the plastic MR materials are not 'deadly' trapped in the polymer matrix; thus, the MR plastomers exhibit higher MR effects and lower sedimentation. The plastic MR materials have attracted increasing attention, and the relevant fundamental mechanisms and practical applications have been intensively studied due to their unique physical and mechanical properties. In this highlight, we have mainly reviewed the preparation and the rheological properties of the MR plastomers. The formation mechanism of the MR plastomers has also been briefly summarized.

Keywords: magnetorheological; plastomers; MR effects

1. Introduction

As an engineering-applicable smart material, the magnetorheological (MR) material is usually prepared by dispersing magnetic particles into a carrying matrix with high viscosity. Applying under an external magnetic field, the dispersing particles are magnetized and the carrying matrix is strengthened by the as-induced magnetic dipole–dipole interactions (MR effects). These functionalized smart materials are expected to possess both high stability and MR effect to meet the requirements of practical applications. In consideration of the mismatch between the heavy magnetic particles and carrying matrix, an optimum host medium is desirable for developing the high-performance MR materials. The MR fluids (MRFs) have been developed as the first kind of MR materials [1–12] and these have been applied in a wide range of fields such as vibration absorbers, mass dampers, brake, sensors, actuators, base isolator, and mount [13–16]. Unfortunately, the sedimentation of the MRFs is believed to be the most awful point, which limits their practical applications. In the past decades, MR elastomers (MREs) have been developed to defeat the worse stability of the MRFs [17–30]. The entrapment of the magnetic particles within elastic polymer matrixes gives rise to stable mechanical properties of the MREs. Much effort has been made to widen the applications of the MREs [31–35], and several structural models [36–42] have been proposed to investigate the MR mechanism. However, the MREs still face many challenges in the practical applications due to the weak MR effects induced by the hard nature of the polymer matrix.

*Corresponding author. Email: gongxl@ustc.edu.cn

Recently, a novel plastic MR material consisted of magnetic particles and soft polymer matrixes has been developed [43–47]. For this type of material, the magnetic particles are highly mobile due to the low cross-linking density of the carrying matrix and the aggregated structures can thus be changed by applying the external magnetic field. Therefore, it exhibits much higher MR effects than the MREs. Similar to the plasticine, this kind of MR material named as MR plastomer (MRP) has the plastic characteristic that enables it to be molded into various shapes [48]. The carrying matrix is critical for the MRP. In the case of highly cross-linking matrix, the magnetic particles were "deadly" anchored in the elastomer (MRE); thus, the high starting shear modulus and low magneto-induced modulus lead to the weak MR effects [49]. If the magnetic particles are highly mobile within the cross-linking polymer matrix, the obtained elastic MR material would exhibit strong MR effects [50]. The MRP is resulted with the further decrease in cross-linking density of polymer matrix. In this case, the magnetic particles could form chain-like structures under the magnetic field, resulting in a very high storage modulus of the material. In comparison with the traditional MR materials, the MRPs are multifunctional. The thermosensitive MRPs can be stored in solid state while working as MRFs under the fluid model [51]. These novel features of the MRPs will find promising applications in vibration dissipation, damping, artificial muscles, drug delivery, valves, and actuators [52–55].

In this highlight, after a brief introduction of the origination of the plastic MR materials and a short summary of the synthetic methods of MRP_S, we will demonstrate how the mechanical properties of MRPs are dependent on the detailed structure of polymer matrix and how the strong MR effects are generated by the strong chain-like structures. Finally, we will give discussion on the mechanism of the magnetic field-induced improvement of mechanical properties and the potential application of MRPs in various fields.

2. Magnetorheological plastomers

Since Shiga et al. reported the silicone gels containing dispersed magnetic particles in 1995 [56], the plastic MR materials started to emerge in the scope of relevant research areas. According to the characteristic of the carrying matrix, the plastic MR materials can be classified into three catalogues, i.e., hydrogels, swollen polymer gels, and pure polymer gels (MRP). The magnetic hydrogels have always been prepared by using water-swollen polymer gels as the dispersing phase. Due to the high viscosity of the polymer matrix, the MR hydrogels exhibit higher stability than the MRFs. Similarly, if the polymers are swollen by high boiling point solvent, the as-obtained MR gels (MRGs) are belonged to the swollen polymer gels. Moreover, there are also many kinds of low cross-linking polymers which possess fluidic characteristics. After incorporating the magnetic particles into the above polymer matrixes, the plastic products are obtained and they are defined as the pure polymer gel-based MR materials (MRP).

At the early stage, various water-swollen polymer gels such as the carrageenan [45], gelatin [57], and guar gum [58] are introduced to the MRF to produce the jellylike MR materials. By use of the gel network as support, the settlement problem of the MRFs has been solved. However, a sharp decrease in MR effect is usually observed in these materials because the network can be destroyed by reversibly applying the magnetic field [59]. Although this giant negative MR effect can be conquered by optimizing the experimental parameters, the unexpected degradation of the carbonyl iron (CI) particles is emerged in water [50]. Recently, a kind of novel hydrogel cross-linked by magnetic particles has been reported [57,60,61]. These magnetic particles are permanently linked to the polymer network; thus, the stability of the hydrogels is substantially improved. The magnetic particles can assemble along the supermolecular hydrogelators to form hybrid nanofibers [57]. Interestingly, as soon as the magnetic field is applied, the viscosity of the magnetic gel is sharply decreased. This work provides a new method to control the mechanical properties of soft materials.

Solvent-swollen polymer matrix has also been considered as an important candidate to improve the MR performances [62–67]. By adjusting the ratio of high boiling point diluents to reactants in the mixing polyurethane-silicone polymer gels, the MR materials can be altered from a liquid state to a solid state [62]. The silicone-based MRGs have a lower viscosity and a higher yield shear stress while polyurethane (PU)-based MRGs exhibit a better sedimentation stability [63]. Moreover, the high mechanical performances are also successfully achieved in the solvent-swollen cross-linked copolyimide (CCPI) [64] and superamolecular polymer systems [65]. Recently, An et al. have reported a new type of physically cross-linked MRGs in which the triblock copolymer polystyrene-block-poly(ethylene-stat-butadiene)-block-polystyrene (SEBS) swollen in paraffin oil was employed as the carrying matrix [46,68–70]. Due to the highly assembled magnetic particles, the as-formed MRGs exhibit a relative MR effect as high as 6000% [46]. Similarly, the MRG has also been prepared by introducing high boiling point solvents to the high cross-linked PU matrix. The dynamic and static yield shear stress is controllable by adjusting the swollen ratio of the PU matrix [66]. Very recently, it has been found that when the swollen solvent exceeds to a critical point (~25 wt%), the final MR gel becomes unstable and the sedimentation appears under a relatively long time (3 months) [67]. Therefore, the swollen ratio of the PU matrix should be controlled within a safe range.

The polymer gels with a low cross-linking density often exhibit a fluidic characteristic. By using such kind of polymers as the carrying matrix, the MR plastomers can be prepared. The first high-performance MRP was prepared by using the low cross-linking PU as the matrix, where the PU matrix was synthesized using toluene diisocyanate (TDI) and poly(propylene glycol) (PPG-1000) as the starting materials (Figure 1(a) and (b)) [48]. The mechanical properties of the PU matrix can be adjusted by varying the molar ratio of hard segment (TDI) to soft segment (PPG-1000); thus, both MRPs and MREs are available in this system. Applying under a magnetic field to the isotropic MRPs (Figure 1(c)), the magnetic forces have driven the CI particles defeat the polymer obstacle to form chain structures, thereby producing the anisotropic MRPs (Figure 1(d)). The mechanical properties of the MR materials are crucially dependent on the inner CI particles assembled structures. The *in situ* observation of the structure evolution opens up a possibility to study the origination of the MR effects. Different from the other two kinds of plastic MR materials, the MRPs have several unique characteristics. First, the matrix of solid-like magnetic gels is not elastic but plastic (Figure 1(a)). Second, the polymer barriers can be easily overcame by the magnetic field-induced torques due to the low cross-linking density of the matrix. Furthermore, a strong MR effect can be achieved since the strong CI chains can be readily formed in the MRP. The last point, the low molecular weight polymer chains can be tangled together and are easily untangled under small disturbance; thus, the self-healing phenomenon is usually observed with the help of magnetic field-induced dipole–dipole interactions [71].

Later, we reported the magnetic plasticine, in which the CI particles were used to substitute the traditional calcium salts dispersed within the paraffin wax/petroleum jelly mixing matrix [51]. Because of the low molecular weight of the paraffin wax and gel-like

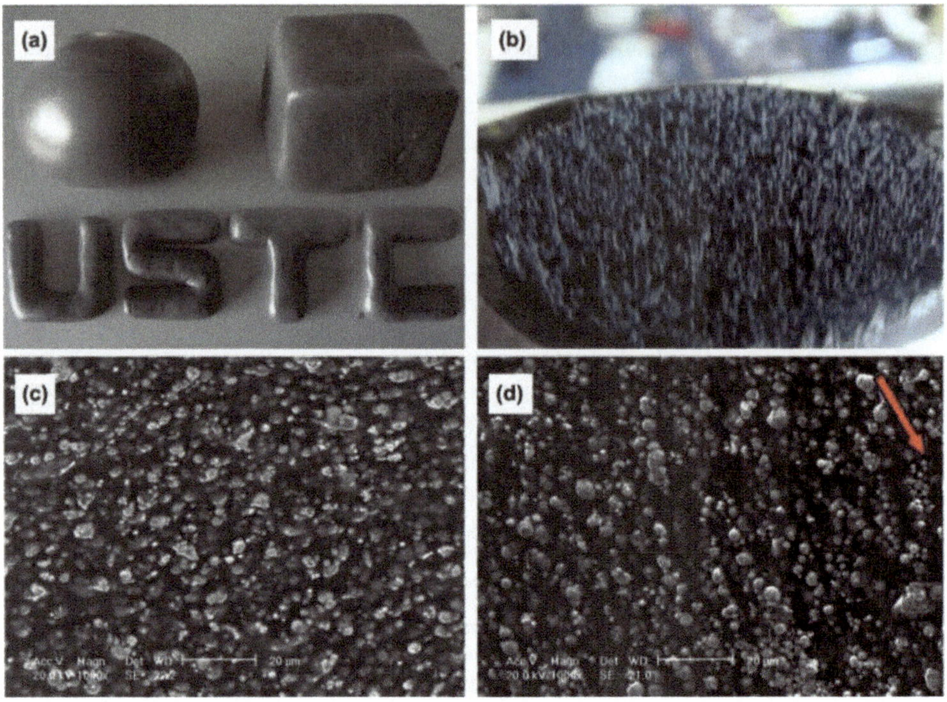

Figure 1. (a) The photograph of MRP with various shapes and (b) the image of the MRP under the magnetic field. The SEM images of (c) isotropic MRP and (d) anisotropic MRP [48]. The scale bar on the SEM images is 20 μm and the red arrow presents the direction of the applied magnetic field.

structure of the petroleum jelly, the as-prepared product possesses all the advantages of plasticine such as the easily shape-molding and good self-healing ability. Similar to the above MRP, the magnetic plasticine exhibits a high MR effect of 430% when the CI content is only 60 wt%. Since the paraffin wax is a phase-changing matrix, this product can be transformed into the MRF when the environmental temperature is higher than 50° C. This feature makes them to be applied reversibly between MRP and MRF by controlling the working temperature. A similar phase-changing MR material based on the thermoresponsive triblock copolymer (Pluronics F127) has also reported [72]. Due to the thermally driven 'liquid to solid' transition nature of the carrying matrix, such product behaves like a high-performance MRF when the suspending medium is in the 'liquid' phase and then performs as the MRE when the matrix is in the 'solid' phase. Clearly, this special kind of thermal sensitive characteristic not only reduces the sentimental defects of the traditional MRFs but also supplies a feasible way for multifunctional MR materials. Actually, the previously reported grease-based MR materials also exhibits such a thermal responsive MR behavior [73,74]. Although the physical state of the MR grease has not been demonstrated in the manuscript, it was estimated to be a kind of MRP with a small starting storage modulus, because the authors have claimed that this product had a strong solid-like structure rather than a fluid-like structure under external magnetic field.

Multifunctional MRP is also achievable by varying the carrying matrix. Shear hardening is a very interesting phenomenon in polymer composite. The storage modulus of the shear hardening polymer increases sharply by increasing the applied shear stress or impact. Introducing the CI particles to the shear hardening polymer makes the novel

MRP possess both high MR effects and excellent shear hardening property [75]. Interestingly, the shock absorbance property of the functional MRP can be tuned by varying the external magnetic field owing to the MR effects. The analysis indicates that the 'cross bonds' between the polymer chains and particle–particle forces are attributed to the multifunctional stimulus-responsive properties.

Rheological tests indicate the MRP is different from the MRE or MRF but inherits a part of their advantages. Applying under the magnetic field, the mechanical properties of the polymer matrixes enhanced due to the magnetic dipole–dipole forces. Because the magnetic particles are mobile, they continuously aggregate to form the particle chains until an equilibrium state is reached. The relatively strong chain structure and soft polymer matrix cause the MRPs to have a higher MR effect than the MREs. In comparison with the anisotropic one, the isotropic MRP has a smaller starting storage modulus induced by the randomly dispersed magnetic particles. Different from the MRGs and MRF, the Bingham viscoplastic model is not suitable to describe the mechanical behaviors of the MRPs. Owing to their soft solid-like characteristic, oscillatory shear rheometry has been proven to be a powerful tool to study their mechanical properties [60,65,66,74–77]. In the linear viscoelastic range, the highest magneto-induced storage modulus of the PU-based MRP reaches as high as 6.54 MPa, corresponding to a relative MR effect of 532% (Figure 2(a)). Under applying a constant magnetic field, a noticeable time-dependent increment of the storage modulus is observed for the isotropic MRPs, indicating the

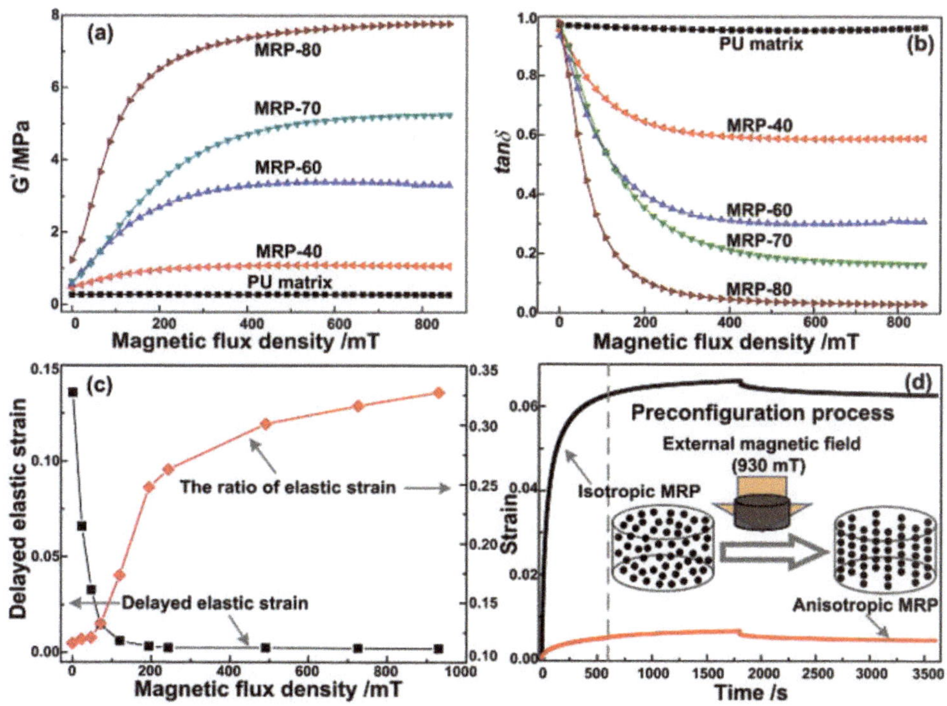

Figure 2. (a) The storage modulus and (b) loss factor of anisotropic MRP under different magnetic fields (The MRP-x denotes the weight percentage of the CI particles) [48]; (c) delayed elastic strain and the ratio of elastic strain to the total strain of the anisotropic MRP under different magnetic fields and (d) the creep and recovery curves of isotropic and anisotropic MRP under a 930 mT magnetic field [81]. (The creep time is 1800 s and the constant stress is set as 2000 Pa.)

structural reorganization processes in these plastic materials [71]. With increase in the magnetic field, the movement of the soft polymer segment decreases as the restriction effect of the CI chains is enhanced, in turn leading to a decrease in the damping factors (Figure 2(b)). As the interfacial slipping between the CI particles and the matrix in the isotropic MRP is larger than the anisotropic MRP, the former exhibits a higher damping factor [48]. Additionally, an obvious magnetostriction effect is observed by monitoring the normal force of MRPs along with the direction of the magnetic field. Therefore, the particle arrangement process and magnetic field strength also generate a great effect on the normal force of MRPs [78,79]. Moreover, the influence of stepwise magnetic field on the storage modulus of MRPs has also been systematically studied. These results have demonstrated that the MR effect and magnetic response velocity are highly dependent on the aggregation structure of the magnetic particles [45,50,67,80].

The MRP exhibits unique nonlinear dynamic properties by loading harmonic strain. The critical points between the linear and nonlinear areas are both found in strain amplitude and frequency dependency testing. Interestingly, although the magnetic field strongly affects the dynamic properties of the MRP, it almost has no influence on the strain- and frequency-induced nonlinearity [71]. By increasing the magnetic field from 0 to 930 mT, the ratio of delayed elastic strain to the total creep strain increases from 11.6% to 32.7% (Figure 2(c)) because of the increment of the restriction effects to the slipping of soft chain segments. Clearly, the MRP transforms from a viscoplastic to a elastoplastic state with increasing magnetic field. In addition, only a small decrease in the modulus is observed in the PU-based MRP with increasing temperature, which is attributed to the low-temperature sensitivity of the PU matrix. The creep strain decreases with increasing temperature under applying a 930 mT magnetic field, while they exhibit a reversible characteristic in the absence of the magnetic field. Although this difference is originated from the magnetic interactions among the CI particles and the coupling effects between the CI particular structures and the polymer matrix, the detailed mechanism of this interesting aspect is still unknown and it is deserved for further investigations. The particle distribution also has a significant influence on the creep behavior of MRP (Figure 2(d)). It is found that the creep strain of isotropic MRP increases more sharply than that of anisotropic MRP at the initial 10 minutes, which reflects that the aggregation process of magnetic particle starts from random dispersion to structured orientation [81].

Besides the normal forces, the squeeze flow behavior of the MRP has also been systematically investigated [82,83]. The elastic deformation region, stress relaxation region, and plastic flow region are presented in the compression and tension testing. It is reported that compressive yield stress and the tensile yield stress are sensitive to the magnetic field, particle concentration, and particle distribution. Because the viscosity of polymer matrix is higher than the silicon oil, the yield stress of the MRP is larger than the relative MRF under the squeeze mode [83]. Moreover, the MRP is a typical soft magnetosensitive polymer composite, so it has a large magneto-deforming effect. The quasi-statical shearing testing measurements indicate that the MRP exhibits a high magneto-damping performance, which can also be described by the mageto-enhanced Bingham fluid-like model [84].

The MRP also possesses the typical structural dependent conductivity [85]. As the CI particles are conductive, the electric resistance of the MRP sharply changes once the CI particular structure bridges the two test points. Applying under the magnetic field, both the thickness of the interface layer of electrode–MRP and the gaps between the adjacent particles decrease simultaneously. To this end, the conductivity of the MRP increases since the electron transfer in the particulate structures is improved (Figure 3). Clearly, this

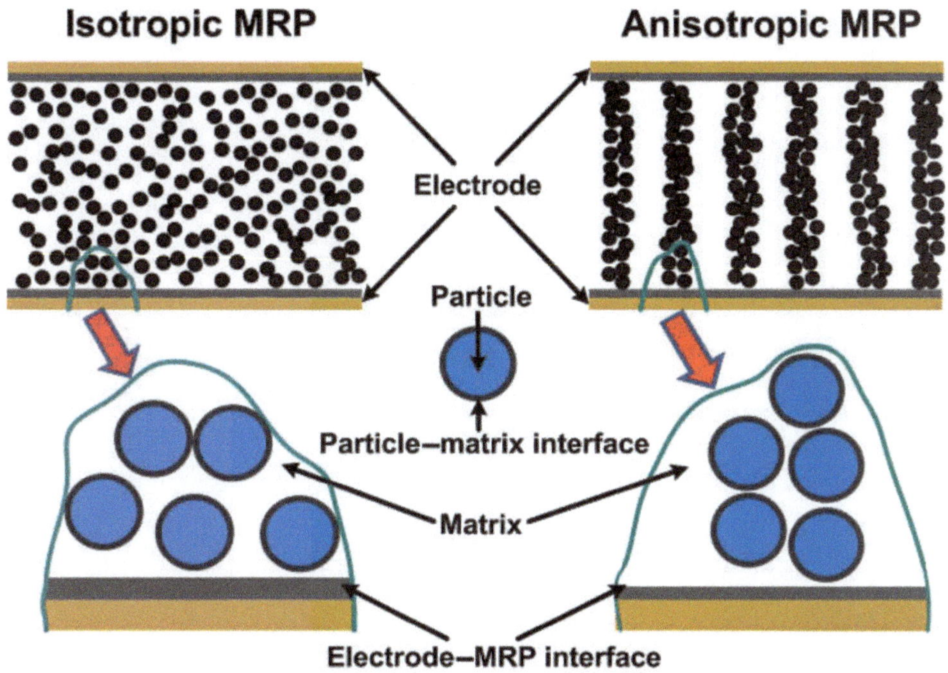

Figure 3. Schematic illustration of the particulate structure of MRP with different particle distribution [85].

work provides a simple method to quantitatively characterize the anisotropy of the MRP, which will help us to understand how the magnetic field affects the interactions between the neighboring CI particles. However, the magnetically dependent conductivity of the MRP has not been carefully explored and more work need to be done to fully understand the exact MR mechanism. Since the MRPs are free-standing and their inner structure is sensitive to the external stimulus, they are expected to be applicable in a number of fields including soft actuator, artificial muscle, and sensors [53–55].

3. Mechanism and simulation

Many interesting/novel properties of the smart MRPs have been reported in recent experimental works. However, the underlying mechanisms of the properties are still mysterious, even though some qualitative mechanisms have already been proposed. Stokesian dynamics theory [86] and computer simulations [87–92] are effective for studying the microstructural formation and evolution processes of MRFs. Therefore, these theories may be useful for describing the unique MR mechanisms of the MRPs due to the similarity of particle movement between the MRPs and MRFs.

With the rapid development of computer science, a large amount of computer-simulation works have been conducted to reveal the magneto-controllable properties of plastic MR materials. Various analytic methods such as the continuous medium theory [93–95], field theory [96], microscopic model [97,98], and Monte Carlo simulations [99] have been developed to describe the magneto-viscoelastic property of ferrogels. Inspired by the aforementioned works, the first simulation work for MRP has already been performed

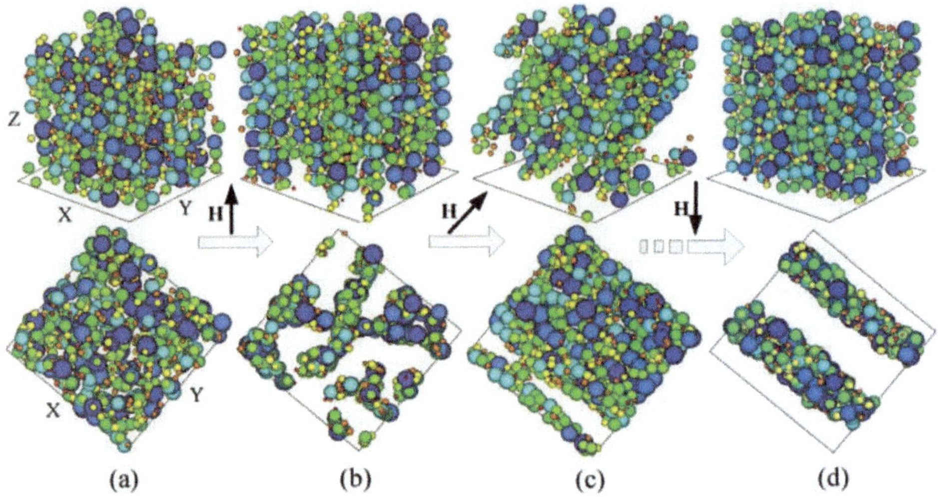

Figure 4. The magneto-induced particulate microstructure of (a) isotropic MRP; (b) anisotropic MRP obtained by applying the magnetic field on the isotropic MRP; (c) and (d) anisotropic MRPs obtained by further rotating the magnetic field. (The upper series shows the axonometric views and the lower series shows the top views. The solid arrows denote the directions of external magnetic field. The particles with the same color indicate the same size) [100].

to demonstrate the magneto-induced rearrangeable characteristics [100]. The size distribution of the CI particles was characterized by a lognormal formula, and a modified dipolar force model was preferably introduced to describe the magnetic interaction of the nearby CI particles. The matrix was treated as a high-viscosity Bingham fluid with a certain yield stress. As an example of the evolution of magneto-induced particulate microstructure in MRPs, Figure 4 illustrates that the initial randomly distributed microstructure (Figure 4(a)) transfers to chain-like/column-like microstructure (Figure 4(b)) after applying an external magnetic field. With the evolution of microstructure, the macroscopic physical or mechanical properties of MRPs vary correspondingly. The chain-like/column-like microstructures can be well remained after removing the external magnetic field, and new microstructures can be formed via the rearrangement of themselves by changing the direction of the magnetic field. Figure 4(c–d) gives an illustration of simulated, magneto-induced rearrangeable microstructure of MRPs from a quasi-steady state to another quasi-steady state in a stepwise rotating magnetic field. Interestingly, the CI particles assemble to planar microstructure in the rotating plane of external magnetic field.

Both the particle–particle model and particle–matrix model are essential for building the entire simulation model. In most of the previous numeric-simulation works, the CI particles were treated as magnetic point dipoles. This treatment leads to some errors in the basic particle model because the CI particles are always aggregated to form large structures in the polymer matrix. Therefore, in the further simulations, proper modification should be conducted on the basic particle model. To this end, an alternative finite-dipole model [101] has been used to solve this problem. Specifically, the magnetization of a particle is represented as a distribution of current density and this method has been proven to be effective for analyzing the dynamic properties of the MR materials. Nevertheless, the current matrix model is mostly based on the model of Newtonian fluid or Bingham fluid, and the motion of magnetic particles in the matrix is treated as Stokesian drag.

These simple models are no longer suitable to deal with the computation of the dynamics properties of MRPs, so these should be improved in future work.

4. Conclusions and perspectives

Plastic MR material has been proven to be a kind of fascinating smart material which possesses both strong MR effects and high stability. The development of novel MRP systems and the optimization of their mechanical properties are considered to be crucial in relevant research areas. In this highlight, we have mainly reviewed the development of the plastic MR materials. The polymer matrix is believed to be the most important parameter for modulating the mechanical properties of the MRPs. Multifunctional MRPs with the combination of MR effects, temperature sensitivity, conductivity sensitivity, and shear thicking/thinning properties certainly deserve future research. The detailed mechanism of the CI aggregation structure transformation, the force transfer during applying the external stimulus, and the formation of the MR effects have remained unclear. Future analysis of the theory model and computer simulation on the structural dependent mechanical properties of the MRP materials will be of high interest.

Due to the unique magnetically dependent mechanical properties, the MRP may find many promising applications in a wealth of areas. Our ongoing project demonstrates that the MRP is effective for the high-performance MR absorber, which should be very useful for intelligent vibration dissipation. The multifunctional MRP and the PU sponge-reinforced MRP can be designed to fabricate the future-generation body armor [75,102]. Furthermore, the magnetic field-dependent conductivity of the MRP also enables them be applied in the field of smart devices such as sensors and on-off switches [103].

Disclosure statement

No potential conflict of interest was reported by the authors.

Funding

This work was supported by the Collaborative Innovation Center of Suzhou Nano Science and Technology. Financial support from the National Natural Science Foundation of China [grant number 11372301], [grant number 11125210]; the Anhui Provincial Natural Science Foundation of China [grant number 1408085QA17];and the National Basic Research Program of China (973 Program) [grant number 2012CB937500] is gratefully acknowledged.

References

[1] E. Lemaire, G. Bossis, and Y. Grasselli, *Yield stress and structuration of magnetorheological suspensions*, J. Magn. Magn. Mater. 122 (1993), pp. 51–52. doi:10.1016/0304-8853(93)91037-8.

[2] W.H. Li, G. Chen, and S.H. Yeo, *Viscoelastic properties of MR fluids*, Smart. Mater. Struct. 8 (1999), pp. 460–468. doi:10.1088/0964-1726/8/4/303.

[3] W.H. Li, H. Du, G. Chen, and S.H. Yeo, *Viscoelastic properties of MR fluids under oscillatory shear*, Smart Struct. Mater. 2001 Damping Isolation. 4331 (2001), pp. 333–342.

[4] J. De Vicente, F. Gonzalez-Caballero, G. Bossis, and O. Volkova, *Normal force study in concentrated carbonyl iron magnetorheological suspensions*, J. Rheol. 46 (2002), pp. 1295–1303.

[5] J.S. Choi, B.J. Park, M.S. Cho, and H.J. Choi, *Preparation and magnetorheological characteristics of polymer coated carbonyl iron suspensions*, J. Magn. Magn. Mater. 304 (2006), pp. E374–E376. doi:10.1016/j.jmmm.2006.02.055.

[6] B. Hu, A. Fuchs, S. Huseyin, F. Gordaninejad, and C. Evrensel, *Atom transfer radical polymerized MR fluids*, Polymer. 47 (2006), pp. 7653–7663. doi:10.1016/j.polymer.2006.08.069.

[7] M.T. Lopez-Lopez, A. Zugaldia, F. Gonzalez-Caballero, and J.D.G. Duran, *Sedimentation and redispersion phenomena in iron-based magnetorheological fluids*, J. Rheol. 50 (2006), pp. 543–560.

[8] F.F. Fang, H.J. Choi, and M.S. Jhon, *Magnetorheology of soft magnetic carbonyl iron suspension with single-walled carbon nanotube additive and its yield stress scaling function*, Colloid Surf. A-Physicochem. Eng. Asp. 351 (2009), pp. 46–51. doi:10.1016/j.colsurfa.2009.09.032.

[9] M.T. Lopez-Lopez, P. Kuzhir, J.D.G. Duran, and G. Bossis, *Normal stresses in a shear flow of magnetorheological suspensions: Viscoelastic versus Maxwell stresses*, J. Rheol. 54 (2010), pp. 1119–1136.

[10] B.J. Park, F.F. Fang, and H.J. Choi, *Magnetorheology: Materials and application*, Soft Matter. 6 (2010), pp. 5246–5253.

[11] J. De Vicente, D.J. Klingenberg, and R. Hidalgo-Alvarez, *Magnetorheological fluids: A review*, Soft Matter. 7 (2011), pp. 3701–3710.

[12] F.F. Fang, Y.D. Liu, H.J. Choi, and Y. Seo, *Core–Shell structured carbonyl iron microspheres prepared via dual-step functionality coatings and their magnetorheological response*, ACS Appl. Mater. Inter. 3 (2011), pp. 3487–3495. doi:10.1021/am200714p.

[13] M.R. Jolly, J.W. Bender, and J.D. Carlson, *Properties and applications of commercial magnetorheological fluids*, J. Intel. Mat. Syst. Str. 10 (1999), pp. 5–13. doi:10.1177/1045389X9901000102.

[14] D.J. Klingenberg, *Magnetorheology: Applications and challenges*, Aiche. J. 47 (2001), pp. 246–249. doi:10.1002/(ISSN)1547-5905.

[15] A.G. Olabi and A. Grunwald, *Design and application of magneto-rheological fluid*, Mater. Design. 28 (2007), pp. 2658–2664. doi:10.1016/j.matdes.2006.10.009.

[16] A. Farjoud, M. Craft, W. Burke, and M. Ahmadian, *Experimental investigation of MR squeeze mounts*, J. Intel. Mat. Syst. Str. 22 (2011), pp. 1645–1652. doi:10.1177/1045389X11411225.

[17] J.M. Ginder, M.E. Nichols, L.D. Elie, and J.L. Tardiff, *Magnetorheological elastomers: Properties and applications*, Smart Struct. Materials 1999 Smart Mat. Technol.. 3675 (1999), pp. 131–138.

[18] C. Bellan and G. Bossis, *Field dependence of viscoelastic properties of MR elastomers*, Int. J. Mod. Phys. B. 16 (2002), pp. 2447–2453. doi:10.1142/S0217979202012499.

[19] Z. Varga, G. Filipcsei, and M. Zrínyi, *Magnetic field sensitive functional elastomers with tuneable elastic modulus*, Polymer. 47 (2006), pp. 227–233. doi:10.1016/j.polymer.2005.10.139.

[20] A. Boczkowska, S.F. Awietjan, and R. Wroblewski, *Microstructure–property relationships of urethane magnetorheological elastomers*, Smart Mater. Struct. 16 (2007), pp. 1924–1930. doi:10.1088/0964-1726/16/5/049.

[21] A. Boczkowska and S.F. Awietjan, *Smart composites of urethane elastomers with carbonyl iron*, J. Mater. Sci. 44 (2009), pp. 4104–4111. doi:10.1007/s10853-009-3592-7.

[22] J.K. Wu, X.G. Gong, L. Chen, H.S. Xia, and Z.G. Hu, Preparation and characterization of isotropic polyurethane magnetorheological elastomer through in situ polymerization, J. Appl. Polym. Sci. 114 (2009), pp. 901–910. doi:10.1002/app.v114:2.

[23] A. Fuchs, J. Sutrisno, F. Gordaninejad, M.B. Caglar, and L. Yanming, *Surface polymerization of iron particles for magnetorheological elastomers*, J. Appl. Polym. Sci. 117 (2010), pp. 934–942. doi:10.1002/app.v117:2.

[24] W.H. Li and X.Z. Zhang, *A study of the magnetorheological effect of bimodal particle based magnetorheological elastomers*, Smart Mater. Struct. 19 (2010), pp. 035002.

[25] M. Yu and S. Wang, *The composite MRE embedded with a copper coil*, Smart Mater. Struct. 19 (2010), pp. 065023.

[26] J. Kaleta, M. Krolewicz, and D. Lewandowski, *Magnetomechanical properties of anisotropic and isotropic magnetorheological composites with thermoplastic elastomer matrices*, Smart Mater. Struct. 20 (2011), pp. 085006.

[27] W.H. Li and M. Nakano, *Fabrication and characterization of PDMS based magnetorheological elastomers*, Smart Mater. Struct. 22 (2013), pp. 055035.

[28] A.M. Figueiredo Neto, M.H. Godinho, T. Toth-Katona, and P. Palffy-Muhoray, *Optical, magnetic and dielectric properties of non-liquid crystalline elastomers doped with magnetic colloids*, Braz. J. Phys. 35 (2005), pp. 184–189. doi:10.1590/S0103-97332005000100016.

[29] C. Sena, M.H. Godinho, and A.M. Figueiredo Neto, *Optical characterization of urethane/urea elastomers undoped and doped with magnetic nanoparticles subjected to a uniaxial strain*, J. Appl. Phys. 102 (2007), pp. 073507.

[30] C. Sena, M.H. Godinho, P.J. Sebastião, D. Sousa, and A.M. Figueiredo Neto, *Free-standing urethane/urea elastomer films undoped and doped with ferro-nano-particles*, Eur. J. Phys. E. 34 (2011), pp. 9. doi:10.1140/epje/i2011-11009-8.

[31] J.D. Carlson and M.R. Jolly, *MR fluid, foam and elastomer devices*, Mechatronics. 10 (2000), pp. 555–569. doi:10.1016/S0957-4158(99)00064-1.

[32] H.-X. Deng, X.-L. Gong, and L.-H. Wang, *Development of an adaptive tuned vibration absorber with magnetorheological elastomer*, Smart Mater. Struct. 15 (2006), pp. N111–N116. doi:10.1088/0964-1726/15/5/N02.

[33] Y.Q. Ni, Z.G. Ying, and Z.H. Chen, *Micro-vibration suppression of equipment supported on a floor incorporating magneto-rheological elastomer core*, J. Sound Vib. 330 (2011), pp. 4369–4383. doi:10.1016/j.jsv.2011.04.020.

[34] X.Z. Zhang and W.H. Li, *Adaptive tuned dynamic vibration absorbers working with MR elastomers*, Smart Struct. Syst. 5 (2009), pp. 517–529. doi:10.12989/sss.2009.5.5.517.

[35] W.H. Li, X.Y. Wang, X.Z. Zhang, and Y. Zhou, *Development and analysis of a variable stiffness damper using an MR bladder*, Smart Mater. Struct. 18 (2009), pp. 074007.

[36] L.C. Davis, *Model of magnetorheological elastomers*, J. Appl. Phys. 85 (1999), pp. 3348–3351.

[37] I.A. Brigadnov and A. Dorfmann, *Mathematical modeling of magneto-sensitive elastomers*, Int. J. Solids. Struct. 40 (2003), pp. 4659–4674. doi:10.1016/S0020-7683(03)00265-8.

[38] A. Dorfmann and R.W. Ogden, *Magnetoelastic modelling of elastomers*, Eur. J. Mech. A-Solid. 22 (2003), pp. 497–507. doi:10.1016/S0997-7538(03)00067-6.

[39] H.M. Yin, L.Z. Sun, and J.S. Chen, *Magneto-elastic modeling of composites containing chain-structured magnetostrictive particles*, J. Mech. Phys. Solids. 54 (2006), pp. 975–1003. doi:10.1016/j.jmps.2005.11.007.

[40] P.P. Castaneda and E. Galipeau, *Homogenization-based constitutive models for magnetorheological elastomers at finite strain*, J. Mech. Phys. Solids. 59 (2011), pp. 194–215. doi:10.1016/j.jmps.2010.11.004.

[41] E. Galipeau and P.P. Castaneda, *A finite-strain constitutive model for magnetorheological elastomers: Magnetic torques and fiber rotations*, J. Mech. Phys. Solids. 61 (2013), pp. 1065–1090. doi:10.1016/j.jmps.2012.11.007.

[42] S. Rudykh and K. Bertoldi, *Stability of anisotropic magnetorheological elastomers in finite deformations: A micromechanical approach*, J. Mech. Phys. Solids. 61 (2013), pp. 949–967. doi:10.1016/j.jmps.2012.12.008.

[43] M. Zrinyi, *Intelligent polymer gels controlled by magnetic fields*, Colloid Polym. Sci. 278 (2000), pp. 98–103. doi:10.1007/s003960050017.

[44] G. Filipcsei, I. Csetneki, A. Szilagyi, and M. Zrinyi, *Magnetic field-responsive smart polymer composites*, Adv. Polym. Sci. 206 (2007), pp. 137–189.

[45] T. Mitsumata and N. Abe, *Magnetic-field sensitive gels with wide modulation of dynamic modulus*, Chem. Lett. 38 (2009), pp. 922–923. doi:10.1246/cl.2009.922.

[46] H.N. An, S.J. Picken, and E. Mendes, *Enhanced hardening of soft self-assembled copolymer gels under homogeneous magnetic fields*, Soft Matter. 6 (2010), pp. 4497–4503.

[47] J.K. Wu, X.L. Gong, Y.C. Fan, and H.S. Xia, *Physically crosslinked poly(vinyl alcohol) hydrogels with magnetic field controlled modulus*, Soft Matter. 7 (2011), pp. 6205–6212.

[48] Y.G. Xu, X.L. Gong, S.H. Xuan, W. Zhang, and Y.C. Fan, *A high-performance magnetorheological material: Preparation, characterization and magnetic-mechanic coupling properties*, Soft Matter. 7 (2011), pp. 5246–5254.

[49] B. Wei, X.L. Gong, and W.Q. Jiang, *Influence of polyurethane properties on mechanical performances of magnetorheological elastomers*, J. Appl. Polym. Sci. 116 (2010), pp. 771–778.

[50] T. Mitsumata and S. Ohori, *Magnetic polyurethane elastomers with wide range modulation of elasticity*, Polym. Chem. 2 (2011), pp. 1063–1067.
[51] S.H. Xuan, Y.L. Zhang, Y.F. Zhou, W.Q. Jiang, and X.L. Gong, *Magnetic plasticine (TM): A versatile magnetorheological material*, J. Mater. Chem. 22 (2012), pp. 13395–13400.
[52] T.-Y. Liu, S.-H. Hu, T.-Y. Liu, D.-M. Liu, and S.-Y. Chen, *Magnetic-sensitive behavior of intelligent ferrogels for controlled release of drug*, Langmuir. 22 (2006), pp. 5974–5978. doi:10.1021/la060371e.
[53] R. Fuhrer, E.K. Athanassiou, N.A. Luechinger, and W.J. Stark, *Crosslinking metal nanoparticles into the polymer backbone of hydrogels enables preparation of soft, magnetic field-driven actuators with muscle-like flexibility*, Small. 5 (2009), pp. 383–388. doi:10.1002/smll.v5:3.
[54] V.Q. Nguyen, A.S. Ahmed, and R.V. Ramanujan, *Morphing soft magnetic composites*, Adv. Mater. 24 (2012), pp. 4041–4054. doi:10.1002/adma.201104994.
[55] Y.H. Li, G.Y. Huang, X.H. Zhang, B.Q. Li, Y.M. Chen, T.L. Lu, T.J. Lu, and F. Xu, *Magnetic hydrogels and their potential biomedical applications*, Adv. Funct. Mater. 23 (2013), pp. 660–672. doi:10.1002/adfm.v23.6.
[56] T. Shiga, A. Okada, and T. Kurauchi, *Magnetoviscoelastic behavior of composite gels*, J. Appl. Polym. Sci. 58 (1995), pp. 787–792. doi:10.1002/app.1995.070580411.
[57] Z.M. Yang, H.W. Gu, J. Du, J.H. Gao, B. Zhang, X.X. Zhang, and B. Xu, *Self-assembled hybrid nanofibers confer a magnetorheological supramolecular hydrogel*, Tetrahedron. 63 (2007), pp. 7349–7357. doi:10.1016/j.tet.2007.02.009.
[58] A. Tiraferri, K.L. Chen, R. Sethi, and M. Elimelech, *Reduced aggregation and sedimentation of zero-valent iron nanoparticles in the presence of guar gum*, J. Colloid. Interf. Sci. 324 (2008), pp. 71–79. doi:10.1016/j.jcis.2008.04.064.
[59] T. Mitsumata, T. Wakabayashi, and T. Okazaki, *Particle dispersibility and giant reduction in dynamic modulus of magnetic gels containing barium ferrite and iron oxide particles*, J. Phys. Chem. B. 112 (2008), pp. 14132–14139. doi:10.1021/jp805955j.
[60] Y. Pan, Y. Gao, J.F. Shi, L. Wang, and B. Xu, *A versatile supramolecular hydrogel of nitrilotriacetic acid (NTA) for binding metal ions and magnetorheological response*, J. Mater. Chem. 21 (2011), pp. 6804–6806.
[61] P. Ilg, *Stimuli-responsive hydrogels cross-linked by magnetic nanoparticles*, Soft Matter. 9 (2013), pp. 3465–3468.
[62] M.J. Wilson, A. Fuchs, and F. Gordaninejad, *Development and characterization of magnetorheological polymer gels*, J. Appl. Polym. Sci. 84 (2002), pp. 2733–2742. doi:10.1002/(ISSN)1097-4628.
[63] A. Fuchs, M. Xin, F. Gordaninejad, X.J. Wang, G.H. Hitchcock, H. Gecol, C. Evrensel, and G. Korol, *Development and characterization of hydrocarbon polyol polyurethane and silicone magnetorheological polymeric gels*, J. Appl. Polym. Sci. 92 (2004), pp. 1176–1182. doi:10.1002/(ISSN)1097-4628.
[64] A. Fuchs, B. Hu, F. Gordaninejad, and C. Evrensel, *Synthesis and characterization of magnetorheological polyimide gels*, J. Appl. Polym. Sci. 98 (2005), pp. 2402–2413. doi:10.1002/(ISSN)1097-4628.
[65] B. Hu, A. Fuchs, S. Huseyin, F. Gordaninejad, and C. Evrensel, *Supramolecular magnetorheological polymer gels*, J. Appl. Polym. Sci. 100 (2006), pp. 2464–2479. doi:10.1002/(ISSN)1097-4628.
[66] B. Wei, X.L. Gong, W.Q. Jiang, L.J. Qin, and Y.C. Fan, *Study on the properties of magnetorheological gel based on polyurethane*, J. Appl. Polym. Sci. 118 (2010), pp. 2765–2771. doi:10.1002/app.32688.
[67] Y.G. Xu, X.L. Gong, and S.H. Xuan, *Soft magnetorheological polymer gels with controllable rheological properties*, Smart Mater. Struct. 22 (2013), pp. 075029.
[68] H.N. An, S.J. Picken, and E. Mendes, *Nonlinear rheological study of magnetoresponsive soft gels*, Polymer. 53 (2012), pp. 4164–4170. doi:10.1016/j.polymer.2012.07.023.
[69] H.N. An, S.J. Picken, and E. Mendes, *Direct observation of particle rearrangement during cyclic stress hardening of magnetorheological gels*, Soft Matter. 8 (2012), pp. 11995–12001.
[70] H.-N. An, B. Sun, S.J. Picken, and E. Mendes, *Long time response of soft magnetorheologicalgels*, J. Phys. Chem. B. 116 (2012), pp. 4702–4711. doi:10.1021/jp301482a.

[71] X.L. Gong, Y.G. Xu, S.H. Xuan, C.Y. Guo, L.H. Zong, and W.Q. Jiang, *The investigation on the nonlinearity of plasticine-like magnetorheological material under oscillatory shear rheometry*, J. Rheol. 56 (2012), pp. 1375–1391.
[72] K. Shahrivar and J. DeVincente, *Thermoresponsive polymer-based magneto-rheological (MR) composites as a bridge between MR fluids and MR elastomers*, Soft Matter. 9 (2013), pp. 11451–11456.
[73] H. Sahin, X.J. Wang, and F. Gordaninejad, *Temperature dependence of magneto-rheological materials*, J. Intell. Mater. Syst. Struct. 20 (2009), pp. 2215–2222. doi:10.1177/1045389X09351608.
[74] H.J. Choi, B.O. Park, B.J. Park, and M.J. Hato, *Soft magnetic carbonyl iron microsphere dispersed in grease and its rheological characteristics under magnetic field*, Colloid Polym. Sci. 289 (2011), pp. 381–386. doi:10.1007/s00396-010-2363-y.
[75] S. Wang, W.Q. Jiang, W.F. Jiang, F. Ye, Y. Mao, S.H. Xuan, and X.L. Gong, *Multifunctional polymer composite with excellent shear stiffening performance and magnetorheological effect*, J. Mater. Chem. C. 2(34) (2014), pp. 7133–7140.
[76] P.V. Rao, S. Maniprakash, S.M. Srinivasan, and A.R. Srinivasa, *Functional behavior of isotropic magnetorheological gels*, Smart Mater. Struct. 19 (2010), pp. 085019.
[77] T. Mitsumata, S. Ohori, A. Honda, and M. Kawai, *Magnetism and viscoelasticity of magnetic elastomers with wide range modulation of dynamic modulus*, Soft Matter. 9 (2013), pp. 904–912. doi:10.1039/C2SM26717A.
[78] B.X. Ju, M. Yu, J. Fu, X. Zheng, and S.Z. Liu, *Magnetic field-dependent normal force of magnetorheological gel*, Ind. Eng. Chem. Res. 52 (2013), pp. 11583–11589. doi:10.1021/ie4013419.
[79] T.X. Liu, Y.G. Xu, X.L. Gong, H.M. Pang, and S.H. Xuan, *Magneto-induced normal stress of magnetorheological plastomer*, Aip. Adv. 3 (2013), pp. 082122.
[80] T. Mitsumata, S. Ohori, N. Chiba, and M. Kawai, *Enhancement of magnetoelastic behavior of bimodal magnetic elastomers by stress transfer via nonmagnetic particles*, Soft. Matter. 9 (2013), pp. 10108–10116.
[81] Y.G. Xu, X.L. Gong, S.H. Xuan, X.F. Li, L.J. Qin, and W.Q. Jiang, *Creep and recovery behaviors of magnetorheological plastomer and its magnetic-dependent properties*, Soft Matter. 8 (2012), pp. 8483–8492.
[82] B.X. Ju, M. Yu, J. Fu, X. Zheng, and S.Z. Liu, *Magnetic field-dependent normal force of magnetorheological gel*, Ind. Eng. Chem. Res. 52 (2013), pp. 11583–11589. doi:10.1021/ie4013419.
[83] Y.G. Xu, X.L. Gong, T.X. Liu, and S.H. Xuan, *Squeeze flow behaviors of magnetorheological plastomers under constant volume*, J. Rheol. 58(3) (2014), pp. 659–679. doi:10.1122/1.4869350.
[84] T.X. Liu, X.L. Gong, Y.G. Xu, H.M. Peng, and S.H. Xuan, *Magneto-induced large deformation and high-damping performance of a magnetorheological plastomer*, Smart Mater. Struct. 23 (2014), pp. 105028.
[85] Y.G. Xu, X.L. Gong, T.X. Liu, and S.H. Xuan, *Magneto-induced microstructure characterization of magnetorheological plastomers using impedance spectroscopy*, Soft Matter. 9 (2013), pp. 7701–7709.
[86] J.F. Brady and G. Bossis, *Stokesian dynamics*, Annu. Rev. Fluid Mech. 20 (1988), pp. 111–157. doi:10.1146/annurev.fl.20.010188.000551.
[87] M. Mohebi, N. Jamasbi, and J. Liu, *Simulation of the formation of nonequilibrium structures in magnetorheological fluids subject to an external magnetic field*, Phys. Rev. E. 54 (1996), pp. 5407–5413. doi:10.1103/PhysRevE.54.5407.
[88] D.J. Klingenberg, C.H. Olk, M.A. Golden, and J.C. Ulicny, *Effects of nonmagnetic interparticle forces on magnetorheological fluids*, J. Phys.-Condes. Matter. 22 (2010), pp. 324101.
[89] D.J. Klingenberg and J.C. Ulicny, *Enhancing magnetorheology*, Int. J. Mod. Phys. B. 25 (2011), pp. 911–917. doi:10.1142/S021797921105847X.
[90] Y.B. Peng, R. Ghanem, and J. Li, *Investigations of microstructured behaviors of magnetorheological suspensions*, J. Intel. Mat. Syst. Str. 23 (2012), pp. 1351–1370. doi:10.1177/1045389X12447288.
[91] G. Bossis, L. Iskakova, V. Kostenko, and A. Zubarev, *Kinetics aggregation of magnetic suspensions*, Physica. A. 390 (2011), pp. 2655–2663. doi:10.1016/j.physa.2011.02.044.

[92] G. Bossis, P. Lançon, A. Meunier, L. Iskakova, V. Kostenko, and A. Zubarev, *Kinetics of internal structures growth in magnetic suspensions*, Physica. A. 392 (2013), pp. 1567–1576. doi:10.1016/j.physa.2012.11.029.
[93] Y.L. Raikher and O.V. Stolbov, *Magnetodeformational effect in ferrogel samples*, J. Magn. Magn. Mater. 258-259 (2003), pp. 477–479. doi:10.1016/S0304-8853(02)01102-2.
[94] Y.L. Raikher and O.V. Stolbov, *Magnetodeformational effect in ferrogel objects*, J. Magn. Magn. Mater. 289 (2005), pp. 62–65. doi:10.1016/j.jmmm.2004.11.018.
[95] Y.L. Raikher and O.V. Stolbov, *Numerical modeling of large field-induced strains in ferroelastic bodies: A continuum approach*, J. Phys.-Condes. Matter. 20 (2008), pp. 204126.
[96] Y. Han, W. Hong, and L. Faidley, *Coupled magnetic field and viscoelasticity of ferrogel*, Int. J. Appl. Mech. 3 (2011), pp. 259–278. doi:10.1142/S175882511100097X.
[97] A.Y. Zubarev, *On the theory of the magnetic deformation of ferrogels*, Soft Matter. 8 (2012), pp. 3174–3179.
[98] A.Y. Zubarev, *Effect of chain-like aggregates on ferrogel magnetodeformation*, Soft Matter. 9 (2013), pp. 4985–4992.
[99] D.S. Wood and P.J. Camp, *Modeling the properties of ferrogels in uniform magnetic fields*, Phys. Rev. E. 83 (2011), pp. 011402. doi:10.1103/PhysRevE.83.011402.
[100] T.X. Liu, X.L. Gong, Y.G. Xu, S.H. Xuan, and W.Q. Jiang, *Simulation of magneto-induced rearrangeable microstructures of magnetorheological plastomers*, Soft Matter. 9 (2013), pp. 10069–10080.
[101] E.E. Keaveny and M.R. Maxey, *Modeling the magnetic interactions between paramagnetic beads in magnetorheological fluids*, J. Comput. Phys. 227 (2008), pp. 9554–9571. doi:10.1016/j.jcp.2008.07.008.
[102] L. Ge, S.H. Xuan, G.J. Liao, T.T. Yin, and X.L. Gong, *Stretchable polyurethane sponge reinforced magnetorheological material with enhanced mechanical properties*, Smart Mater. Struct. 24 (2015), pp. 037001.
[103] H.M. Pang, S.H. Xuan, T.X. Liu, and X.L. Gong, *Magnetic field dependent electro-conductivity of the graphite doped magnetorheological plastomer* (*in revision*).

6

Femtosecond laser irradiation-induced infrared absorption on silicon surfaces

Qinghua Zhu and Mengyan Shen*

Department of Physics and Applied Physics, and Nanomanufacturing Center, University of Massachusetts Lowell, 1 University Avenue, Lowell, MA 01854, USA

The near-infrared (NIR) absorption below band gap energy of crystalline silicon is significantly increased after the silicon is irradiated with femtosecond laser pulses at a simple experimental condition. The absorption increase in the NIR range primarily depends on the femtosecond laser pulse energy, pulse number, and pulse duration. The Raman spectroscopy analysis shows that after the laser irradiation, the silicon surface consists of silicon nanostructure and amorphous silicon. The femtosecond laser irradiation leads to the formation of a composite of nanocrystalline, amorphous, and the crystal silicon substrate surface with microstructures. The composite has an optical absorption enhancement at visible wavelengths as well as at NIR wavelength. The composite may be useful for an NIR detector, for example, for gas sensing because of its large surface area.

Keywords: near-infrared absorption; nanostructured silicon; amorphous silicon

1. Introduction

Silicon has formed an immense global industry – it has a large variety of applications in optoelectronics. However, the large band gap of silicon solid restricts its applications in making optoelectronic devices for utilizing infrared (IR) light [1–6]. If its wavelength is longer than 1.1 μm, i.e., its photon energy is less than the silicon band gap energy (1.1 eV), the light cannot be absorbed as it passes through silicon solid because the photon energy is not large enough to excite an electron to the conduction band from the valence band of the silicon solid. To improve the optical properties of silicon, femtosecond laser irradiation method has been applied to treat silicon surfaces [3,7–10]. In order to increase the silicon absorption in the IR range, the microstructured silicon was formed with femtosecond laser irradiation on silicon surfaces in the presence of the background gas of SF_6. After sulfur atoms were doped into the silicon surface, the band gap of silicon was filled with the sulfur impurity energy states. As a result, the absorption for IR light with photon energy less than 1.1 eV can be obtained [3,7]. The IR absorption has potential applications for the IR optoelectronic devices, such as IR image detector.

Here we report that the IR absorption of silicon can be directly obtained through fabricating a silicon substrate surface with a femtosecond laser irradiation in air without applying any special background gas [11–13]. The scanning electron microscope (SEM)

*Corresponding author. Email: mengyan_shen@uml.edu

images of the fabricated silicon surface have shown arrays of silicon hills in micrometers with smaller nano-scaled structures on the hills' surfaces, which were generated during the laser irradiation. The phonon Raman spectroscopy analysis of the fabricated surface has shown that amorphous silicon as well as silicon nano-crystalline was generated on the silicon surface. The IR optical absorption was largely increased after being processed by the irradiation of femtosecond laser. This absorption depends on several laser-processing parameters such as laser pulse duration, pulse number, and pulse energy. The increment of the IR absorption is originated from the composite of nano-crystalline silicon and amorphous silicon on the silicon substrate surface because amorphous silicon provides electronic energy states in the energy band gap for the IR absorption [14].

On the one hand, amorphous silicon-related papers of laser irradiation mainly concentrated on the application of the laser-annealing effect, i.e., amorphous silicon became silicon crystalline or micro/nanostructured silicon for better optoelectronics applications [15–26]. On the other hand, amorphous silicon and lattice distortions could be formed with the femtosecond laser irradiation [27–31]. The formed amorphous silicon was analyzed with different techniques including transmission electron microscopy, energy dispersive X-ray analysis [27], non-destructive optical imaging method of scanning laser microscopy [28], and phonon Raman spectroscopy [29,30]. However, the thickness of the formed amorphous silicon layer was found to be only tens of nanometers, and its IR optical property has not been noticed and studied because the amount of the amorphous silicon was too small to influence the optical properties of the silicon substrate. In the present work, we have transformed silicon solid smooth surface into a composite of nanocrystalline silicon and amorphous silicon on the silicon solid surface with microstructures using a simple femtosecond laser irradiation in air. We found that the microstructures enhanced the optical absorptions and the amount of the formed amorphous silicon was large enough to considerably influence the IR optical absorption of the irradiated silicon substrates. The composite will be useful for manufacturing silicon-related IR optoelectronic device, such as IR sensors and solar cells.

2. Experimental

A clean silicon wafer (p-Si (100), thickness of 500 μm, resistivity of 20 Ω•m) was placed on the translation stage and was irradiated in air by a 1 kHz train of 35 fs, 6 mJ pulses at 800 nm wavelength from an amplified Ti:sapphire laser system with the incident laser beam perpendicular to the wafer surface. The sample was scanned line by line with the distance of 150 μm between two lines at even speed to make the silicon receive uniform exposure to the laser in an area of 1 cm × 1 cm. Laser beam splitters are used to control the average laser fluence. The pulse duration was controlled by adjusting the laser compressor and measured by an autocorrelator around the sample surface location between a lens and its focal point. The sample translation velocity in the direction perpendicular to the laser beam direction was used to control the average pulse number irradiated on the substrate surface. The average pulse number n was calculated by applying the formula $n = wf/v$, where w is the full width at half maximum (FWHM) of the Gaussian laser beam profile on the sample surface, f is the frequency of the laser pulse train, and v is the sample translation velocity [12]. In this work, w was 100 μm, f was 1000, and v varied from 0.3 mm/s to 2.5 mm/s. After the laser irradiation, the silicon wafer was cleaned by distilled water ultrasonically to remove the silicon dust sprayed out during the laser treatment procedure and then further washed with 10% hydrofluoric acid to remove the silicon oxide layer formed during the laser ablation procedure. The cleaning procedure can eliminate the effect of silicon dust and silicon oxide layer during Raman spectrum test and optical absorptance measurement.

The crystallinity of the sample at room temperature was analyzed with a Raman spectrometer (Optic Senterra, Bruker). A continuous wave laser at the wavelength of 532 nm was used as the excitation source for the Raman spectroscopy measurement. The sample surface structure was also studied by using an SEM (JEOL 7401F). The optical property of the sample was studied with a spectrophotometer (U-4001, HITACHI) equipped with a precise integrating sphere. The reflectance (R) and transmittance (T) were measured with the integrating sphere. The absorptance (A) of the sample was then obtained with the formula $A = 1 - T - R$.

3. Experimental results and discussion

Raman spectroscopy analysis is an effective method to study the silicon surface irradiated by femtosecond laser [29,30]. Figure 1 shows the Raman spectrum of the c-Si after the femtosecond laser irradiation. In the experiment to prepare the sample, the laser pulse duration was 35 fs, the average laser fluence was 6 kJ/m^2, and the pulse number was 250. Figure 1(a) shows the comparison of Raman spectrum of laser processed Si with that of Si crystal. The Raman main peak of silicon crystal is at 520.0 cm^{-1}, while the main peak of the laser-irradiated silicon has shifted to lower energy from 520.0 cm^{-1} to 513.5 cm^{-1} with an asymmetry broadening. Because of the phonon confinement effect, the redshift and the asymmetric broadening of the one phonon Raman signal at 520 cm^{-1} have been used to estimate the average size of small silicon crystals [23,32]. The redshift and the asymmetry broadening of the Raman peak around 520.0 cm^{-1} indicate that silicon nanocrystals were generated during the laser irradiation. The Raman spectrum suggests that the average size of the generated silicon nanocrystals was around 10 nm [33] although the spectrum cannot give the shapes and the size distribution of the silicon nanocrystals. Figure 1(b) shows the Raman spectrum of weak signals from wavenumber of 100 cm^{-1} to 900 cm^{-1}. The broad shoulder near the nanocrystal Si peak of 513.5 cm^{-1} can be observed. As shown in Figure 1(b), we have deconvoluted the spectrum to obtain the peak centered at a wavenumber of 513.5 cm^{-1} which belongs to the nanocrystal Si, and four broad peaks which are the characteristic peaks of amorphous silicon at wavenumber of 480 cm^{-1} (transverse optical mode), 380 cm^{-1} (longitudinal optical mode), 301 cm^{-1} (longitudinal acoustic mode), and 150 cm^{-1} (transverse acoustic mode) [34–39]. The Raman spectrum analysis has revealed that nanostructured silicon and amorphous silicon were generated simultaneously during the laser irradiation.

Figure 2 shows the Si surface SEM images after laser irradiation. Figure 2(a) shows that the structures formed on the silicon surface after the femtosecond laser irradiation are arrays of silicon micro-hills. The micro-hills were formed because of the line-by-line laser scanning. The cross section of each one of the hills is in isosceles triangle-like shape. The bottom side of the triangle is about 80-micrometer long. The height of the triangle is about 100 micrometers. Figure 2(b) shows the silicon micro-hill feature viewed after rotating the substrate at 90° and titling at 45° to the surface normal. Figure 2(c) and 2(d) shows that on the surface of silicon micro-hills, there are smaller structures generated during the laser irradiation process. Figure 2(e) and 2(f) shows the SEM image with larger magnification at different locations of the hills; the images show the same nanostructures or submicrostructures at different locations of the surface of the micro-hill surfaces. The laser wavelength was 532 nm for the excitation source for the

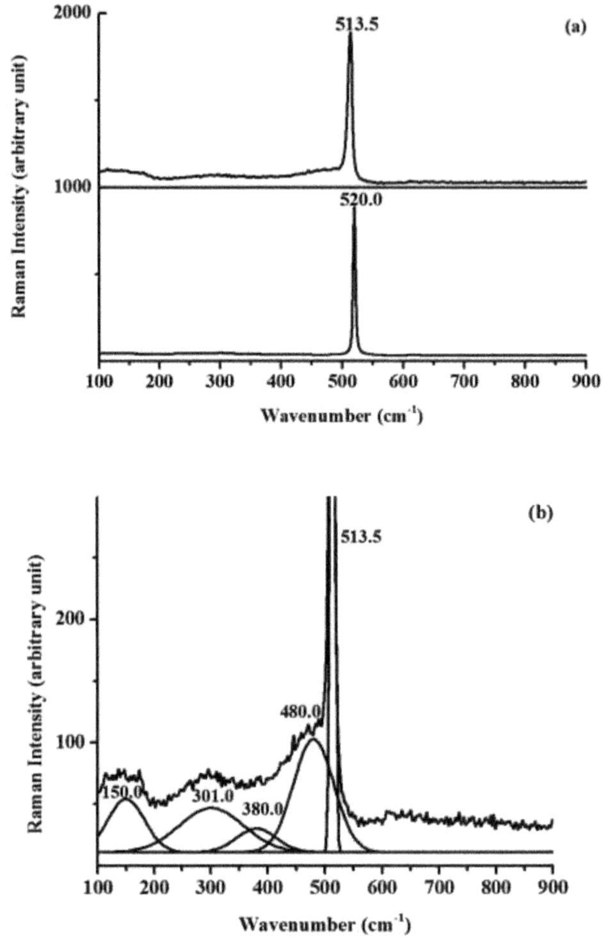

Figure 1. The Raman spectrum of silicon after the irradiation of laser with pulse duration of 35 fs, an average laser fluence of 6 kJ/m^2, and a pulse number of 250. (a) The comparison of the Raman spectrum of the sample (the upper spectrum) with that of crystalline silicon that shows a clean peak at 520 cm^{-1} (in the lower spectrum). (b) The deconvolution of the Raman spectrum between 100 cm^{-1} and 600 cm^{-1}.

Raman spectroscopy measurement. The laser penetration depth, $\lambda/4\pi\kappa$ (where λ and κ are the wavelength and the imaginary part of the complex refractive index of silicon, respectively [40]), is less than 2 μm that is much smaller than the surface structures shown in the images of Figure 2(e) and 2(f). The Raman signals of the sample were from the nanocrystals and amorphous silicon but not from the silicon crystal microstructures beneath the nanocrystals and amorphous silicon. Therefore, we have transformed the crystal silicon into a composite of nanocrystalline, amorphous, and crystal silicon substrate surface with micro-morphologies.

Figure 3 shows the optical absorptance of a laser-irradiated silicon sample between 380 and 2000 nm. The experimental condition used was 35 fs pulse duration, 6 kJ/m^2 average pulse fluence, and 250 laser pulses. The absorptance of regular silicon is also shown in Figure 3 for comparison with the laser-irradiated Si sample. As wavelength

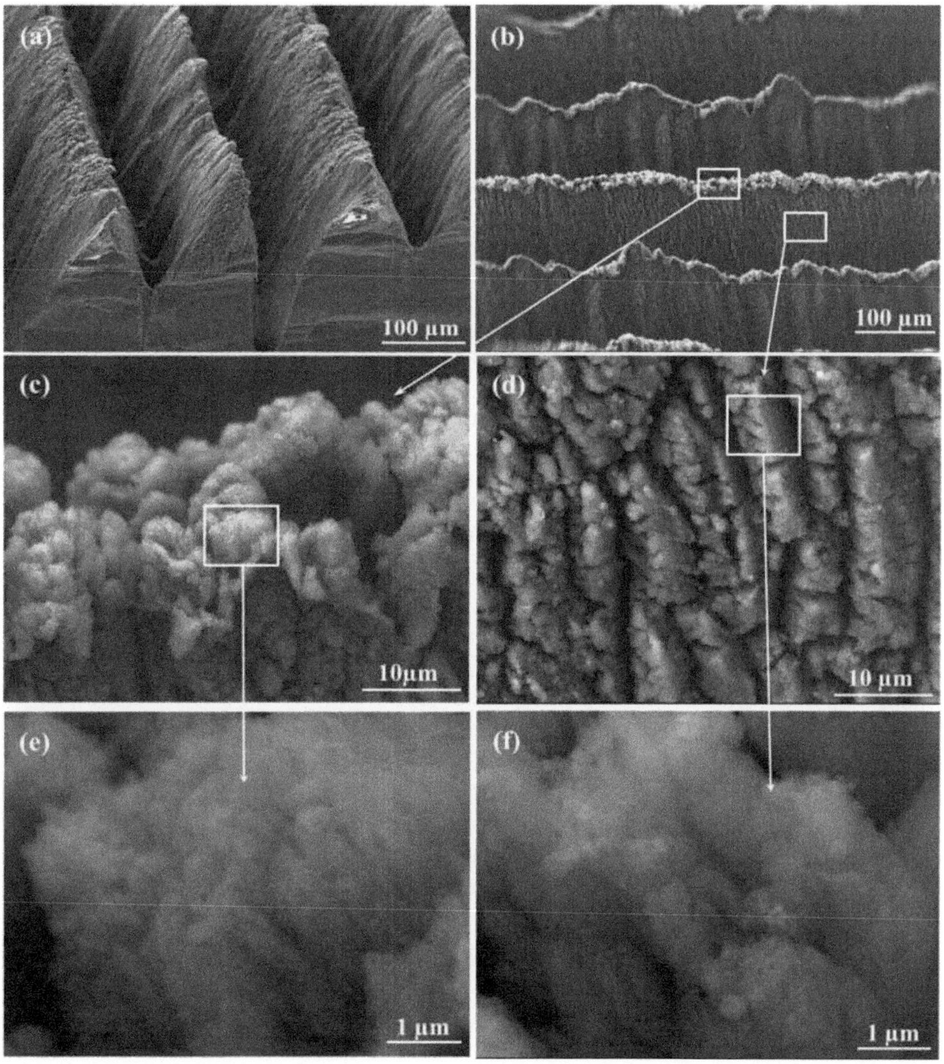

Figure 2. SEM images of the Si surface after irradiation of laser with pulse duration of 35 fs, an average laser fluence of 6 kJ/m^2, and a pulse number of 250. (a) Si micro-hills, (b) Si micro-hills in (a) after rotating of about 90°, (c) structures on the top surface of silicon micro-hills in (b) for higher magnification, (d) structures on the side surface of silicon micro-hills in (b) for higher magnification, (e) magnified image of (c), and (f) magnified image of (d).

increases, the absorptance of the sample decreases. The absorption is almost a linear function of wavelength between 400 nm and 2000 nm. The comparison indicates that the structured silicon's optical absorptance is much higher than that of the regular silicon in the measured wavelength range. The absorptance at wavelength of 1.5 μm of the structured silicon is about 50% that is much higher than that of unstructured silicon substrate. In visible wavelengths part, the absorptance increment is mainly originated from the effect of multiple reflections, i.e., the incident light is reflected multiple times on the micro/nanostructured surface and is absorbed by the silicon [7–11]. In near-infrared (NIR) region, there is an obvious drop in wavelength around 1.1 μm for the unstructured silicon

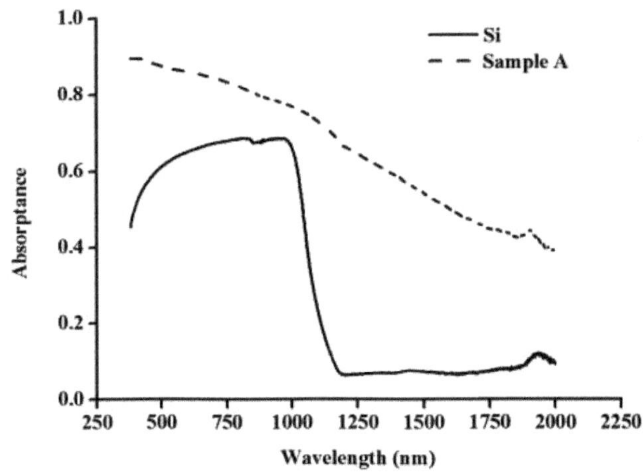

Figure 3. The optical absorption comparison of the femtosecond laser irradiated sample A with crystalline silicon. The silicon substrate has been irradiated with laser with pulse duration of 35 fs, an average laser fluence of 6 kJ/m^2, and a pulse number of 250.

absorptance curve, which is corresponding to the silicon band gap. The light with wavelength longer than 1.1 μm does not have enough photon energy to excite electron from the valence band to the conduction band of silicon solid. As a result, silicon is almost transparent to the light in this range. However, the silicon after laser irradiation has a large IR optical absorptance because the disordered amorphous silicon does not have a defined band gap.

Figure 4 shows laser pulse duration effect on the IR absorptance when keeping the pulse number as 250 and the average pulse fluence as 6 kJ/m^2. As the pulse duration decreased from 100 to 35 fs, the IR absorptance increased in the NIR range although changing the pulse duration did not make a large difference. The IR absorptance change was less than 0.1 or 10%. Meanwhile, the disappeared band gap when the pulse duration

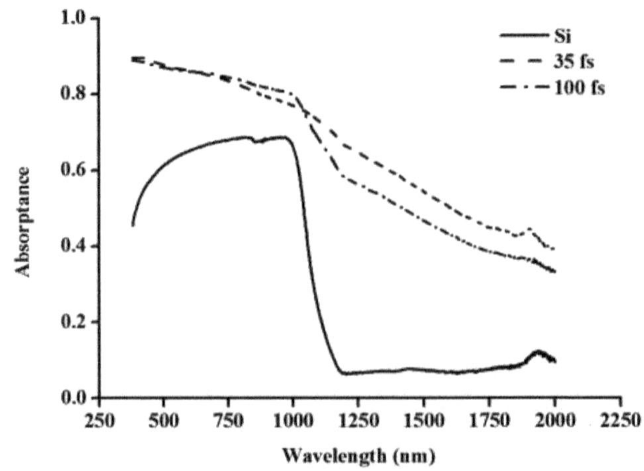

Figure 4. The optical absorptance of silicon after laser irradiation at two different laser pulse durations, 35 fs and 100 fs. The average laser fluence and pulse number are 6 kJ/m^2 and 250, respectively.

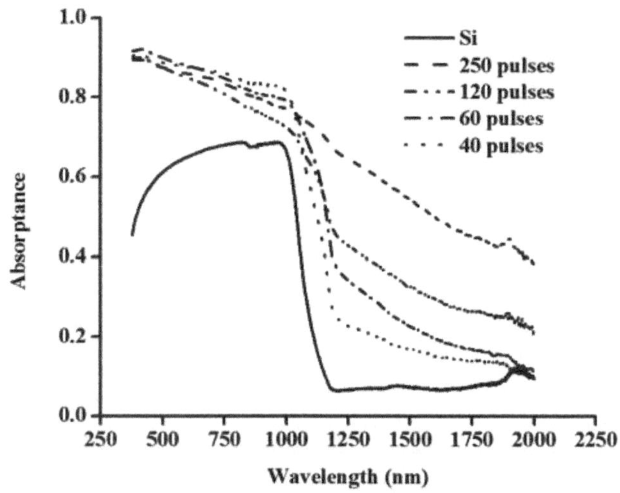

Figure 5. Pulse number dependence of optical absorptance of silicon after laser irradiation. The pulse duration and the average laser fluence are 35 fs and 6 kJ/m^2, respectively.

was 35 fs around 1.1 μm was observable when the pulse duration was adjusted to 100 fs. When irradiating with longer pulse duration, we got less amorphous silicon, and thus less IR absorptance.

Figure 5 shows the laser pulse number dependence of the optical absorptance while keeping the average laser fluence as 6 kJ/m^2 and the pulse duration as 35 fs. When the pulse number increased from 40 to 250, the absorptance changed less than 10% for the wavelength less than 1000 nm, but increased significantly between 1000 and 2000 nm. For the small pulse number, the optical absorptance approached to that of unstructured silicon, i.e., although a small number of femtosecond pulse irradiation changed the surface structures, there was not enough amorphous silicon to change in the NIR optical property of the silicon substrate as a whole system. When the pulse number is 250, NIR absorptance reached the maximum. As the pulse number was larger than 250, the induced NIR absorptance became smaller, which will be further analyzed in the following text.

Figure 6 shows the laser fluence dependence of optical absorptance when keeping the pulse number as 250 laser pulses and the pulse duration as 35 fs. For the wavelength between 1250 and 2000 nm, the absorptance increases as the laser fluence increases. For the case of average laser fluence of 1.5 and 0.7 kJ/m^2, the characteristic Si band gap can still be observed around the wavelength of 1.1 μm. While as the fluence increases to 6 kJ/m^2, the band gap cannot be found on the absorptance spectrum. However, we also found that when the fluence became stronger than 6 kJ/m^2, the NIR absorptance became smaller and similar to that of Si irradiated with lower laser fluences as shown in Figure 6 for the fluence of 7 kJ/m^2. This is opposite to the fluence dependence of optical absorptance when the fluence was weaker than 6 kJ/m^2.

In Figure 7(a), we compare Sample A in Figures 1 and 3 with the samples with more irradiation, i.e., more accumulated irradiation energy absorbed by the substrate. Sample B was made with the irradiation of the laser 35 fs pulse duration, fluence of 7 kJ/m^2, and 250 laser pulses; and Sample C at 35 fs pulse duration, fluence of 6 kJ/m^2, and 300 laser pulses. Figure 7(b) shows the corresponding Raman spectra of Samples A, B, and C. From the Raman spectra, we know that the crystalline peak position of Sample A was at

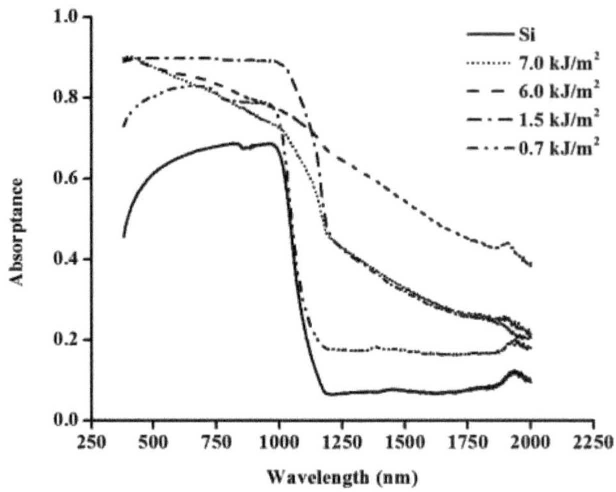

Figure 6. Laser fluence dependence of optical absorptance of silicon after laser irradiation. The pulse duration and the pulse number are 35 fs and 250, respectively.

513.5 cm^{-1}, Sample B at 516.0 cm^{-1}, and Sample C at 518.0 cm^{-1}. The amorphous broad band at 480 cm^{-1} in Samples B and C became weak relative to the crystalline peak in comparison with that in Sample A. Thus, the amorphous silicon became crystalline silicon when more laser energy was irradiated on the amorphous silicon, and the IR absorptance decreased as shown in Figure 7(a). Therefore, as the pulse number or pulse energy increases, the NIR absorptance decreases after reaching a maximum value. The silicon amorphous structure goes back to silicon crystalline structure under the influence of laser irradiation [15–26]. The silicon structure can be changed back and forth between crystalline and amorphous by controlling the experimental conditions.

The Raman spectrum of the silicon after the laser irradiation shows that the crystalline silicon can change to amorphous silicon under the effect of laser irradiation. From the obvious IR absorption, we know that there is a significant amount of amorphous silicon formed. The sharp and redshifted main Raman peak indicates that some of the silicon in the surface is in the form of nanocrystalline. The silicon atoms of a nanocrystalline surface are usually in a deformed lattice that is different from that in the bulk. The surface atoms can easily join with the surrounding amorphous silicon. The amorphous silicon fills in the space surrounding silicon nanocrystals and the crystal silicon substrate surface with microstructures. This forms a composite that absorbs NIR light significantly. The composite may be useful for an NIR detector, for example, for gas sensing because of the enhancement in its surface area.

4. Summary

The femtosecond laser irradiation on silicon solid leads to the formation of a composite of nanocrystalline, amorphous, and silicon substrate surface with microstructures. The composite has an optical absorption enhancement both at the visible wavelengths and NIR wavelengths. The transformations between crystalline silicon and amorphous silicon with the femtosecond pulse laser irradiation may provide a new method to make low cost IR detectors with silicon.

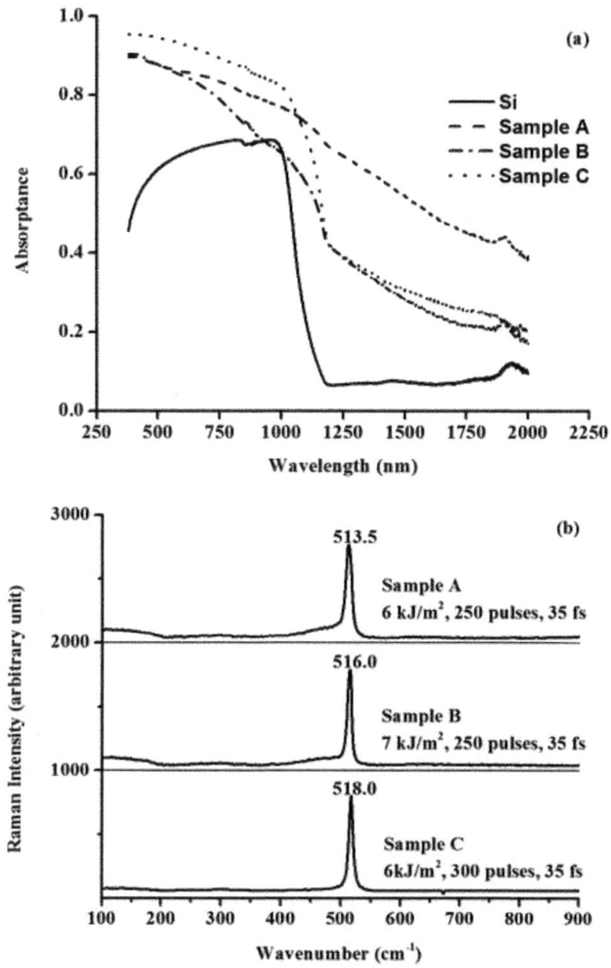

Figure 7. The radiation effect on the silicon. Sample A is made with the irradiation of the laser 35 fs pulse duration, fluence of 6 kJ/m^2, and 250 laser pulses; Sample B at 35 fs pulse duration, fluence of 7 kJ/m^2, and 250 laser pulses; and sample C at 35 fs pulse duration, fluence of 6 kJ/m^2, and 300 laser pulses. (a) The optical absorptance of silicon after different laser irradiations. (b) The Raman spectra of silicon samples after different laser irradiations.

Acknowledgments

We thank Dr H. Huo, Dr C. Wang, Mr H. Ren, and Mr Y. Li for experimental assistances, and Dr E. Ada and Dr L. Li for helpful discussions.

Disclosure statement

No potential conflict of interest was reported by the authors.

Funding

This research was supported by the National Science Foundation [grant number CMMI-1031111].

References

[1] H. Sai, Y. Kanamori, and M. Kondo, *Flattened light-scattering substrate in thin film silicon solar cells for improved infrared response*, Appl. Phys. Lett. 98 (2011), pp. 113502.

[2] D.E. Carlson and C.R. Wronski, *Amorphous silicon solar cell*, Appl. Phys. Lett. 28 (1976), pp. 671.

[3] J.E. Carey, C.H. Crouch, M. Shen, and E. Mazur, *Visible and near-infrared responsivity of femtosecond-laser microstructured silicon photodiodes*, Opt. Lett. 30 (2005), pp. 14.

[4] H. Zhou, G. Fang, L. Yuan, C. Wang, X. Yang, H. Huang, and X. Zhao, *Deep ultraviolet and near infrared photodiode based on n-ZnO/p-silicon nanowire heterojunction fabricated at low temperature*, Appl. Phys. Lett. 94 (2009), pp. 1.

[5] H. Zogg, A. Fach, J. John, J. Masek, P. Mueller, C. Paglino, and S. Blunier, *Photovoltaic IV-VI on Si infrared sensor arrays for thermal imaging*, Opt. Eng. 34 (1995), pp. 1964.

[6] H. Zogg, C. Maissen, J. Masek, and S. Blunier, *Photovoltaic lead-chalcogenide on silicon infrared sensor arrays*, Opt. Eng. 33 (1994), pp. 1440.

[7] T.H. Her, R.J. Finlay, C. Wu, S. Deliwala, and E. Mazur, *Microstructuring of silicon with femtosecond laser pulses*, Appl. Phys. Lett. 73 (1998), pp. 1673.

[8] A.J. Pedraza, J.D. Fowlkes, and D.H. Lowndes, *Surface nanostructuring of silicon*, Appl. Phys. Lett. 77 (2000), pp. 3018.

[9] M.Y. Shen, C.H. Crouch, J.E. Carey, and E. Mazur, *Femtosecond laser-induced formation of submicrometer spikes on silicon in water*, Appl. Phys. Lett. 85 (2004), pp. 5694.

[10] M. Shen, J.E. Carey, C.H. Crouch, M. Kandyla, H.A. Stone, and E. Mazur, *High-density regular arrays of nanometer-scale rods formed on silicon surfaces via femtosecond laser irradiation in water*, Nano. Lett. 8 (2008), pp. 2087–2091. doi:10.1021/nl080291q

[11] T.-H. Her, R.J. Finlay, C. Wu, and E. Mazur, *Femtosecond laser-induced formation of spikes on silicon*, Appl. Phys. A. 70 (2000), pp. 383–385. doi:10.1007/s003390051052

[12] C.H. Crouch, J.E. Carey, M. Shen, E. Mazur, and F.Y. Genin, *Infrared absorption by sulfur-doped silicon formed by femtosecond laser irradiation*, Appl. Phys. A. 79 (2004), pp. 1635. doi:10.1007/s00339-004-2676-0.

[13] C. Wu, C.H. Crouch, L. Zhao, J.E. Carey, R. Younkin, J.A. Levinson, E. Mazur, R.M. Farrell, P. Gothoskar, and A. Karger, *Near-unity below-band-gap absorption by microstructured silicon*, Appl. Phys. Lett. 78 (2001), pp. 13.

[14] J. Zhu, Z. Yu, G.F. Burkhard, C. Hsu, S.T. Connor, Y. Xu, Q. Wang, M. McGehee, S. Fan, and Y. Cui, *Optical absorption enhancement in amorphous silicon nanowire and nanocone arrays*, Nano Lett. 9 (2009), pp. 279–282. doi:10.1021/nl802886y

[15] J.-M. Shieh, Z.-H. Chen, B.-T. Dai, Y.-C. Wang, A. Zaitsev, and C.-L. Pan, *Near-infrared femtosecond laser-induced crystallization of amorphous silicon*, Appl. Phys. Lett. 85 (2004), pp. 1232.

[16] G.J. Lee, S.H. Song, Y.P. Lee, H. Cheong, C.S. Yoon, Y.D. Son, and J. Jang, *Arbitrary surface structuring of amorphous silicon films based on femtosecond-laser-induced crystallization*, Appl. Phys. Lett. 89 (2006), pp. 151907.

[17] M.-J. Sher, K. Hammond, L. Christakis, and E. Mazur, *The photovoltaic potential of femtosecond-laser textured amorphous silicon*, Laser-based Micro- and Nanopackaging and Assembly VII, Proc. SPIE. 8608 (2013), pp. 86080R.

[18] A. Gat, J.F. Gibbons, T.J. Magee, J. Peng, V.R. Deline, P. Williams, and C.A. Evans Jr, *Physical and electrical properties of laser-annealed ion-implanted silicon*, Appl. Phys. Lett. 32 (1978), pp. 276.

[19] A. Gat, L. Gerzberg, J.F. Gibbons, T.J. Magee, J. Peng, and J.D. Hong, *cw laser anneal of polycrystalline silicon: Crystalline structure, electrical properties*, Appl. Phys. Lett. 33 (1978), pp. 775.

[20] R.T. Young, C.W. White, G.J. Clark, J. Narayan, W.H. Christie, M. Murakami, P.W. King, and S.D. Kramer, *Laser annealing of boron-implanted silicon*, Appl. Phys. Lett. 32 (1978), pp. 139.

[21] M. Miyasaka and J. Stoemenos, *Excimer laser annealing of amorphous and solid-phase crystallized silicon films*, J. Appl. Phys. 86 (1999), pp. 5556.

[22] D. Murley, N. Young, M. Trainor, and D. McCulloch, *An investigation of laser annealed and metal-induced crystallized polycrystalline silicon thin-film transistors, Electron Devices*, IEEE Trans. 48 (2001), pp. 1145–1151. doi:10.1109/16.925240.

[23] M.O. Thompson, G.J. Galvin, J.W. Mayer, P.S. Peercy, J.M. Poate, D.C. Jacobson, A.G. Cullis, and N.G. Chew, *Melting temperature and explosive crystallization of amorphous silicon during pulsed laser irradiation*, Phys. Rev. Lett. 52 (1984), pp. 2360–2363. doi:10.1103/PhysRevLett.52.2360

[24] J. Bonse, S. Baudach, J. Krüger, W. Kautek, and M. Lenzner, *Femtosecond laser ablation of silicon–modification thresholds and morphology*, Appl. Phys. A. 74 (2002), pp. 19–25. doi:10.1007/s003390100893

[25] A.E. Bell, *Review and analysis of laser annealing*, RCA Review 40 (1979), pp. 295.

[26] S. Kawamura, J. Sakurai, M. Nakano, and M. Takagi, *Recrystallization of Si on amorphous substrates by doughnut-shaped cw Ar laser beam*, Appl. Phys. Lett. 40 (1982), pp. 394.

[27] J. Jia, M. Li, and C.V. Thompson, *Amorphization of silicon by femtosecond laser pulses*, Appl. Phys. Lett. 84 (2004), pp. 3205.

[28] J. Bonse, *All-optical characterization of single femtosecond laser-pulse-induced amorphization in silicon*, Appl. Phys. A. 84 (2006), pp. 63–66. doi:10.1007/s00339-006-3583-3

[29] F. Costache, S. Kouteva-Arguirova, and J. Reif, *Sub-damage-threshold femtosecond laser ablation from crystalline Si: Surface nanostructures and phase transformation*, Appl. Phys. A. 79 (2004), pp. 1429. doi:10.1007/s00339-004-2803-y

[30] M. Schade, O. Varlamova, J. Reif, H. Blumtritt, W. Erfurth, and H.S. Leipner, *High-resolution investigations of ripple structures formed by femtosecond laser irradiation of silicon*, Anal. Bioanal. Chem. 396 (2010), pp. 1905–1911. doi:10.1007/s00216-009-3342-3

[31] M. Straub, M. Schüle, M. Afshar, D. Feili, H. Seidel, and K. König, *Sub-15 femtosecond laser-induced nanostructures emerging on Si(100) surfaces immersed in water: Analysis of structural phases*, Appl. Phys. A. 115 (2014), pp. 221–228. doi:10.1007/s00339-013-7980-0

[32] H. Richter, Z.P. Wang, and L. Ley, *The one phonon Raman spectrum in microcrystalline silicon*, Solid State Commun. 39 (1981), pp. 625–629. doi:10.1016/0038-1098(81)90337-9

[33] M. Shen, *Raman phonons of silicon wires*, Phys. Lett. A 180 (1993), pp. 295–298. doi:10.1016/0375-9601(93)90714-B

[34] A.T. Voutsas, M.K. Hatalis, J. Boyce, and A. Chiang, *Raman spectroscopy of amorphous and microcrystalline silicon films deposited by low-pressure chemical vapor deposition*, J. Appl. Phys. 78 (1995), pp. 6999.

[35] B. Li, D. Yu, and S. Zhang, *Raman spectral study of silicon nanowires*, Phys. Rev. B 59 (1999), pp. 1645–1648. doi:10.1103/PhysRevB.59.1645

[36] Z. Iqbal and S. Veprek, *Raman scattering from hydrogenated microcrystalline and amorphous silicon*, J. Solid State Phys. 15 (1982), pp. 377–392. doi:10.1088/0022-3719/15/2/019

[37] X.L. Wu, G.G. Siu, S. Tong, X.N. Liu, F. Yan, S.S. Jiang, X.K. Zhang, and D. Feng, *Raman scattering of alternating nanocrystalline silicon/amorphous silicon multilayers*, Appl. Phys. Lett. 69 (1996), pp. 523.

[38] R. Saleh and N.H. Nickel, *Raman spectroscopy of B-doped microcrystalline silicon films*, Thin Solid Films. 427 (2003), pp. 266–269. doi:10.1016/S0040-6090(02)01203-8

[39] X.C. Wang, H.Y. Zheng, C.W. Tan, F. Wang, H.Y. Yu, and K.L. Pey, *Femtosecond laser induced surface nanostructuring and simultaneous crystallization of amorphous thin silicon film*, Opt. Express 18 (2010), pp. 18.

[40] G. Ghosh, *Handbook of Optical Constants of Solids*, Academic Press, New York, 1998.

7

Sustainable shape memory polymers based on epoxidized natural rubber cured by zinc ferulate via oxa-Michael reaction

Xuhui Zhang, Tengfei Lin, Zhenghai Tang, Kailun Sun and Baochun Guo*

Department of Polymer Materials and Engineering, South China University of Technology, Guangzhou 510640, P. R. China

Although various shape memory polymers (SMPs) or diverse applications have been widely reported, the SMPs based on rubbers have been rarely realized due to the low triggering temperature of rubbers. In another aspect, the SMPs based on sustainable substances are highly desired for the growing shortage in fossil resources. In the present study, we accordingly developed the sustainable SMPs with tunable triggering temperature, based on natural rubber (NR) and ferulic acid (FA) as the raw materials. Specifically, the SMPs are based on a crosslinked network of epoxidized natural rubber (ENR) crosslinked by *in situ* formed zinc ferulate (ZDF) via oxa-Michael reaction. The excellent shape memory effect (SME) is found in these SMPs, as evidenced by the high fixity/recovery ratio and the tunable triggering temperature. With the incorporation of natural halloysite nanotubes (HNTs), the stress and recovery rate of the SMPs are found to be tunable, which widens the application of this kind of SMPs. The combination of adoption of sustainable raw materials, and the excellent and tunable SME makes these SMPs potentially useful in many applications, such as various actuators and heat-shrinkable package materials.

Keywords: shape memory polymers; epoxidized natural rubber; ferulic acid; oxa-Michael reaction; sustainable

1. Introduction

SMPs are smart materials that can change shape and memorize their original shape when exposed to an external stimulus, such as temperature, light, electricity and solvent [1 –5]. Heating is the most common way to trigger the shape memory effect (SME) [6,7]. Generally, SMPs possess a fixing phase provided by a crosslinked network as well as a transition phase supported by the glass transition and/or crystalline melting transition [8]. SMPs are deformed at a temperature above the transition temperature (T_{trans}) and the deformed sample is cooled to a temperature below T_{trans} under stress to fix the temporary shape. When reheated above T_{trans} without stress, SMPs recover their permanent shape due to entropy elasticity of polymer chains [9]. SMPs have attracted significant attention due to their diverse applications such as medical devices, sensors, switches and self-deployable structures [10 –13].

Although vitrification is utilized as the transition mechanism in many SMPs, it is seldom used in elastomer-based SMPs because T_g of rubber is too low for practical applications [14].

*Corresponding author. Email: psbcguo@scut.edu.cn

Crystalline compounds with relatively high melting point, such as fatty acid, are introduced into elastomers to act as transition phase. Weiss *et al.* [15] mixed zinc stearate with a sulfonated ethylene-propylene-diene monomer (EPDM) (EPDM with pendent sulfonate groups) to prepare SMPs with a transition phase supported by zinc stearate. The resultant blends exhibited good SME and the length recovery was ~92%. It is conceivable that if T_g of rubber is able to approach to room temperature or above, rubber should also be a suitable choice for SMP fabrication. By this way, we may also tune the transition temperature by simply changing the amount of the curing agent. Chang *et al.* [16] revealed that epoxidized natural rubber (ENR) crosslinked by 3-amino-1,2,4-triazole showed high T_g and exhibited excellent SME based on the vitrification mechanism.

Recently, extensive interest has focused on 'sustainable' polymers with biobased raw materials instead of petroleum-based resources due to the increasing consumption of limited fossil feedstock [17 –19]. Sustainable materials are those whose production is supported inexhaustibly by nature. A variety of sustainable sources have been explored for polymers, including vegetable oils, plant sugars, rosins and lignin [20 –23]. However, the main challenge of 'sustainable' polymer fabrication, namely physicochemical properties of products is hardly comparable or superior to their counterparts prepared from fossil resources, is urgent to be solved.

In our previous studies, the efficient crosslinking between ENR and zinc acrylate (ZDA) via oxa-Michael reaction has been revealed and SME based on glass transition has been demonstrated [24,25]. Very recently, the oxa-Michael reaction between zinc ferulate (ZDF) and ENR has also been revealed [26]. The ZDF-cured ENR should be potentially ideal SMPs in virtue of tunable T_g and the utilization of sustainable raw materials including NR and FA. In ENR crosslinked by ZDF, the crosslinked networks serve as the permanent shape and the glass transition serves as the shape triggering mechanism. It should be emphasized that FA exhibits potential antioxidant capability due to a stabilized phenoxy radical arising from the phenolic nucleus and an extended side chain conjugation [27,28]. Besides, FA shows unique features of nontoxicity and pharmacological functions [29,30]. More importantly, FA is a sustainable material as a ubiquitous plant constituent of metabolism [31]. Combining the sustainability of NR, the precursor of ENR, raw materials for the SMPs fabrication in the present work are sustainable and essentially nontoxic.

In the present study, the SMPs based on ENR/ZDF networks were fabricated. The transition behaviors and the SME were investigated. To better conform the practical applications, halloysite nanotubes (HNTs), a kind of naturally occurring nanotubular clay, were incorporated into the ENR/ZDF network to regulate the mechanical properties and SME of the SMPs. The effects of HNT incorporation on the mechanical properties and SME were evaluated and correlated with the structures of the composites.

2. Experimental methods

2.1. Raw materials

ENR with an epoxidization degree of 50% was produced by the Agricultural Products Processing Research Institute, Chinese Academy of Tropical Agricultural Science, Zhanjiang, P. R. China. Chemically pure FA was purchased from Aladdin Industrial Corporation, China, and zinc oxide (ZnO) is industrially available products and used as received. HNTs were mined from Hubei Province, China, and purified accordingly to the previously reported procedure [32].

2.2. *Preparation of ENR cured by ZDF*

ENR, FA, ZnO, and HNTs were mixed in an open two-roll mill and compositions of the rubber compounds are listed in Table 1. Due to the formation of zinc ferulate by *in situ* reaction between ferulic acid and ZnO, molar ratios between FA and ZnO are set as 2:1. For all samples, the well mixed compounds were hot pressed at 160 °C with optimized curing time (Tc90) to cure the compounds. As shown in the below Figure 1, Tc90 is determined as the time when the torque is equal to $[M_L+(M_H-M_L)*90\%]$.

2.3. *Characterizations*

Curing behaviors of ENR and ENR/HNT composites cured by ZDF were determined at 160 °C using a U-CAN UR-2030 vulcameter (Taiwan). Tensile tests of composites at room temperature were measured accordingly to ISO 37-2005, using a UCAN UT-2060 machine (Taiwan). Tensile tests at elevated temperature were determined by a TA Q800 dynamic mechanical analyzer (USA). After equilibrium at the testing temperature, samples with thickness of ~0.5 mm were stretched with an extension rate of 0.5 N/min until the stress/strain limit reached.

X-ray diffraction (XRD) data were collected at ambient temperature on a Rigaku Dmax/III diffractometer, using Cu Kα radiation (λ = 1.54 Å) with an accelerating voltage and current of 40 kV and 30 mA, respectively. Samples were scanned from 4° to 60°, with a step length of 0.02° at 24 °C.

Dynamic mechanical analysis (DMA) was also performed with a TA Q800 dynamic mechanical analyzer (USA). The test was carried out at the tension condition with a frequency of 1 Hz. In each case, a rectangular sample (typical dimensions = 8 × 6 × 0.5 mm) was loaded under tension and an oscillatory deformation with an amplitude of 0.5% strain and a force track of 125% was applied. The scanning temperature was ranged from −50 to 100 °C at a heating rate of 3 °C/min.

The cryogenically fractured surfaces of the ENR/HNT composites were observed by field emission scanning electron microscopy (FESEM, Nova Nano SEM 430). Energy dispersive X-ray spectra (EDS) of Zn in the ZDA-cured ENR samples were collected using a Horiba EX-250 spectrometer.

Shape memory behaviors of ZDF-cured ENR and ENR/HNT nanocomposites were characterized by the thermomechanical cycle experiments and measured with a TA Q800 instrument. The test was carried out under controlled force mode. Prior to deformation, samples with dimension of 8 × 6 × 0.5 mm were heated to 30 °C above T_g, and equilibrated for 10 min. In step 1 (deformation step), samples were deformed with an extension rate of 0.5 N/min, from a preloaded value of 0.001 N to a designated value with a setting strain (S_s). In step 2 (cooling step), samples were cooled to 40 °C below T_g (T_g-40 °C) at a rate of 3 °C/min under a constant force to prevent the recovery. In step 3 (unloading and shape fixing step), the force imposed on the samples was unloaded to the preloaded value of 0.001 N at a rate of 0.5 N/min. An additional 20 min isothermal step was followed to ensure shape fixing at T_g-40 °C. Upon unloading, part of the strain (S_s-S_f) was instantaneously recovered, leaving a fixing strain (S_f). In the final step (recovery step), samples were reheated to 30 °C above T_g at a rate of 3 °C/min and held for another 10 min to recover any possible residual strain. Then a permanent strain ($S_{p(n-1)}$) was left. This four-step thermomechanical cycle was repeated three times for each sample.

Table 1. Compositions of ENR and ENR/HNT composites cured by ZDF.

Sample	EFeZn5	EFeZn10	EFeZn15	EFeZn25	EFeZn40	EFeZn25H10	EFeZn25H20	EFeZn25H40	EFeZn25H60
ENR	100	100	100	100	100	100	100	100	100
FA	5	10	15	25	40	25	25	25	25
ZnO	1.05	2.10	3.15	5.25	8.40	5.25	5.25	5.25	5.25
HNTs						10	20	40	60

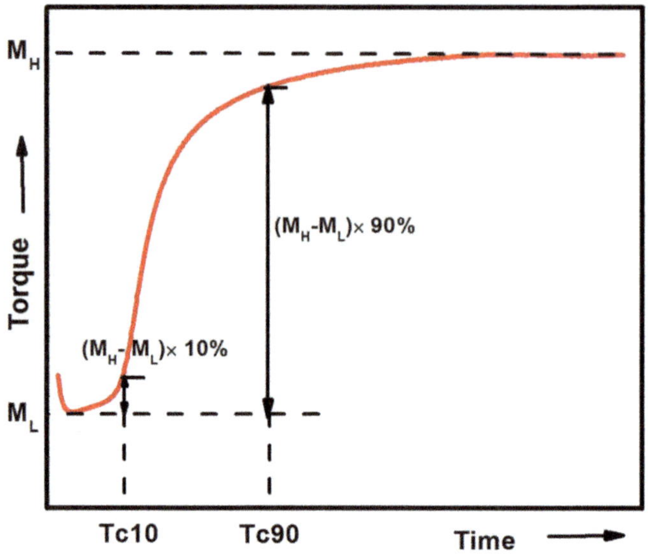

Figure 1. Typical curing curve for rubber compound.

The fixing ratios (R_f) and recovery ratios (R_r) were defined as:

$$R_f(\%) = \frac{S_f - S_{p(n-1)}}{S_s - S_{p(n-1)}} \times 100, \quad (1)$$

$$R_r(\%) = \frac{S_s - S_{p(n)}}{S_s - S_{p(n-1)}} \times 100. \quad (2)$$

3. Results and discussion

3.1. SME of ENR/ZDF network

To verify to the reaction between ZnO and FeA and the formation of ZDF, XRD data was collected (Figure 2). In the experiment, ZnO and FeA were incorporated into SBR, and the resulting FeA/ZnO/SBR compound was treated at 160 °C for different time. It should be noted that SBR is chosen as the rubber matrix to replace ENR, with the aim to avoid the consumption of the yielded ZDF. The samples are named as SFeZn25-xmin, where S represents SBR to distinguish with ENR. XRD patterns show that the peaks of SFeZn25-0 min are the superposition of peaks of FeA and ZnO. The heat treated samples (SFeZn25-10 min, SFeZn25-20 min, SFeZn25-30 min) exhibit a new peak at 6.2°, and the intensity increase monotonously with the increasing of curing time. What's more, the intensity of peaks belonging to FeA and ZnO reduce greatly along the curing time. Both the appearance of a new peak and the reduction of peak intensity of FeA and ZnO provide convincing evidences for the formation of ZDF.

According to previous work, the crosslinking mechanism between ENR and ZDF is related to oxa-Michael reaction. Scheme 1 shows the formula of the crosslinking reaction. Due to ring-opening of epoxy group in ENR, hydroxyl formed under the catalysis of Zn^{2+}. Then the conjugated C=C in ZDF is attacked by the generated –OH, resulting in

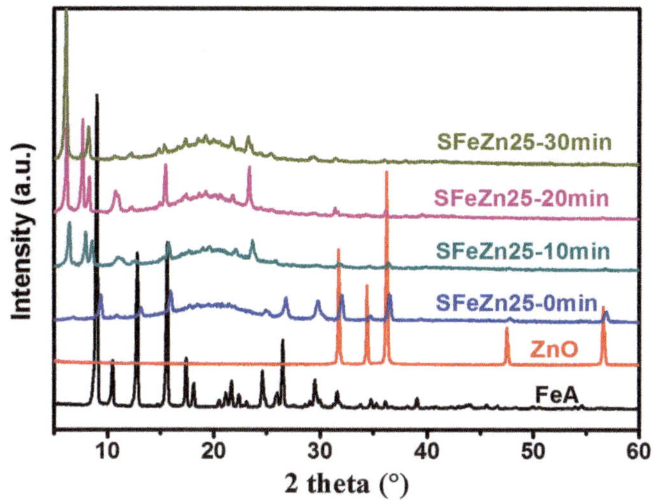

Figure 2. XRD patterns of FeA, ZnO, FeA/ZnO/SBR compound (SFeZn25-0 min) and heat treated FeA/ZnO/SBR compounds with different curing time (SFeZn25-10 min, SFeZn25-20 min, SFeZn25-30 min).

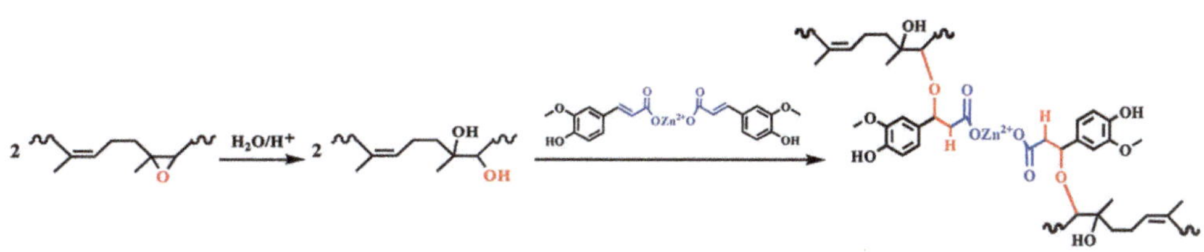

Scheme 1. Formula of the crosslinking reaction of ENR cured by ZDF.

formation of ether bond as the crosslinks. The detailed molecular mechanism of this crosslinking reaction has been disclosed via the characterizations of model substance [26]. To investigate the dispersion of ZDF, SEM and EDS spectra for EFeZn40 were recorded (Figure 3). From the SEM photo (Figure 3(a)), one can observe that, even though high content of FA and ZnO are added, the sample exhibits smooth fracture surface and no aggregated particles are found. The observed ridges are actually caused by the cryogenic fracture process. The dispersion of ZDF could further be evaluated by the zinc element mapping. From the EDS spectrum (Figure 3(b)), it is clearly that zinc element is uniformly dispersed in the whole sample. The uniform dispersion of ZDF is realized by the conversion of inorganic ZnO particles into reactive ZDF molecules and the subsequent crosslinking of ENR by ZDF through oxa-Michael reaction.

DMA measurement was conducted to study the influence of ZDF content on T_g of ENR. Figure 4 presents curves of tan δ of ZDF-cured ENR with various ZDF content. As revealed, with the increasing of ZDF content, peak values of tan δ decrease continuously. For example, when ZDF content increases from 5 phr to 40 phr, tan δ decreases by 33.3%. This decrease should be ascribed to the consistently increased confinement of chain motion arising from the increase in crosslink density. Regarding to T_g, along the increasing of curing agents, T_g exhibits an increasing trend (inset in Figure 4). The T_gs for all the

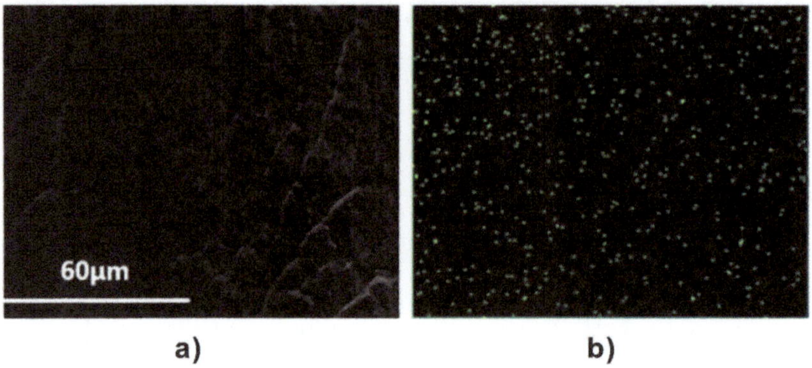

Figure 3. SEM image (left) and EDS spectrum of zinc element (right) for EFeZn40.

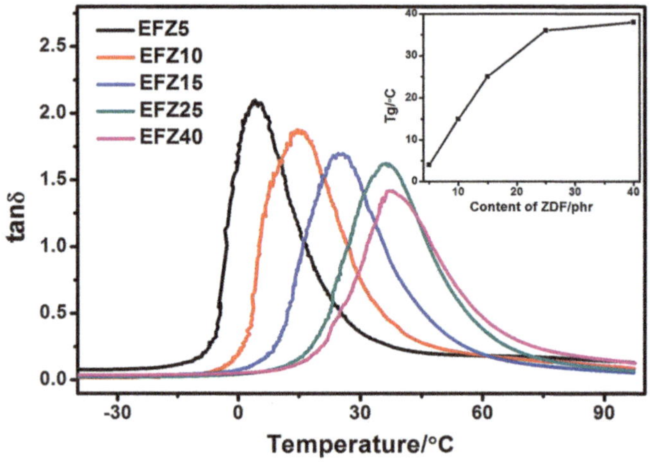

Figure 4. Tan δ of ENR cured by ZDF with variable ZDF content.

samples are above 0 °C, which are quite high values for crosslinked diene-based rubbers. The increment in T_g is also ascribed to the strong confinement of segment motion due to the increased crosslink density. From these results, we may come to a conclusion that T_g of ZDF-cured ENR can be tuned in a wide range from 4 °C to 38 °C, just by changing curing agent concentration. Accordingly, these ZDF-cured ENR exhibits potential as SMPs.

To evaluate the SME of ZDF-cured ENR, the thermomechanical cycling was performed under controlled force mode. The thermomechanical cycling for every sample is repeated for three times. As revealed in Figure 5, the strain change during unloading (step 3) is minimal, indicating excellent fixity of temporary shape while the starting point is almost overlapped with the ending point, suggesting good recovery of the original shape. Although the first cycle shows a little difference from the latter two, the second and third cycles are completely overlapped, indicating that shape memory performances of both EFeZn15 and EFeZn25 show excellent repeatability. The fixing ratio (R_f) and recovery ratio (R_r) were calculated and compared in Figure 6. Both R_f and R_r of EFeZn15 and

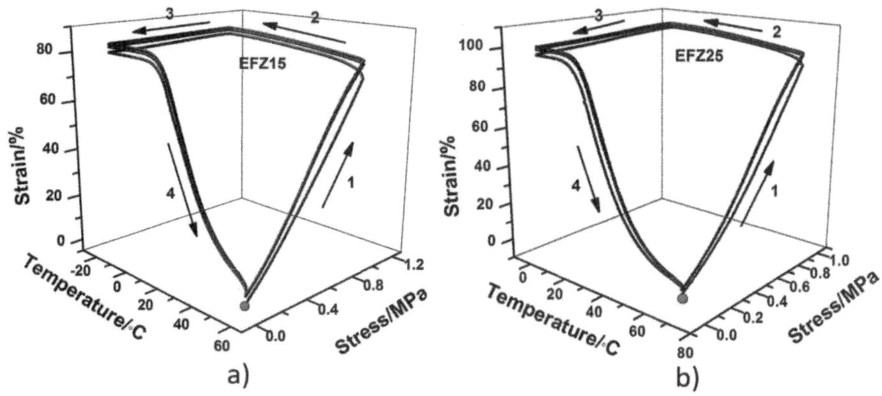

Figure 5. Thermomechanical cycles of EFeZn15 (a) and EFeZn25 (b) to characterize the shape memory behaviors. The red points indicate the starting point of the test.

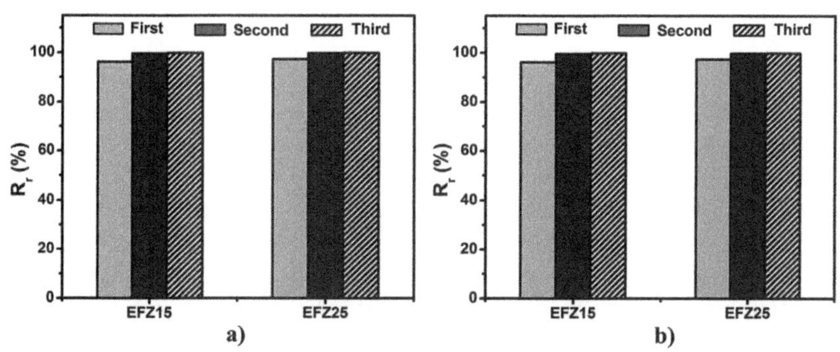

Figure 6. Histogram of R_f (a) and R_r (b) of EFeZn15 and EFeZn25.

EFeZn25 are more than 93%, suggesting that changes in ZDF concentration have a minimal effect on shape memory performance. It is notable that the R_f and R_r are improved to some degree with the increase in cycling times. The relatively poorer R_f and R_r of first cycle should be ascribed to the unrelaxed stress remained during processing. Besides, the slippage of excessive crystalline aggregates of ZDF may be another origin for the imperfect R_f and R_r in the first cycle. As to the following cycling, much better shape memory performance is realized due to the relaxation of the residual stress and rearrangement of the excessive ZDF aggregates during the first cycling. Similar behaviors have been found in other systems filled with small crystalline substance or inorganic filler [25,33].

3.2. Regulation of SME of ZDF-cured ENR by HNTs

HNTs are naturally occurring nanotubular clay. Compared to the more commonly used 2D clay such as montmorillonite, HNTs can be easily dispersed via conventional processing facilities [34]. In virtue of the favorable interfacial interaction, such as hydrogen bonding, HNTs have been demonstrated to be very effective in reinforcing polar rubber without sophisticated treatment [35]. In the present study, to further

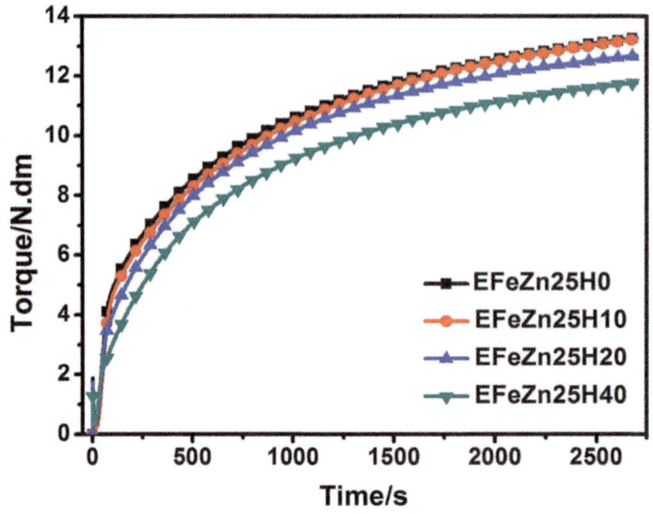

Figure 7. Curing curves of ENR and ENR/HNT nanocomposites cured by ZDF.

regulate the SME of ZDF-cured ENR, HNTs are incorporated as reinforcement for ZDF-cured ENR.

Curing curve measured using a vulcameter is a convenient method to monitor the curing process and reflect curing characteristics, such as scorch time, optimal curing time, maximum torque (M_H), and curing rate. Generally the maximum torque is indicative for crosslink density. Figure 7 depicts curing profiles of ENR and ENR/HNT nanocomposites cured by ZDF. As shown, the M_H and curing rate decrease continuously with the increment in HNT content, indicating the detrimental effect of HNTs on crosslinking of ZDF-cured ENR. This observation may be attributed to the partial consumption of ZDF by HNTs. As was reported, Zn^{2+} can catalyze the oxa-Michael reaction between ENR and ZDA. For one thing, ZDA can promote nucleophilic attack between –OH in ENR and –C=C– in ZDA due to formation of carbonyl-metal ion complexation [36]. For the other thing, ZDA can act as nucleophilic cation as well as Lewis acid to catalyze the ring-opening reaction of ENR [37,38]. Similarly, the *in situ* formed ZDF also acts as catalyst and curing agent. However, with addition of plenty of HNTs, ZDF are partially adsorbed on the surface of HNTs due to hydrogen bonds and complexation interaction, which consequently suppresses the catalytic and crosslinking efficiencies. Therefore, although HNTs may improve torque via the hydromechanics effect, the resultant M_H is unexpectedly decreased due to the suppressed crosslinking of the matrix.

The dispersion state of HNTs in ZDF-cured ENR is evaluated by morphological observations. Figure 8 displays SEM images of ENR/HNT nanocomposites cured by ZDF with 10 phr and 40 phr HNTs, respectively. As shown, even though 40 phr HNTs are added, HNTs disperse uniformly and most of the tubes are individualized in the ENR matrix. Considering that existence of plenty of hydroxyls in the ZDF-cured ENR network, which could form hydrogen bonds with surface hydroxyls on the surface of HNTs, so the uniform dispersion of HNTs should be attributed to the hydrogen bonding between ENR and HNTs. In addition, the residual epoxy groups also make a contribution to the well dispersion due to the potential hydrogen bonds between epoxy groups and surface hydroxyls on HNTs. Similar interaction between carboxylated styrene-butadiene rubber

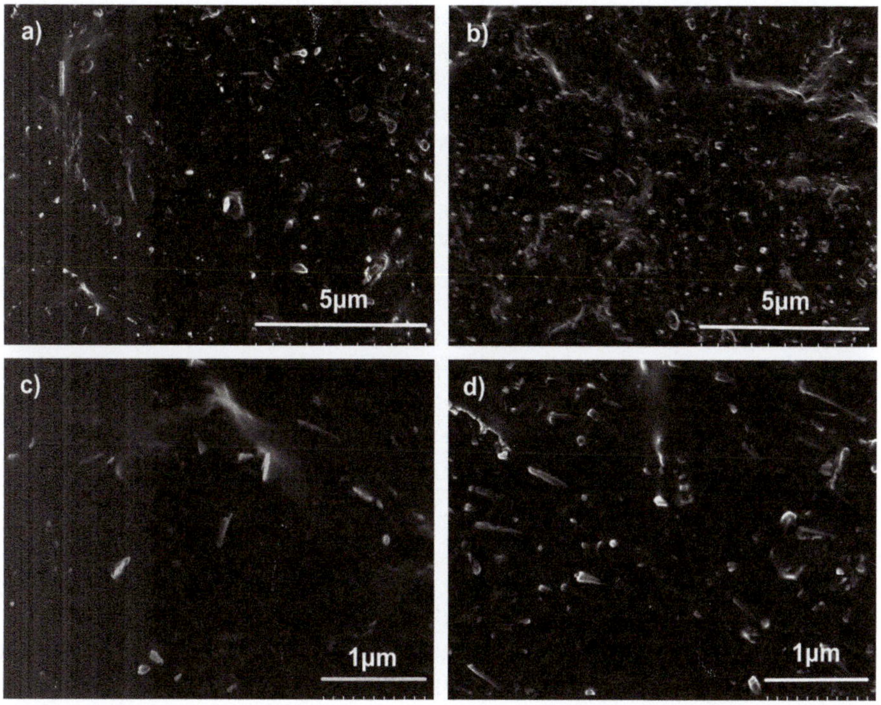

Figure 8. SEM images of ZDF-cured ENR/HNT composites: EFeZn25H10 (a, c) and EFeZn25H40 (b, d).

and HNTs has been reported [35]. As shown in Figure 8, blurry interfaces are exhibited, which further suggests strong interfacial interaction.

Figure 9 depicts stress-strain curves of ZDF-cured ENR/HNT nanocomposites with variable HNT loading at temperatures below and above T_g (at 70 °C). As indicated, with the increasing of HNT content, tensile strength and modulus increase continuously. Compared with neat ENR, modulus (at 100% strain) of EFeZn25H60 increases by

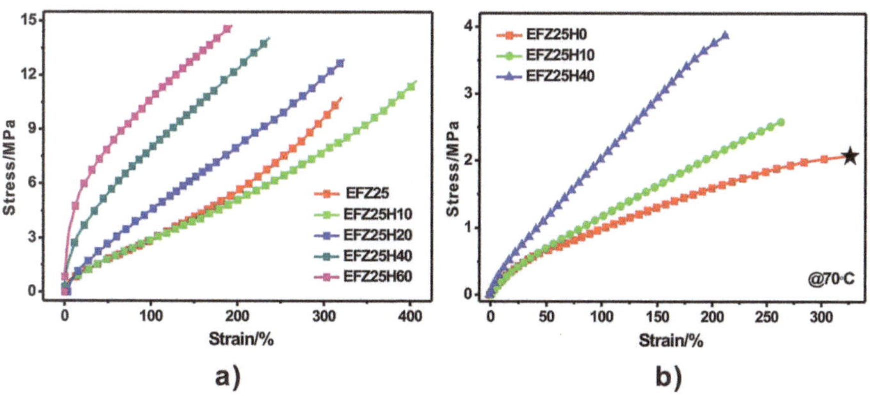

Figure 9. Stress-strain curves of ENR/HNT nanocomposites cured by ZDF at room temperature (a) and at 70 °C (b). The black asterisk is used to illustrate that the sample is not fractured but the test is stopped due to reaching the elongation limit of DMA instruments.

250% (from 3.0 to 10.6 MPa). The striking enhancements in moduli should originate from the strong interfacial adhesion and excellent compatibility between ENR and HNTs, which is consistent with the morphological observations. Meanwhile, elongation at break decreases sharply with increment of HNTs. The phenomenon should be ascribed to the inevitable aggregation of partial HNTs when high content of HNT is incorporated. These aggregated HNTs act as regions of stress concentration and results in the decrease in elongation at break. It is essential to evaluate the tensile properties at temperature above T_g since forces will be applied to samples at high temperature when thermomechanical cycling of SMPs are performed. Figure 9(b) displayed stress-stain curves of ZDF-cured ENR and ENR/HNT composites at 70 °C. An obviously enhanced tensile strength and modulus are exhibited at elevated temperature with the inclusion of HNTs. For example, compared with neat ENR, modulus (at 100% strain) of EFeZn25H40 increases by one-fold. In summary, the incorporation of HNTs effectively improves modulus of ZDF-cured ENR both at room temperature and elevated temperature.

To investigate the effect of HNT content on viscoelasticity of ENR/HNT composites, DMA measurement was performed. The dependences of storage modulus (E') and loss factor (tan δ) on temperature for neat ENR and ENR/HNT composites are show in Figure 10(a) and (b), respectively. All the samples exhibit glassy state with high modulus at low temperature and rubbery state with low modulus at high temperature. The sharp decrease in E' around 30 °C is due to the glass transition. The transition favors the deformation at elevated temperature and fixity of temporary shape at lower temperature, ensuring the shape memory performance. Via the inclusion of HNTs, E' of composite in both glassy region and rubbery region improves strikingly. For example, compared with neat ENR, E' values for EFeZn25H40 at 0 °C and 70 °C increase by about one-fold and seven-fold, respectively. The dramatic reinforcement of HNTs toward ENR is ascribed to the uniform dispersion of HNTs facilitated by strong interfacial adhesion between HNTs and ENR, which guarantees efficient force transferring between HNTs and ENR. Theoretically, higher modulus in glassy state is beneficial to R_f. As is well known, temporary shape of SMPs are fixed at glassy state. For sample with higher modulus in glassy state, larger stress is required for the relaxation and creep of polymer chain, which results in a reduced retraction during fixing and an increased R_f. On the contrary, higher modulus in rubbery state is detrimental to shape recovery. The drive force for shape recovery is the entropy elasticity, which is regulated by temperature. For sample with

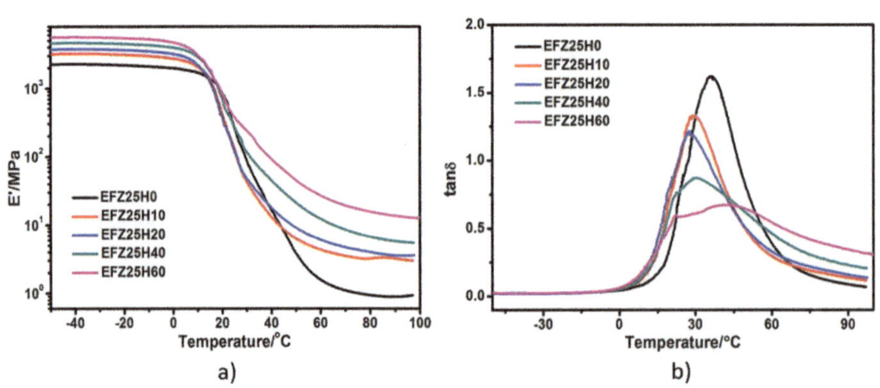

Figure 10. Storage modulus (a) and tan δ (b) of neat ENR and ENR/HNT composites as a function of temperature.

higher modulus in rubbery state, larger recovery stress and higher entropy elasticity are required during shape recovery. Accordingly, under the same recovery temperature, lower R_r and recovery rate are exhibited. The peak value of tan δ decreases consistently with the increasing filler loading, indicating the continuous decrease of segment mobility caused by HNTs. This phenomenon is attributed to the formation of large interface region and strong interfacial interaction. Moreover, it is notable that the starting temperature of T_g transition is declined along the increasing of HNT loading, indicating a decreased crosslink degree of ENR, which is consistent with the results of curing curves (Figure 5). As is discussed above, the partial absorption of ZDF by HNTs should be responsible for this phenomenon.

Thermomechanical cycles of ENR and ENR/HNT composites including variable filler contents under force controlled mode were conducted to investigate the effect of HNTs on shape memory performance. The setting strains are all similar value (approximately 100%) to more accurately study the effect. As shown in Figure 11, all samples exhibit excellent shape fixing and recovery properties on the whole, as evidenced by little strain change during unloading (step 3) and the overlapping of the starting point and ending point of thermomechanical cycles. It is notable that with almost the same strain, compared with neat ENR, the corresponding stress (step 2) for EFeZn25H40 increases greatly by 138%. Therefore, the incorporation of inorganic nanotubes seems to be an efficient way to tune stress of the SMPs. Figure 12 summarizes R_f and R_r of ZDF-cured ENR and ENR/HNT composites. In Figure 12(a), all samples exhibit a high R_f value and the R_f value is increasing with the increment of HNT content. By incorporating HNTs, interfacial interaction between HNTs and ENR confines the motion of ENR chain and retards the relaxation of polymer chain. Therefore, retraction during stretching and cooling is less and larger R_f is found for ENR/HNT composites. Figure 12(b) depicts the effect of HNT content on the R_r value for ENR/HNT composites, which shows a decreasing trend in the first cycle along the increase in HNT content. This unexpected result is related to the inevitable aggregate of nanotubes as well as unexpected slippage during stretching. On one hand, when HNTs are incorporated, excess HNTs tend to aggregate, which is

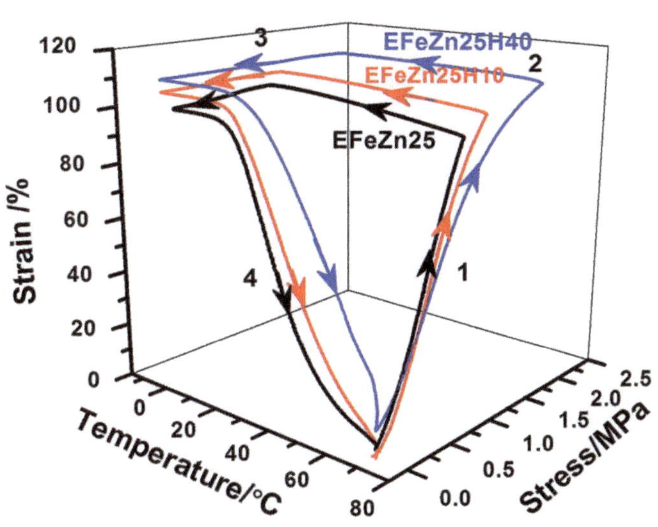

Figure 11. Thermomechanical cycles of ENR and ENR/HNT composites.

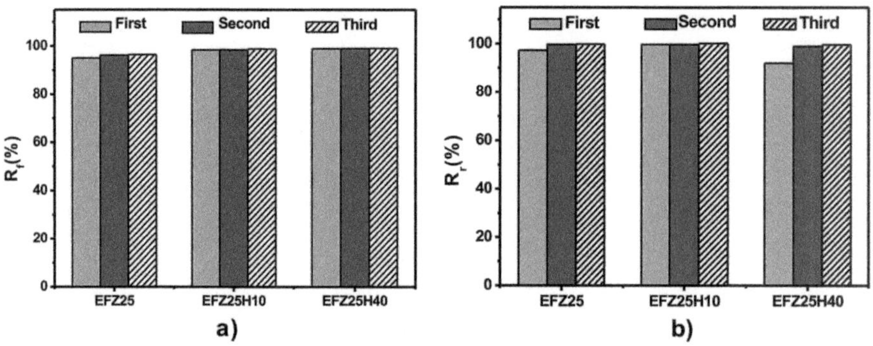

Figure 12. R_f (a) and R_r (b) of ENR and ENR/HNT composites.

detrimental to elasticity of the composites. On the other hand, the slippage between ENR chain and HNTs during stretching is inevitable and the corresponding deformation is permanent, which is detrimental to shape recovery. The two phenomena become apparent at higher HNT content. Therefore, with increasing HNT content, the R_r value decreases consistently. Moreover, the improvement of modulus caused by HNT incorporation discussed above also makes a contribution to the trend of R_f and R_r change. Notably, R_r of ENR/HNT composites could be tuned to high value by increasing the cycling times. Such increase in the R_r value along increasing cycling numbers is due to that the rearrangement of the fillers occur during thermomechanical cycles, which decreases the deformation caused by slipping between ENR chain and HNTs.

To demonstrate the shape memory process visually, shape evolution photos of EFeZn25 and EFeZn25H40 from a temporary spiral to permanent linear shape are taken and arranged in Figure 13. The recovery process for EFeZn25 and EFeZn25H40 at 40 °C are displayed in Figure 13(a) and (b), respectively. One can observe that when 40 phr HNTs are incorporated, the recovery rate declines obviously under the same recovery temperature. For example, the recovery time of EFeZn25 is only 7 s while for EFeZn25H40, the recovery time reaches 20 s. However, when the recovery temperature of EFeZn25H40 is elevated to 60 °C (Figure 13(c)), the recovery rate could be tuned to that achieved in EFeZn25, indicating the recovery rate is conveniently tuned. According to the reports by Kim *et al.* [39,40], the sharp transition from glassy state to rubbery state is crucial to the material for use as SMPs. A high elasticity ratio between glass-state modulus and rubbery-state modulus (E_g'/E_r') allows easy shaping at $T > T_g$ and good resistance to deformation at $T < T_g$. For samples used to take evolution photos, the order of value of E_g'/E_r' is (c) > (a) > (b) (the values are 320, 246 and 131, respectively, calculated from corresponding DMA data), which is coincident with recovery rates displayed in Figure 13. Therefore, we can simply regulate the recovery rate by changing elasticity ratio, which is controllable via changing filler content and recovery temperature. Considering the tunable T_g in a wide range by changing ZDF loading, tunable stress by changing HNT content and tunable recovery rate, this rubber-based SMP is highly controllable and is potential in extensive applications.

At last, it is necessary to emphasize that the relative low triggering temperature may cause the question that the temporary shape of such SMPs cannot be readily fixed at room temperature, which is a widespread question for SMPs with relative low transition temperature. However, a cold condition for the storage of those SMPs is easy to achieve.

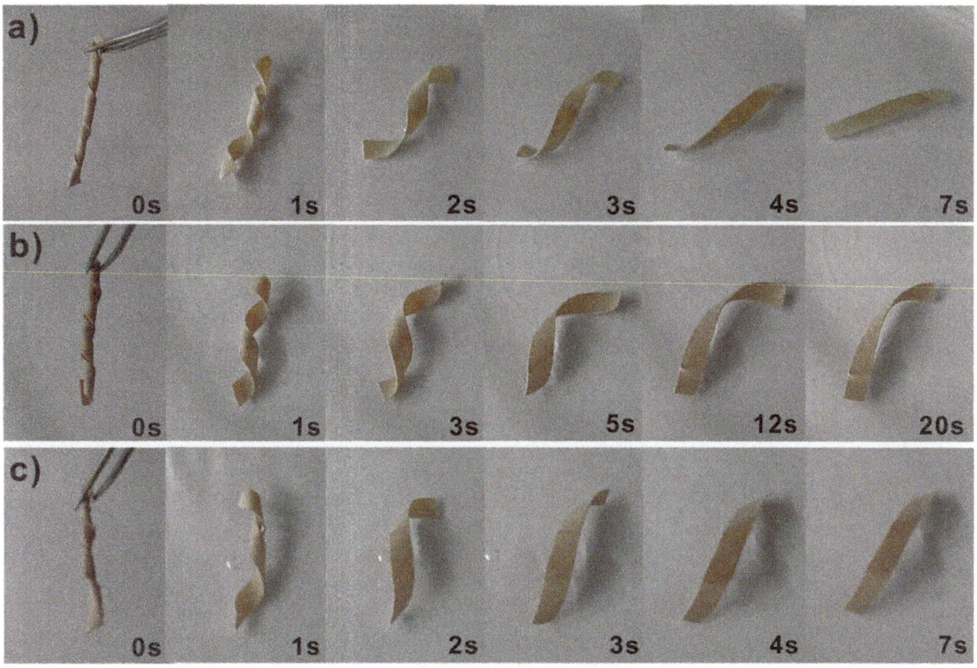

Figure 13. Evolution of (a) EFeZn25 at 45 °C, (b) EFeZn25H40 at 45 °C and (c) EFeZn25H40 at 60 °C from a temporary spiral to permanent linear shape.

On the other hand, SMPs with low transition temperature is also desired in some specific cases. For example, human tissues are soft and vulnerable, and body temperature, namely the triggering temperature for biological SMPs, is about 37°C, which requires low triggering temperatures and low moduli of SMPs when those materials enter into human body.

4. Conclusion

The viability of a kind of rubber-based SMPs fabricated with ZDF-cured ENR is demonstrated in utility temperature range. Excellent shape memory performance is achieved as evidenced by high R_f and R_r. The triggering temperature can be tuned in a wide range from 4 °C to 38 °C just by changing curing agent concentration. HNTs are incorporated to further regulate the shape memory effect. With the incorporation of HNTs, the stress and recovery rate of the SMPs are found to be tunable, which widens the application of this kind of SMPs. The main raw materials for the developed SMPs (NR, FA, and HNTs) are inexpensive and sustainable, endowing extra significance to this kind of SMPs. The combination of adoption of sustainable raw materials, and the excellent and tunable SME makes these SMPs potentially useful in many applications, such as various actuators and heat-shrinkable package materials.

Disclosure statement

No potential conflict of interest was reported by the authors.

Funding

This work was supported by the National Natural Science Foundation of China [grant number 51222301], [grant number 51473050], [grant number 51333003]; Research Fund for the Doctoral Program of Higher Education of China [grant number 20130172110001]; Fundamental Research Funds for the Central Universities [grant number 2014ZG0001].

References

[1] E. Gil and S. Hudson, *Stimuli-reponsive polymers and their bioconjugates*, Prog. Polym. Sci. 29 (2004), pp. 1173–1222. doi:10.1016/j.progpolymsci.2004.08.003

[2] C. Zeng, H. Seino, J. Ren, and N. Yoshie, *Polymers with multishape memory controlled by local glass transition temperature*, ACS Appl. Mater. Interfaces 6 (2014), pp. 2753–2758. doi:10.1021/am405287p

[3] Z. Tang, D. Sun, D. Yang, B. Guo, L. Zhang, and D. Jia, *Vapor grown carbon nanofiber reinforced bio-based polyester for electroactive shape memory performance*, Compos. Sci. Technol. 75 (2013), pp. 15–21. doi:10.1016/j.compscitech.2012.11.019

[4] A. Lendlein, H. Jiang, O. Jünger, and R. Langer, *Light-induced shape-memory polymers*, Nature 434 (2005), pp. 879–882. doi:10.1038/nature03496

[5] W.M. Huang, B. Yang, L. An, C. Li, and Y.S. Chan, *Water-driven programmable polyurethane shape memory polymer: Demonstration and mechanism*, Appl. Phys. Lett. 86 (2005), pp. 114105. doi:10.1063/1.1880448

[6] F. Li, X. Zhang, J. Hou, M. Xu, X. Luo, D. Ma, and B.K. Kim, *Studies on thermally stimulated shape memory effect of segmented polyurethanes*, J. Appl. Polym. Sci. 64 (1997), pp. 1511–1516. doi:10.1002/(ISSN)1097-4628

[7] J. Hao and R. Weiss, *Mechanically tough, thermally activated shape memory hydrogels*, ACS Macro. Lett. 2 (2013), pp. 86–89. doi:10.1021/mz3006389

[8] A. Lendlein and S. Kelch, *Shape-memory polymers*, Angew. Chem. Int. Edit. 41 (2002), pp. 2034–2057. doi:10.1002/1521-3773(20020617)41:12<2034::AID-ANIE2034>3.0.CO;2-M

[9] M. Behl, J. Zotzmann, and A. Lendlein, *Shape-memory polymers and shape-changing polymers*, in *Shape-Memory Polymers*, A. Lendlein, eds., Vol. 226, Springer, Berlin, 2010, pp. 1–40.

[10] J. Kunzelman, T. Chung, P.T. Mather, and C. Weder, *Shape memory polymers with built-in threshold temperature sensors*, J. Mater. Chem. 18 (2008), pp. 1082–1086. doi:10.1039/b718445j

[11] A. Lendlein and R.S. Langer, *Biodegradable Shape Memory Polymeric Sutures*, United States Patent US8834522, 2012.

[12] C. Liu, H. Qin, and P. Mather, *Review of progress in shape-memory polymers*, J. Mater. Chem. 17 (2007), pp. 1543–1558. doi:10.1039/b615954k

[13] M.C. Serrano, L. Carbajal, and G.A. Ameer, *Novel biodegradable shape-memory elastomers with drug-releasing capabilities*, Adv. Mater. 23 (2011), pp. 2211–2215. doi:10.1002/adma.v23.19

[14] H. Zhang, H. Wang, W. Zhong, and Q. Du, *A novel type of shape memory polymer blend and the shape memory mechanism*, Polymer 50 (2009), pp. 1596–1601. doi:10.1016/j.polymer.2009.01.011

[15] R. Weiss, E. Izzo, and S. Mandelbaum, *New design of shape memory polymers: Mixtures of an elastomeric ionomer and low molar mass fatty acids and their salts*, Macromolecules 41 (2008), pp. 2978–2980. doi:10.1021/ma8001774

[16] Y.-W. Chang, J.K. Mishra, J.-H. Cheong, and D.-K. Kim, *Thermomechanical properties and shape memory effect of epoxidized natural rubber crosslinked by 3-amino-1,2,4-triazole*, Polym. Int. 56 (2007), pp. 694–698. doi:10.1002/(ISSN)1097-0126

[17] H. Kang, M. Li, Z. Tang, J. Xue, X. Hu, L. Zhang, and B. Guo, *Synthesis and characterization of biobased isosorbide-containing copolyesters as shape memory polymers for biomedical applications*, J. Mater. Chem. B 2 (2014), pp. 7877–7886. doi:10.1039/C4TB01304B

[18] Y. Liu, K. Yao, X. Chen, J. Wang, Z. Wang, H.J. Ploehn, C. Wang, F. Chu, and C. Tang, *Sustainable thermoplastic elastomers derived from renewable cellulose, rosin and fatty acids*, Polym. Chem. 5 (2014), pp. 3170–3181. doi:10.1039/c3py01260c

[19] S. Wang, S. Vajjala Kesava, E.D. Gomez, and M.L. Robertson, *Sustainable thermoplastic elastomers derived from fatty acids*, Macromolecules 46 (2013), pp. 7202–7212. doi:10.1021/ma4011846

[20] Y. Xia and R.C. Larock, *Vegetable oil-based polymeric materials: Synthesis, properties, and applications*, Green Chem. 12 (2010), pp. 1893–1909. doi:10.1039/c0gc00264j

[21] R. Bhardwaj and A.K. Mohanty, *Advances in the properties of polylactides based materials: A review*, J. Biobased Mater. Bioenergy 1 (2007), pp. 191–209. doi:10.1166/jbmb.2007.023

[22] P.A. Wilbon, F. Chu, and C. Tang, *Progress in renewable polymers from natural terpenes, terpenoids, and rosin*, Macromol. Rapid Commun. 34 (2013), pp. 8–37. doi:10.1002/marc.v34.1

[23] A. Gandini, *The irruption of polymers from renewable resources on the scene of macromolecular science and technology*, Green Chem. 13 (2011), pp. 1061–1083. doi:10.1039/c0gc00789g

[24] T. Lin, S. Ma, Y. Lu, and B. Guo, *New design of shape memory polymers based on natural rubber crosslinked via oxa-Michael reaction*, ACS Appl. Mater. Interfaces 6 (2014), pp. 5695–5703. doi:10.1021/am500236w

[25] T. Lin and B. Guo, *Curing of rubber via oxa-Michael reaction toward significantly increased aging resistance*, Ind. Eng. Chem. Res. 52 (2013), pp. 18123–18130. doi:10.1021/ie403485e

[26] T. Lin, X. Zhang, Z. Tang, and B. Guo, *Renewable conjugated acids as curatives for high-performance rubber/silica composites*, Green Chem. 17 (2015), pp. 3301–3305. doi:10.1039/C5GC00834D

[27] D.N. Krishnan, N. Prasanna, E.P. Sabina, and M. Rasool, *Hepatoprotective and antioxidant potential of ferulic acid against acetaminophen-induced liver damage in mice*, Comp. Clin. Path 22 (2012), pp. 1177–1181. doi:10.1007/s00580-012-1546-y

[28] N. Kumar and V. Pruthi, *Potential applications of ferulic acid from natural sources*, Biotechnol. Rep. 4 (2014), pp. 86–93. doi:10.1016/j.btre.2014.09.002

[29] A. Saija, A. Tomaino, R.L. Cascio, D. Trombetta, A. Proteggente, A. De Pasquale, N. Uccella, and F. Bonina, *Ferulic and caffeic acids as potential protective agents against photooxidative skin damage*, J. Sci. Food Agric. 79 (1999), pp. 476–480. doi:10.1002/(ISSN)1097-0010

[30] C. Mancuso and R. Santangelo, *Ferulic acid: Pharmacological and toxicological aspects*, Food Chem. Toxicol. 65 (2014), pp. 185–195. doi:10.1016/j.fct.2013.12.024

[31] E. Graf, *Antioxidant potential of ferulic acid*, Free Radic. Biol. Med. 13 (1992), pp. 435–448. doi:10.1016/0891-5849(92)90184-I

[32] D.G. Shchukin, G.B. Sukhorukov, R.R. Price, and Y.M. Lvov, *Halloysite nanotubes as biomimetic nanoreactors*, Small 1 (2005), pp. 510–513. doi:10.1002/(ISSN)1613-6829

[33] W.S. Guo, H.L. Kang, Y.W. Chen, B.C. Guo, and L.Q. Zhang, *Stronger and faster degradable biobased poly(propylene sebacate) as shape memory polymer by incorporating boehmite nanoplatelets*, ACS Appl. Mater Interfaces 4 (2012), pp. 4006–4014. doi:10.1021/am300828u

[34] M.L. Du, B.C. Guo, and D.M. Jia, *Newly emerging applications of halloysite nanotubes: A review*, Polym. Int.. 59 (2010), pp. 574–582.

[35] M. Du, B. Guo, Y. Lei, M. Liu, and D. Jia, *Carboxylated butadiene–styrene rubber/halloysite nanotube nanocomposites: Interfacial interaction and performance*, Polymer 49 (2008), pp. 4871–4876. doi:10.1016/j.polymer.2008.08.042

[36] Y.-M. Hong, Z.-L. Shen, X.-Q. Hu, W.-M. Mo, X.-F. He, B.-X. Hu, and N. Sun, *Acid-catalyzed intramolecular oxa-Michael addition reactions under solvent-free and microwave irradiation conditions*, Arkivoc. 14 (2009), pp. 146–155.

[37] M. Chini, P. Crotti, L.A. Flippin, F. Macchia, and M. Pineschi, *Regiochemical control of the ring-opening of 1, 2-epoxides by means of chelating processes. 2. Synthesis and reactions of the cis-and trans-oxides of 4-[(benzyloxy) methyl] cyclohexene, 3-cyclohexenemethanol, and methyl 3-cyclohexenecarboxylate*, J. Org. Chem. 57 (1992), pp. 1405–1412. doi:10.1021/jo00031a018

[38] I. Paterson and D.J. Berrisford, *Meso epoxides in asymmetric synthesis: Enantioselective opening by nucleophiles in the presence of chiral Lewis acids*, Angew. Chem. Int. Ed. Engl. 31 (1992), pp. 1179–1180. doi:10.1002/(ISSN)1521-3773

[39] B.K. Kim, S.Y. Lee, and M. Xu, *Polyurethanes having shape memory effects*, Polymer 37 (1996), pp. 5781–5793. doi:10.1016/S0032-3861(96)00442-9

[40] B.K. Kim, S.Y. Lee, J.S. Lee, S.H. Baek, Y.J. Choi, J.O. Lee, and M. Xu, *Polyurethane ionomers having shape memory effects*, Polymer 39 (1998), pp. 2803–2808. doi:10.1016/S0032-3861(97)00616-2

8

Natural rubber/nitrile butadiene rubber/hindered phenol composites with high-damping properties

Xiuying Zhao[a,b], Jingna Yang[a], Detao Zhao[a], Yonglai Lu[a,b], Wencai Wang[b]*, Liqun Zhang[a]* and Toshio Nishi[c]

[a]Key Laboratory of Beijing City on Preparation and Processing of Novel Polymer Materials, Beijing University of Chemical Technology, Beijing 100029, P. R. China; [b]Beijing Engineering Research Center of Advanced Elastomers, Beijing University of Chemical Technology, Beijing 100029, P. R. China; [c]Department of Organic and Polymeric Materials, Graduate School of Science and Engineering, Tokyo Institute of Technology, 2-12-1 O-okayama, Meguro-ku, Tokyo 152-8552, Japan

New natural rubber (NR)/nitrile butadiene rubber (NBR)/hindered phenol (AO-80) composites with high-damping properties were prepared in this study. The morphological, structural, and mechanical properties were characterized by atomic force microscopy (AFM), polarized Fourier transform infrared spectrometer (FTIR), dynamic mechanical thermal analyzer (DMTA), and a tensile tester. Each composite consisted of two phases: the NR phase and the NBR/AO-80 phase. There was partial compatibility between the NR phase and the NBR/AO-80 phase, and the NR/NBR/AO-80 (50/50/20) composite exhibited a co-continuous morphology. Strain-induced crystallization occurred in the NR phase at strains higher than 200%, and strain-induced orientation appeared in the NBR/AO-80 phase with the increase of strain from 100% to 500%. The composites had a special stress–strain behavior and mechanical properties because of the simultaneous strain-induced orientation and strain-induced crystallization. In the working temperature range of a seismic isolation bearing, the composites (especially the NR/NBR/AO-80 (50/50/20) composite) presented a high loss factor, high area of loss peak (TA), and high hysteresis energy. Therefore, the NR/NBR/AO-80 rubber composites are expected to have important application as a high-performance damping material for rubber bearing.

Keywords: natural rubber; nitrile butadiene rubber; damping properties; strain-induced orientation; rubber bearing

1. Introduction

As highly unexpected natural disasters, earthquakes seriously do damage to human lives and properties. Destruction of buildings is the most direct cause of economic losses and casualties. Rubber bearings, as the most commonly used seismic isolators for buildings, are composed of alternating layers of natural or synthetic rubber bonded to intermediate steel plates [1–4]. Usually, the isolators are set between the foundation and the building structure [5]. It needs not only to carry the weight of the whole

*Corresponding authors. Email: wangnus@126.com (Wencai Wang); zhanglq@mail.buct.edu.cn (Liqun Zhang)

building, but also to have a long life (normally more than 70 years). Besides having good damping properties, the rubbers used for isolators must have high flexibility, high elasticity, and high strength [2,5].

Nowadays, natural rubber (NR) is the most widely used material for seismic isolation bearings [6]. NR can bring rubber bearings enough flexibility, which can decrease the earthquake response of buildings. NR also has the advantages of high elasticity, low compression set, and high strength (because of strain-induced crystallization) [7]. Strain-induced crystallization is an important character of NR used in bearings because it can not only prevent the propagation of cracks, but also lead to an increase in the shear stiffness of bearings at high strains, which limit lateral displacements. However, the damping properties of NR cannot meet the requirements for seismic isolation bearings [8]. The horizontal force–displacement relationship for NR bearings is almost linear, and the equivalent damping ratio is only 2%–4% at 100% shear strain, much lower than the 10%–20% for high-damping rubber (HDR) bearing applications. Generally, the loss factor of the rubbers used in seismic isolation bearings should be higher than 0.1, which was tested under the conditions corresponded to the seismic bearing service environment. To enhance the energy dissipation capability of the low damping NR bearings, lead–rubber (LR) bearings are widely used [9]. A LR bearing is similar to an NR bearing, but with an added lead core inserted into a central hole of the bearing. The lead core undergoes plastic deformation and strong energy dissipation during horizontal shaking. However, LR bearings also have some disadvantages; for example, their high-damping properties are useful only for a narrow earthquake frequency range and are not so effective for large deformations. HDR bearings are a new type of rubber bearings in which a new rubber compound, instead of a lead core, is used to provide energy dissipation [10]. Conventional HDR composites contain large amounts of reinforcing fillers such as carbon black and silica. Their hardness and shear modulus are high; thus, high levels of damping are difficult to achieve. No single conventional rubber composite can satisfy all the requirements for HDR bearings. Therefore, the development of a new elastomeric system with high damping, high modulus, and high elasticity is highly desirable.

Because of their excellent damping performance, organic hybrid elastomer composites have become the focus of many studies in recent years [11–13]. In our previous studies [14–16], we prepared a kind of organic hybrid materials with excellent damping properties by adding hindered phenol to the rubber matrix. In particular, the AO-80/NBR rubber composites exhibited high dynamic mechanical losses and high glass transition temperatures depending on the AO-80 content [16]. For the AO-80/NBR (40/100) composite, the temperature ranges in which $\tan\delta > 0.3$ and $\tan\delta > 0.1$ are 3°C to 35°C and −3°C to 50°C, respectively. At the same time, the nitrile butadiene rubber (NBR) matrix in the composites without filler reinforcement can be oriented under stretching, and the tensile strengths of the composites are above 17 MPa. The NBR/AO-80 composites meet the needs, such as high strength, high damping and high flexibility, of HDR composites for HDR bearings. However, the elasticity of the NBR/AO-80 composites cannot meet the needs of the isolator. As stated above, NR is a material with high elasticity and strain-induced crystallization [17]. If NR and NBR/AO-80 composites are combined, the two components could complement each other to form new composites with properties needed by HDR bearings. In this work, NR/NBR/AO-80 composites with high-damping properties were prepared. The microstructures, crystallization, and orientation under stretching, and damping properties of the new rubber composites were investigated.

2. Experimental

2.1. Materials

NR (smoked sheet #1) was obtained from Nanjing Shengdong Chemical Co., Ltd. (Nanjing, China). NBR (N220S) with an acrylonitrile mass fraction of 41% was provided by Japan Synthetic Rubber Co., Ltd. (Tokyo, Japan). AO-80 (ADK-ATAB-AO-80) in the form of crystalline powder was provided by Asahi Denka Co., Ltd. (Tokyo, Japan). Other chemicals and ingredients were purchased in China. All materials were used without further purification.

2.2. Sample preparation

The NR/NBR/AO-80 rubber composites were prepared according to the following procedures:

2.2.1. NBR/AO-80 mixture

After the as-received NBR was kneaded on a Φ 152.4 mm two-roll mill at room temperature for 3 min, AO-80 was added in the NBR/AO-80 mass ratio of 100/40. The mixture was then kneaded at room temperature for 5 min, further kneaded on the two-roll mill at 135°C for 5 min, and cooled to room temperature. The mixture was then blended at room temperature with compounding and crosslinking additives, consisting of 5.0 phr of zinc oxide, 2.0 phr of stearic acid, 0.5 phr of diphenyl guanidine, 0.5 phr of dibenzothiazole disulfide, 0.2 phr of tetramethylthiuram disulfide, and 2.0 phr of sulfur. Finally, the mixture was kneaded on the two-roll mill at room temperature for 5 min to form the NBR/AO-80 mixture.

2.2.2. NR mixture

The as-received NR was kneaded on a Φ 152.4 mm two-roll mill at room temperature for 3 min. The mixture was then blended with compounding and crosslinking additives, including 3.0 phr of zinc oxide, 1.0 phr of stearic acid, 1.2 phr of *N*-cyclohexyl-2-benzothiazole sulfenamide, 0.2 phr of tetramethylthiuram disulfide, 1 phr of sulfur, and 2 phr of Antioxidant 4010NA. Finally, the mixture was kneaded on the two-roll mill at room temperature for 5 min to form the NR mixture.

2.2.3. NR/NBR/AO-80 rubber composites

The NR mixture and the NBR/AO-80 mixture in the NR/NBR/AO-80 mass ratios of 100/0/0, 75/25/10, 50/50/20, 25/75/30, and 0/100/40 were kneaded on a Φ 152.4 mm two-roll mill at room temperature for 5 min to form NR/NBR/AO-80 mixtures. These mixtures were then hot pressed and vulcanized at 145°C under the pressure of 15 MPa for various periods of time, and then naturally cooled down to room temperature to form NR/NBR/AO-80 composites. The vulcanization time was pre-determined for each composite by using a disk rheometer (P355C2) purchased from Huanfeng Chemical Technology and Experimental Machine Co. (Beijing, China).

2.3. *Analysis and characterization*

The morphological, structural, and mechanical properties of the NR/NBR/AO-80 composites were evaluated by using atomic force microscopy (AFM), polarized Fourier transform infrared spectrometer (FTIR), dynamic mechanical thermal analyzer (DMTA), and tensile testing.

The AFM images of representative surfaces of the composites were taken by an AFM (Multimode 8, Bruker, Germany) in the tapping mode. This measurement provided the modulus of the surface depicted as DMT modulus images. The samples were polished at −130°C by using a cryo-ultramicrotome (EM UC7, Leica, Germany) before observation.

Polarized FTIR measurements were conducted on a TENSOR 27 FTIR (Bruker, Germany). The polarized FTIR spectra were acquired by scanning the specimens in the wavenumber range of 400–4000 cm^{-1} for 32 times with a resolution of 2 cm^{-1} at room temperature, using the infrared transmittance spectra technique. The FTIR spectra of the NR/NBR/AO-80 rubber composites were acquired from ultra-thin film specimens subjected to uniaxial elongation on a stretching machine mounted in the sample compartment of the spectrometer. During the uniaxial elongation, the interferograms were acquired in small strain intervals with light alternately polarized parallel and perpendicular to the stretching direction.

Tensile tests of the NR/NBR/AO-80 composites were conducted according to ASTM standard (D412: dumbbell-shaped) on a tensile tester (Sans, Ltd., China). The hysteresis losses under stretching of the composites were also measured with the same machine. The specimens (100 mm long, 10 mm wide, and 2 mm thick) were stretched and retracted at a constant rate of 20 mm min^{-1}. At the same time, the stress–strain curves were recorded [18]. The measurements were continued up to four cycles to eliminate the Mullins effect. The hysteresis loss was defined as the area enclosed by the stretch curve and retraction curve.

Dynamic viscoelasticity measurements were carried out on a dynamic mechanical thermal analyzer (DMTA VA3000, 01dB-Metravib Co., Ltd., France). The DMTA specimens were 20 mm long, 6 mm wide, and 2 mm thick. The temperature dependence of the dynamic tensile modulus was measured in the temperature range −100 to 100°C at a frequency of 10 Hz and a heating rate of 3°C°min^{-1}.

3. Results and discussion

3.1. *Phase morphology of NR/NBR/AO-80 composites*

The morphology of the NR/NBR/AO-80 composites was observed by AFM. As shown in Figure 1, the lighter regions represent the NBR/AO-80 phase and the darker regions represent the NR phase. When the mass ratio of NR/NBR/AO-80 is 25/75/30, NR is the dispersed phase and NBR/AO-80 is the continuous phase (see Figure 1(a)). However, at an NR/NBR mass ratio of 50/50, the composite exhibits a co-continuous morphology (see Figure 1(b)). The NR phase becomes the continuous phase with further increases in the amount of NR, as shown in Figure 1(c). The particle size of the dispersed phase is a few microns. The AFM micrographs show that the NR/NBR/AO-80 composites have a two-phase structure, consistent with the DMTA results shown in Figure 2. The plot of tanδ versus temperature for each NR/NBR/AO-80 composite presents two distinct loss peaks. The first loss peak at the lower temperature is the glass transition peak of the NR phase; the second loss peak at the higher temperature is the glass transition peak of the NBR/AO-80 phase. The second loss peak shifts slightly to lower temperatures with increasing amount of NR in the composites, probably because there is a partial compatibility between

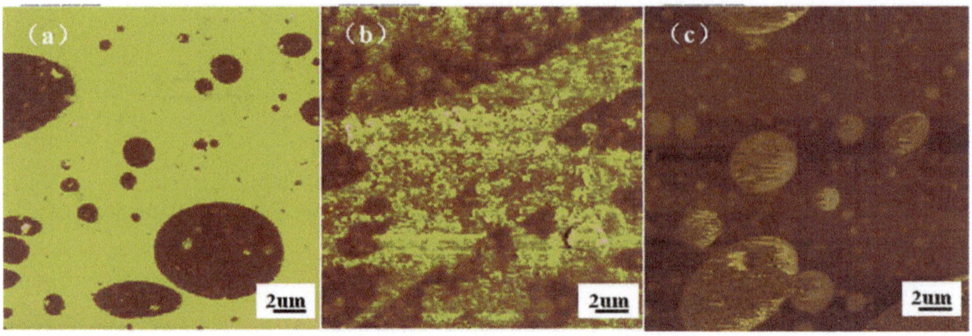

Figure 1. AFM micrographs of NR/NBR/AO-80 composites with various mass ratios NR/NBR/AO-80: (a) 25/75/30, (b) 50/50/20, and (c) 75/25/10.

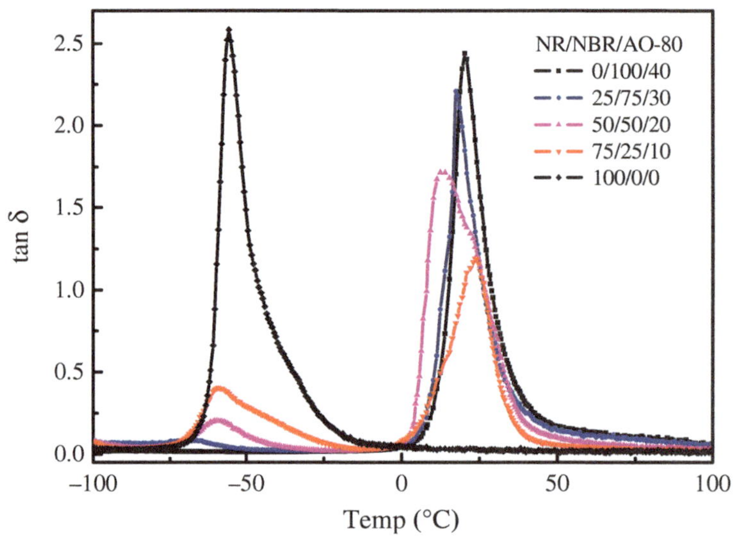

Figure 2. Temperature dependence of loss tangent (tanδ) for NR/NBR/AO-80 composites with various NR/NBR/AO-80 mass ratios.

the NR phase and the NBR/AO-80 phase and most of the AO-80 is accumulated in the NBR phase, regardless of the ratio of the NR phase to the NBR/AO-80 phase. The temperature of the loss peak of the NBR/AO-80 phase is the lowest at an NR/NBR/AO-80 ratio of 50/50/20. In other words, the NR/NBR/AO-80 (50/50/20) composite has the best compatibility between the NR phase and the NBR/AO-80 phase, consistent with the AFM results (Figure 1). The damping properties of the NR/NBR/AO-80 composites will be further discussed later in this article.

3.2. Strain-induced orientation and strain-induced crystallization of NR/NBR/AO-80 composites

Polarized FTIR is a valuable method to study the strain-induced orientation/crystallization of crosslinked rubber [19–22]. The FTIR spectrum of the NR/NBR/AO-80 (75/25/10)

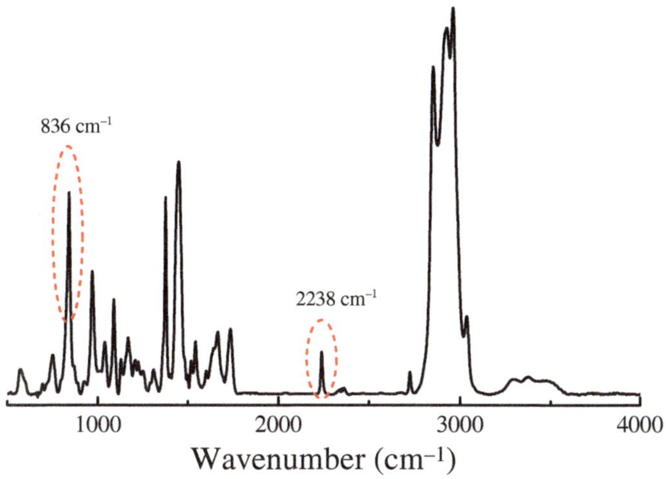

Figure 3. FTIR spectrum of NR/NBR/AO-80 (75/25/10) composite.

composite is shown in Figure 3. We selected the absorption bands at 836 and 2238 cm^{-1} to monitor the orientation and crystallization of the NR/NBR/AO-80 composites under strain. The absorption band at 836 cm^{-1} is attributed to C—H out of plane bending vibration of cis-1, 4-structure of NR; the absorption band at 2238 cm^{-1}, which is the most characteristic absorption bank of NBR, is assigned to the cyano group of NBR [23,24].

Figure 4 shows the FTIR spectra taken with light alternately polarized perpendicular and parallel to the stretching direction under uniaxial elongations as high as 500%. At zero strain, the 'perpendicular' spectra and the 'parallel' spectra in Figure 4(a) are exactly the same, indicating that the composite is isotropic. However, with increasing strain (from 100% to 500%), the peaks at 836 cm^{-1} become higher in the 'perpendicular' spectrum than in the 'parallel' spectrum, indicating that the functional groups concerned have been oriented. Meanwhile, the peaks shift to 844 cm^{-1} at the strain of 300% because of the strain-induced crystallization of NR [21].

Similarly, as shown in Figure 4(b), as the strain increases from 0% to 500%, the peaks at 2238 cm^{-1} in the 'perpendicular' spectra are higher than the corresponding peaks in the 'parallel' spectra, indicating that the –CN groups are oriented at high strains. That is, strain-induced orientation occurs in the NBR/AO-80 phase of the NR/NBR/AO-80 composites at high strains. Both strain-induced orientation and strain-induced crystallization occur simultaneously in the NR/NBR/AO-80 composites. Moreover, with increasing strain, the bands at 836 and 2238 cm^{-1} becomes higher in both the 'perpendicular' spectra and the 'parallel' spectra, and both bands are much lower in the 'parallel' spectra than in the 'perpendicular' spectra; i.e., both bands are vertical bands [21].

The dichroic function (DF) is often used to evaluate the orientation of composites [25,26]. The DF can be estimated according to the following equation:

$$\mathrm{DF} = \frac{(A_{\parallel}/A_{\perp})-1}{(A_{\parallel}/A_{\perp})+2} \tag{1}$$

where $A_{\parallel}$ and $A_{\perp}$ are the areas of the band in the 'parallel' spectrum and 'perpendicular' spectrum at a strain, respectively. Figure 5 shows the relationship between the DF and

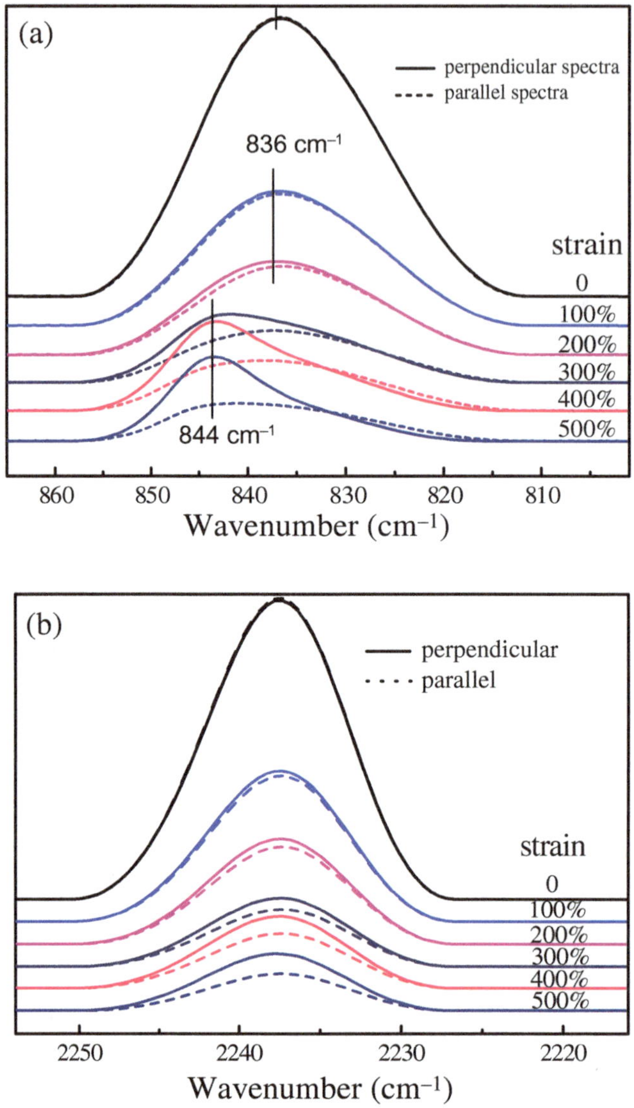

Figure 4. FTIR polarization spectra of NR/NBR/AO-80 (75/25/10) composite under different uniaxial elongations at (a) 836 cm^{-1} and (b) 2238 cm^{-1}.

strain of the NR/NBR/AO-80 (75/25/10) composite. As discussed above, the bands at 836 and 2238 cm^{-1} represent the functional groups of the NR phase and NBR/AO-80 phase of the NR/NBR/AO-80 composites, respectively. The DF of the NBR phase (the band at 2238 cm^{-1}) increases linearly when strain increase from 0% to 500%. However, the DF of the NR phase (the band at 836 cm^{-1}) is not linear. It means that the orientation of NBR molecules began under a small strain. But NR molecules have no significantly change at strains lower than 200%, and the orientation of the NR molecules increase linearly with increasing strain (higher than 200%).

As is well known, the molecular chains of the continuous phase, NR, are very flexible in the NR/NBR/AO-80 composites; however, the molecular chains of the dispersed phase,

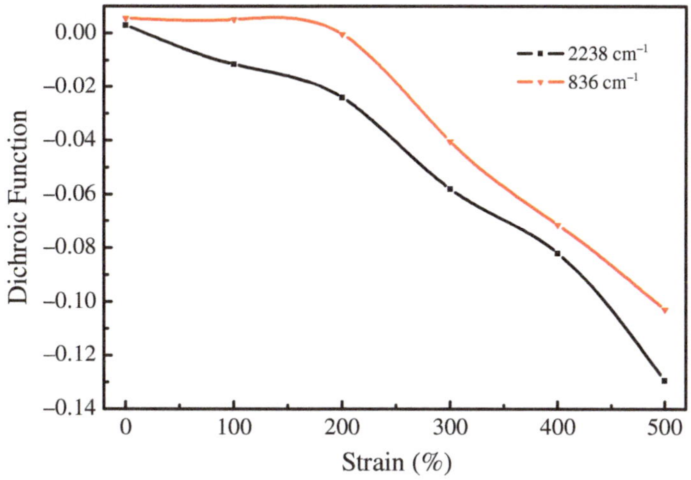

Figure 5. Strain dependence of DF for absorption bands at 836 and 2238 cm^{-1} of FTIR spectrum of NR/NBR/AO-80 (75/25/10) composite.

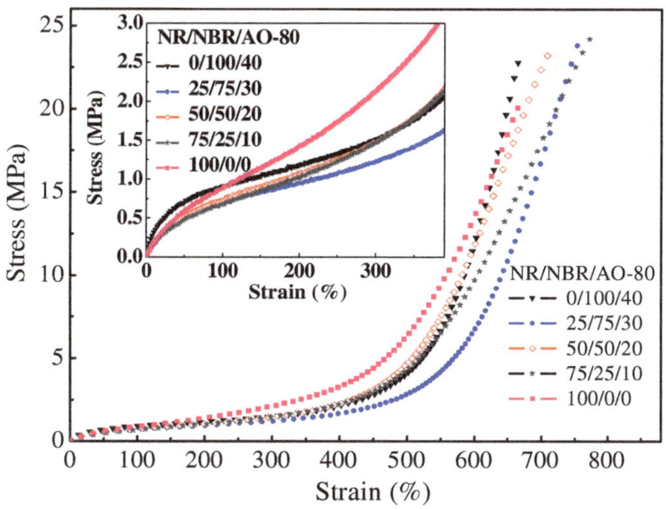

Figure 6. Stress–strain curves of NR/NBR/AO-80 composites with different mass ratios of NR/NBR/AO-80.

NBR/AO-80, are rigid because of the strong polarity and hydrogen bonding between NBR and AO-80 [14]. When faced with a low strain, the flexible chains can quickly compensate for the change in internal stress by adjusting the molecular conformation. But the rigid chains can be deformed only by high strain or appropriate molecular chain orientation. That is, the NBR/AO-80 composites are easily aligned and oriented because of the hydrogen bond network. Moreover, the values of DF for both bands decrease with increasing strain, indicating a good stress transfer between the continuous phase and the dispersed phase of the composites. It also can be seen later from the stress–strain curves of the NR/NBR/AO-80 composites (Figure 6).

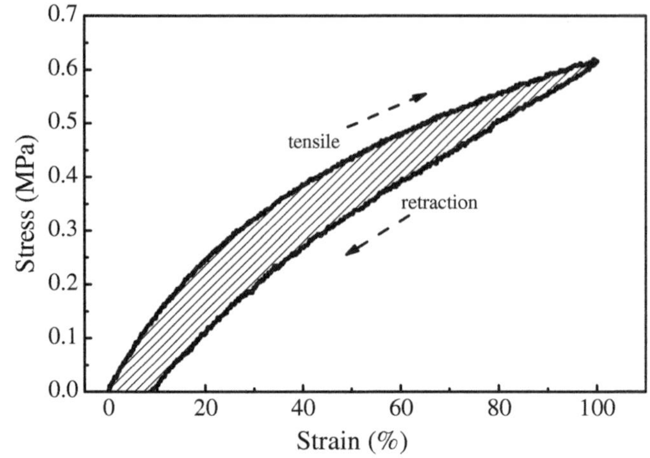

Figure 7. Schematic diagram of hysteresis loop of rubber composite.

Figure 6 shows the tensile stress–strain curves of the NR/NBR/AO-80 composites with various NR/NBR/AO-80 mass ratios. All the curves show the same trend: the samples had a large deformation (up to 400%) under small stress (about 2 MPa); the stress of the samples increased rapidly when the strain was greater than a certain value (about 500%). It is important that the tensile strength and elongation at break of all the composites are above 20 MPa and 650%, respectively, because the rubber composites used in isolators require the performance of large deformation and low modulus [5,27]. NR has excellent mechanical properties, mainly because of its strain-induced crystallization. As for NBR/AO-80, there is a good compatibility and strong intermolecular interactions between NBR and AO-80, and both of these properties lead to strain-induced orientation [14]. Furthermore, all the NR/NBR/AO-80 composites have similar tensile strengths and elongations at break. We have thus reconfirmed that strain-induced orientation and crystallization occur simultaneously in the two phases of the composites: the NR phase and NBR/AO-80 phase.

3.3. Damping properties of NR/NBR/AO-80 composites

To characterize the energy dissipation and the hysteresis performance of the NR/NBR/AO-80 composites, the hysteresis loop of the tensile stress–strain curve was examined. As shown in Figure 7, the shaded area surrounded by the tension curve and retraction curve is noted as the hysteresis energy density (HED), representing the consumed energy of the composites during one draw-reply cycle [28].

Table 1 describes the relationship between the hysteresis energy and the strain of the NR/NBR/AO-80 composites. It can be seen that the hysteresis energy of all composites gradually grows with the strain increasing from 50% to 200%, an indication that the hysteresis loss as well as the damping properties increase. On the other hand, for the same strain, the hysteresis energy increases with increasing content of NBR/AO-80, confirming that the NBR/AO-80 phase improves the damping properties of NR/NBR/AO-80 significantly because of the formation of strong hydrogen bonds between the polar rubber NBR and the organic filler AO-80. The intermolecular interactions greatly increase the frictional damping of molecular motion; as a result, the damping properties of the composites improve [28].

Table 1. Relationship between HED and strain of NR/NBR/AO-80 composites.

	HED of NR/NBR/AO-80 composites (MPa)				
Strain (%)	(0/100/40)	(25/75/30)	(50/50/20)	(75/25/10)	(100/0/0)
50	3.34	2.73	3.11	1.67	1.65
100	8.82	6.83	7.74	4.30	4.76
150	16.05	14.42	13.63	9.00	10.02
200	25.41	18.13	15.23	11.42	14.24

The loss factor (tanδ) reflects the external and the internal friction and represents the ratio of the energy dispersed to the energy accumulated in one deformation cycle. Figure 2 shows the temperature dependence of tanδ for the NR/NBR/AO-80 composites with various mass ratios of NR/NBR/AO-80. It indicates that NR has only one tanδ peak with a peak value of 2.6 at about −55°C. For the NBR/AO-80 composites, there is also a single tanδ peak with a peak value of 2.5 at about 20°C. But the NR/NBR/AO-80 composites present two distinct tanδ peaks, corresponding to the NR phase and NBR/AO-80 phase. The working temperature of isolators is the ambient temperature under the building, i.e., 0–40°C. Therefore, the damping property of an NR bearing is poor. The NBR/AO-80 composites have better damping properties than NR in the above working temperature range. With increasing amount of NBR/AO-80, the NR/NBR/AO-80 composites show two tanδ peaks. All the NR/NBR/AO-80 composites show a tanδ peak with a high peak value within the working temperature range.

Table 2 shows the damping properties of the NBR/AO-80 phase of the NR/NBR/AO-80 composites. With increasing NBR/AO-80 content, the tanδ_{max} of the NBR/AO-80 phase of the NR/NBR/AO-80 composites increases from 1.2 to 2.2, and the temperature of the tanδ_{max} (peak value) is around 20°C. At a ratio of NR/NBR/AO-80 of 50/50/20, the temperature range in which tan $\delta > 0.1$ is 2.8–47.1°C, and the temperature range in which tan $\delta > 0.3$ is 4.4–36.8°C. These temperature ranges are consistent with the working temperature range of seismic isolators [5]. These results show that the NR/NBR/AO-80 composites are better materials than NR for seismic isolators. The TA value, the area under the tanδ versus temperature curve, is commonly used to characterize the damping performance of rubbers. Table 2 shows that the TA value of NR/NBR/AO-80 composites reaches 38.6°C in the temperature range of −10°C to 50°C with increasing NBR/AO-80 content. The larger the TA

Table 2. Damping properties of NR/NBR/AO-80 composites with various mass ratios of NR/NBR/AO-80.

	NR/NBR/AO-80 composites				
Properties[a]	(0/100/40)	(25/75/30)	(50/50/20)	(75/25/10)	(100/0/0)
Tanδ_{max}	2.5	2.2	1.7	1.2	–
Temperature of tanδ_{max} (°C)	20.6	17.5	13.4	23.7	–
Temperature range in which tan$\delta > 0.3$ (°C)	10.7–41.1	8.8–39.0	4.4–36.7	8.8–34.6	0
Temperature range in which tan$\delta > 0.1$ (°C)	7.0–64.7	4.9–55.9	2.8–47.1	3.3–36.3	0
TA[b] from −10 to 50°C (°C)	37.0	33.5	38.6	22.5	1.9

[a]The damping properties of the NBR/AO-80 phase.
[b]TA denotes the area under the tanδ versus temperature curve.

value, the better the damping performance of the NR/NBR/AO-80 composite. The NR/NBR/AO-80 (50/50/20) composite has excellent damping properties and mechanical properties in the rubber bearing working temperature range and could be an ideal material for the seismic isolators. For HDR bearings, the equivalent damping ratio or tan δ should be higher than 0.1 in the working temperature [1]. The NR/NBR/AO-80 composites have a broad temperature range in which tan $\delta > 0.1$ and can be good candidates for HDR bearings.

4. Conclusions

NR/NBR/AO-80 rubber composites were successfully prepared in this study. Each composite consisted of two phases: the NR phase and the NBR/AO-80 phase. There was a partial compatibility between the NR phase and the NBR/AO-80 phase, and the NR/NBR/AO-80 (50/50/20) composite exhibited a co-continuous morphology. Strain-induced crystallization occurred in the NR phase at strains higher than 200%, and strain-induced orientation appeared in the NBR/AO-80 phase with the increase of strain from 100% to 500%. The composites had a special stress–strain behavior and mechanical properties because of the simultaneous strain-induced orientation and strain-induced crystallization. In the working temperature range of a seismic isolation bearing, the composites (especially the NR/NBR/AO-80 (50/50/20) composite) presented a high loss factor, high area of loss peak, and high hysteresis energy. Therefore, the NR/NBR/AO-80 rubber composites are expected to have important application as a high-performance damping material for rubber bearing.

Acknowledgments

We would like to thank Dr Tung W. Chan from the Department of Materials Science and Engineering at Virginia Polytechnic Institute and State University for improving the written English.

Disclosure statement

No potential conflict of interest was reported by the authors.

Funding

This work was supported by the National Natural Science Foundation of China [grant number 51320105012], [grant number 51221002], [grant number 51103006], [grant number 51373010].

References

[1] T. Nishi and N. Murota, *Elastomeric seismic-protection isolators for buildings and bridges*, Chin. J. Polym. Sci. 31 (2013), pp. 50–57. doi:10.1007/s10118-013-1217-8

[2] A. Islam, M. Jameel, and M.Z. Jumaat, *Seismic isolation in buildings to be a practical reality: behavior of structure and installation technique*, J. Eng. Technol. Res. 3 (2011), pp. 99.

[3] J. Su, L.H. Wen, S. Hei, and F.L. Zhou, *Recent advances in building isolated laminated rubber bearings*, Low Temp. Archit. Technol. 29 (2010), pp. 29–31.

[4] G.P. Warn and K.L. Ryan, *A review of seismic isolation for buildings: Historical development and research needs*, Buildings 2 (2012), pp. 300–325. doi:10.3390/buildings2030300

[5] A. Marioni, *Portuguese conference on structural engineering*, LNEC, Lisbon, Nov. 25–28, 1998.

[6] A.W. Taylor, A.N. Lin, and J.W. Martin, *Performance of elastomers in isolation bearings: A literature review*, Earthq. Spectra 8 (1992), pp. 279–303. doi:10.1193/1.1585682

[7] M. Kluppel, W. Grellmann, and G. Heinrich, *In-situ structural characterization of rubber during deformation and fracture*, in *Fracture Mechanics and Statistical Mechanics of*

Reinforced Elastomeric Blends, in K. Brüning, K. Schneider, and G. Heinrich, eds., Springer, Berlin, 2013, pp. 43–80.

[8] J.C. Ramallo, E.A. Johnson, and B.F. Spencer, *"Smart" base isolation systems*, J. Eng. Mech. 128 (2002), pp. 1088–1099. doi:10.1061/(ASCE)0733-9399(2002)128:10(1088)

[9] J. Weisman and G.P. Warn, *Stability of elastomeric and lead-rubber seismic isolation bearings*, J. Struct. Eng. 138 (2011), pp. 215–223. doi:10.1061/(ASCE)ST.1943-541X.0000459

[10] K. Fuller, J. Gough, T. Pond, and H. Ahmadi, *High damping natural rubber seismic isolators*, J. Struct. Cont. 4 (1997), pp. 19–40. doi:10.1002/(ISSN)1538-523X

[11] C. Wu, T. Yamagishi, Y. Nakamoto, S. Ishida, K. Nitta, and S. Kubota, *Organic hybrid of chlorinated polyethylene and hindered phenol. I. Dynamic mechanical properties*, J. Polym. Sci., Part B: Polym. Phys. 38 (17) (2000), pp. 2285–2295. doi:10.1002/(ISSN)1099-0488

[12] C. Wu, T. Yamagishi, Y. Nakamoto, S. Ishida, S. Kubota, and K. Nitta, *Organic hybrid of chlorinated polyethylene and hindered phenol. II. Influence of the chemical structure of small molecules on viscoelastic properties*, J. Polym. Sci., Part. B: Polym. Phys. 38 (11) (2000), pp. 1496–1503. doi:10.1002/(ISSN)1099-0488

[13] C. Wu, K. Mori, Y. Otani, N. Namiki, and H. Emi, *Effects of molecule aggregation state on dynamic mechanical properties of chlorinated polyethylene/hindered phenol blends*, Polymer 42 (19) (2000), pp. 8289–8295. doi:10.1016/S0032-3861(01)00203-8

[14] B. Qiao, X.Y. Zhao, D.M. Yue, L.H. Zhang, and S.Z. Wu, *A combined experiment and molecular dynamics simulation study of hydrogen bonds and free volume in nitrile-butadiene rubber/hindered phenol damping mixtures*, J. Mater. Chem. 22 (2012), pp. 12339. doi:10.1039/c2jm31716h

[15] X.-Y. Zhao, Y.-J. Cao, H. Zou, J. Li, and L.-Q. Zhang, *Structure and dynamic properties of nitrile-butadiene rubber/hindered phenol composites*, J. Appl. Polym. Sci. 123 (2012), pp. 3696–3702. doi:10.1002/app.v123.6

[16] X.-Y. Zhao, P. Xiang, M. Tian, H. Fong, R. Jin, and L.-Q. Zhang, *Nitrile butadiene rubber/hindered phenol nanocomposites with improved strength and high damping performance*, Polymer 48 (2007), pp. 6056–6063. doi:10.1016/j.polymer.2007.08.011

[17] S. Toki and B.S. Hsiao, *Nature of strain-induced structures in natural and synthetic rubbers under stretching*, Macromolecules 36 (2003), pp. 5915–5917. doi:10.1021/ma034729e

[18] A.M. Kucherskii, *Hysteresis losses in carbon-black-filled rubbers under small and large elongations*, Polym. Test. 24 (2005), pp. 733–738. doi:10.1016/j.polymertesting.2005.04.005

[19] H.W. Siesler, *Rheo-optical Fourier transform infrared spectroscopy of polymers. 10: Strain-induced crystallization of sulfur-crosslinked natural rubber during cyclic deformation*, Appl. Spectrosc. 39 (1985), pp. 761–765. doi:10.1366/0003702854249853

[20] H.W. Siesler, *Rheo-optical Fourier transform IR (FTIR) spectroscopy of polyurethane elastomers*, Polym. Bull. 9 (1983), pp. 557–562. doi:10.1007/BF00307878

[21] H.W. Siesler, *Rheo-optical Fourier-transform infrared (FTIR) spectroscopy of polymers*, Colloid. Polym. Sci. 262 (1984), pp. 223–229. doi:10.1007/BF01458965

[22] C. Yan, H. Li, J. Zhang, Y. Ozaki, D. Shen, D. Yan, A.-C. Shi, and S. Yan, *Surface-induced anisotropic chain ordering of polycarprolactone on oriented polyethylene substrate: Epitaxy and soft epitaxy*, Macromolecules 39 (2006), pp. 8041–8048. doi:10.1021/ma061188v

[23] S. Joly, G. Garnaud, R. Ollitrault, L. Bokobza, and J.E. Mark, *Organically modified layered silicates as reinforcing fillers for natural rubber*, Chem. Mater. 39 (2002), pp. 8041.

[24] R. Sreeja, S. Najidha, S.R. Jayan, P. Predeep, M. Mazur, and P.D. Sharma, *Electro-optic materials from co-polymeric elastomer–acrylonitrile butadiene rubber (NBR)*, Polymer 47 (2006), pp. 617–623. doi:10.1016/j.polymer.2005.09.024

[25] L. Bokobza, L. Ladouce, Y. Bomal, and B. Amram, *Infrared dichroism and birefringence studies of silica-filled styrene-butadiene rubbers*, J. Appl. Polym. Sci. 82 (2001), pp. 1006–1012. doi:10.1002/(ISSN)1097-4628

[26] S. Besbes, L. Bokobza, and H. Siesler, *Rheo-optical Fourier-transform infrared spectroscopy of polymers 16 orientation and strain-induced conformational regularity of poly(dimethylsiloxane) networks during cyclic elongation and recovery*, Makromol. Chem. Macromol. Symp. 72 (1993), pp. 191–199. doi:10.1002/(ISSN)1521-3900a

[27] P. Tan and F.L. Zhou, *Research and application of seismic isolation system for civil engineering structures*, Constr. Technol. 37 (2008), pp. 5.

[28] Z. Huang, *Fatigue failure resistance and mechanism of different fillers/carbon black/styrene-butadiene rubber composites*. MS Thesis, Beijing University of Chemical Technology, Beijing, May 2013.

9

Water presence detection in a concrete crack using smart aggregates

Qingzhao Kong[a], Qian Feng[b,c] and Gangbing Song[a,d]*

[a]Department of Mechanical Engineering, University of Houston, Houston, TX 77204, USA; [b]Hubei Key Laboratory of Earthquake Early Warning, Institute of Seismology, China Earthquake Administration, Wuhan 430071, China; [c]Wuhan Institute of Earthquake Engineering, Wuhan 430071, China; [d]School of Civil Engineering, Dalian University of Technology, Dalian, Liaoning, China

Liquid migrating into existing concrete cracks is a serious problem for the reliability of concrete structures and can sometimes induce full concrete structural failures. In this paper, the authors present recent research on water presence detection in concrete cracks using piezoceramic-based smart aggregate (SA) transducers. The active sensing approach, in which one piezoceramic transducer is used to generate stress waves and others are used to detect the stress wave responses, is adopted in this research. Cracks formed in concrete structures act as stress reliefs, which attenuate the energy of the signals received by the SAs. In case of a crack being filled with liquid, which changes the wave impedance, the piezoceramic transducers will report higher received energy levels. A wavelet packet-based approach is developed to provide calculated energy values of the received signal. These different values can help detect the liquid presence in a concrete crack. A concrete beam specimen with three embedded SAs was fabricated and tested. Experimental results verified that the SA-based active sensing approach can detect a concrete crack and further detect the liquid presence in the concrete crack.

Keywords: water detection; concrete crack; smart aggregate; active sensing; piezoceramic sensors and actuators2

1. Introduction

Aging infrastructure all around the world demands significant focus and expenditure. Most aging infrastructure is concrete structure. Health monitoring of concrete structure has been studied for many years. Although the study has been ongoing for many years, the detection of liquid presence in concrete cracks rarely attracts enough attention. However, the detection of liquid presence in concrete cracks is significant for the health of the concrete. It is well known that cracks in concrete structures act as conduits, which could cause liquid migration. In the case of decommissioned nuclear structures, it is of great importance to detect the radioactive liquid leaking through cracks in concrete walls to the ground. Additionally, the detection of liquid presence in concrete cracks is also crucial for concrete dams. If water with particular chemical compositions remains in concrete cracks for a long time, the erosion of the concrete interface and embedded rebar may cause a sustained damage and can even lead to complete structural failure of

*Corresponding author. Email: gsong@uh.edu

the concrete dam. Therefore, the detection of liquid presence in concrete cracks is essential and meaningful to the concrete structures.

Crack detection for concrete structures is one of the most popular topics for concrete health evaluation and has been studied in the recent past two decades. Image-based crack detection is a traditional method for concrete crack monitoring. Ito et al. [1] presented integrated image processing techniques with a high-resolution camera to extract crack regions. Fujita et al. [2] demonstrated two methods of preprocessing in order to eliminate the noise from the crack images. Consequently, Yamaguchi and Hashimoto [3,4] developed a highly accurate and efficient method for concrete crack detection using percolation-image processing. Ultrasonic-based concrete structure health monitoring is also an effective method. Tanaka [5] presented the single and multiple inspections of concrete cracks using ultrasonic sensors. Antonaci et al. [6] described the mechanical characteristics of the concrete with discontinuity surfaces under loads using a nonlinear ultrasonic non-destructive technique. The impedance-based crack detection process is a real-time, qualitative damage detection method [7,8]. The impedance-based approach has been applied to the health monitoring of trusses [9], reinforced concrete bridges [10], reinforced concrete specimens [11,12], reinforced concrete shear walls [13] and composite reinforced concrete/masonry walls [14].

Moisture and water detection in concrete structures is another field that attracted concerns in recent years. Multon and Toutlemonde [15] presented some quantitative results to describe the moisture transfers in a concrete beam using weighing and gamma densitometry. Yeo et al. [16] and Ho et al. [17] successfully developed Fiber Bragg Grating (FBG) sensors with moisture-sensitive polymer coatings to measure the moisture content of different concrete samples. Ong et al. [18] applied the wireless, passive embedded sensor to monitor the water content in real time. Water in concrete can be detected by analyzing the propagation velocity of ultrasonic waves. The disadvantages of this method are probably the high cost of bulky equipment and lack of robustness.

Although crack and water detection in concrete have been studied for many years, not enough attention is concentrated on the detection of liquid presence in concrete cracks. In this paper, the authors utilize the SA-based active sensing approach to detect water presence in a concrete crack. When a concrete crack is filled with water, the water changes the wave impedance and influences the received signal energy from sensors. Therefore, the water presence in a concrete crack can be detected by comparing the received energy levels before and after the water migrates into the concrete crack.

2. Piezoceramic-based smart aggregate transducer

Piezoelectricity is the major property of piezoceramic materials. When a mechanical stress is applied to such the material, an electric field proportional to the magnitude of the stress is produced. Conversely, when an electric field is applied to the piezoelectric material, a mechanical stress develops and may produce a shape change of the piezoelectric material.

Lead zirconium titanate (PZT) is one of the most commonly used piezoceramic materials. In order to protect the fragile PZT patch, smart aggregate (SA) is designed by sandwiching a waterproof piezoelectric patch with lead wires between two mating marble blocks as shown in Figure 1 [19]. The fabrication of the SA with a BNC connector is shown in Figure 2. The dimension of the PZT patch is 15 mm × 15 mm, and the

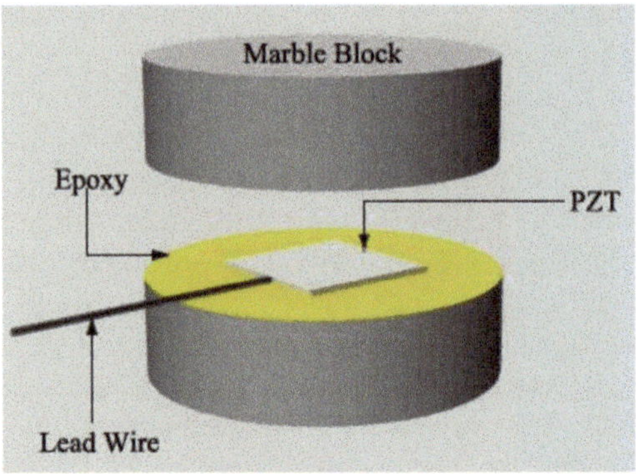

Figure 1. Structure of the SA.

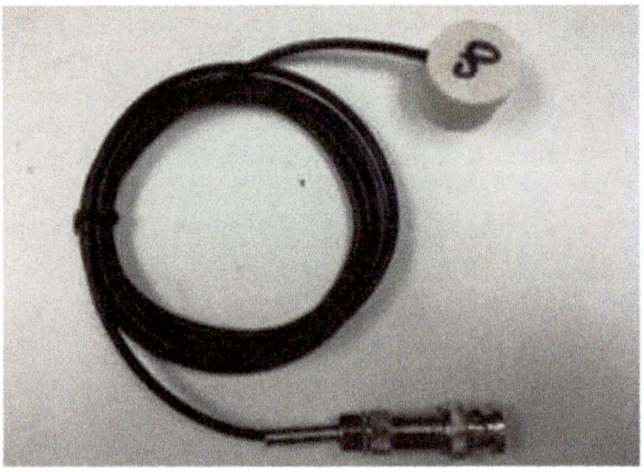

Figure 2. A photo of an SA transducer.

thickness is 0.3 mm. The diameter and height of the SA are 25 mm and 20 mm, respectively. Owing to the great stability of physical and chemical properties of marble, distributed SAs functioning as actuators and sensors can be embedded into concrete structures, which can sustainably monitor the health of structures. In recent years, SAs have been deployed for the structural health monitoring of various fields, including early-age concrete hydration monitoring [20,21], soil freeze-thaw monitoring [22], bridge impact detection [23], and concrete crack monitoring [24].

3. Principles

3.1. Active sensing approach

Water detection using an active sensing approach relies on the change of received stress wave energy. Figure 3 is a diagram of stress wave propagation through a crack between SAs. One SA is utilized as an actuator to generate the desired stress waves that propagate in the concrete. The other SA is utilized as a sensor to detect the stress wave response. When a crack occurs in the concrete, the crack functions as a stress relief, which reduces the received stress wave energy. In this case, very little stress wave energy can be detected from the SA sensor.

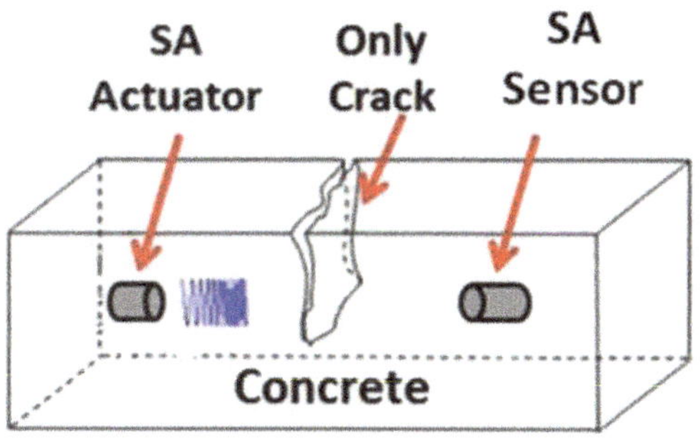

Figure 3. The stress wave propagation through a concrete crack.

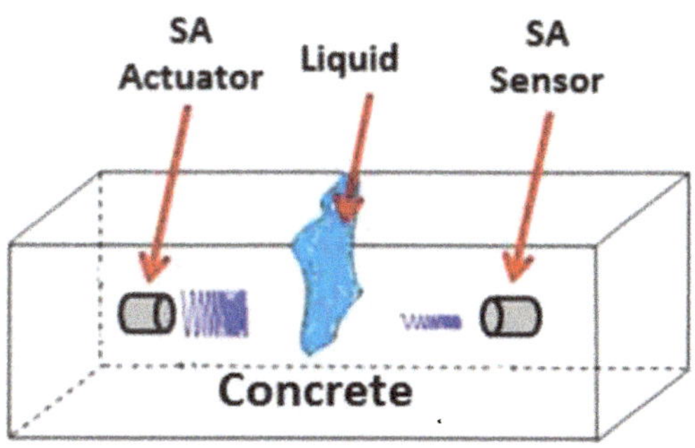

Figure 4. The stress wave propagation through water.

Subsequently, liquid migrating into the concrete crack changes the wave impedance when a stress wave propagates through the crack. Stress wave energy partly recovers the wave propagation, as shown in Figure 4. The whole process of water presence detection in a concrete crack is presented as the 'interruption' of the stress wave transition when cracks form and the 'recovery' when water migrates into the concrete crack. Since the wave impedance changes by both crack and water, the energy level of the received stress wave can be utilized to detect the crack and the further water migration in the crack.

3.2. Wavelet packet-based analysis

The collected sensor energy is an important indicator to evaluate the different conditions of crack and water presence in concrete. To quantitatively determine the case when water migrates into a concrete crack, wavelet packet-based analysis is applied to offer the corresponding values of the received energy due to different cases.

The sensor signal X can be decomposed by n-level wavelet packet decomposition into 2^n signal sets $\{X_1, X_2, \ldots, X_{2^n}\}$. $E_{i,j}$ is the energy of the decomposed signal, where i is the time index and j is the frequency band ($j = 1 \ldots 2^n$). X_j can be expressed in Equation (1).

$$X_j = \left[X_{j,1}, X_{j,2}, \ldots, X_{j,m}\right] \tag{1}$$

where m is the amount of sampling data. Additionally, the energy of the decomposed signal is defined in Equation (2):

$$E_{i,j} = ||X_j||^2 = {x_{j,1}}^2 + {x_{j,2}}^2 + \cdots + {x_{j,m}}^2 \tag{2}$$

The energy vector of the signal at time index i is defined in Equation (3):

$$E_i = \left[E_{i,1}, E_{i,2}, \ldots, E_i, 2^n\right] \tag{3}$$

Based on the definition of energy vectors (E_i), the total sensor energy associated with time index i can be obtained by the sum of energy vector values. The total sensor energy E is given in Equation (4):

$$E = \sum_{k=1}^{2^n} \boldsymbol{E}_{i,k} \tag{4}$$

3.3. Prior experimental verifications

3.3.1. Water presence detection

In this verification, two SAs were fixed by clips with a 1 mm gap between them, as shown in Figure 5. SA-1 was used as an actuator connected to a function generator (Agilent 33120A) and SA-2 was used as a sensor connected to an oscilloscope (Waverunner LT342). The SA actuator generated a swept sine wave signal via the function generator. The parameters of the swept sine wave are shown in Table 1. In the meanwhile, SA-2 was used to detect the wave response. The amplitude of the wave was recorded by an oscilloscope.

The test was operated in two cases described as follows:

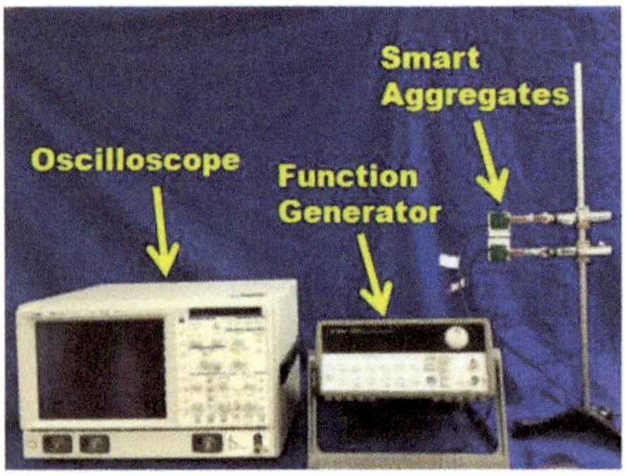

(a)

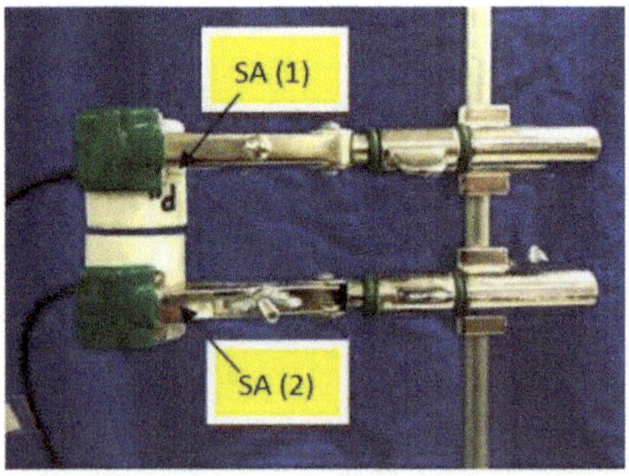

(b)

Figure 5. Test setup: (a) test setup photo; (b) close-up photo of two SAs.

Table 1. Detailed parameters of the swept sine wave signal.

Start frequency	1 Hz
Stop frequency	1 MHz
Amplitude	10 V

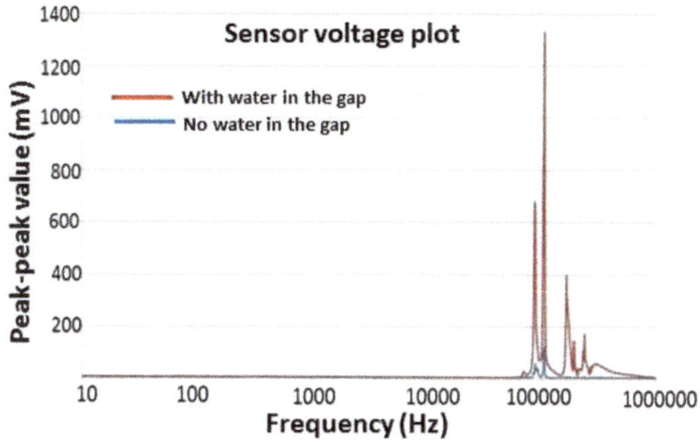

Figure 6. Amplitudes of the sensor signal with and without water in the gap.

Case 1: The 1 mm gap was open to air. SA-1 generated a swept sine wave from 1 Hz to 1 MHz and the signal amplitude detected by the SA sensor was measured by the oscilloscope.

Case 2: The 1 mm gap was filled with a few water drops. The rest of the process was all repeated as in Case 1.

Figure 6 presents the amplitudes of the received sensor signal in two cases: (1) no water in the gap and (2) with water in the gap. The X-axis is a logarithmic coordinate of the frequency from 1 Hz to 1 MHz. The blue curve represents the sensor signal amplitude in case of no water. The red curve depicts the sensor signal amplitude in case of water presence in the gap. When there was no water in the gap, the stress wave could hardly propagate through the air gap from the SA actuator to the SA sensor. Therefore, the amplitude of the received signal was very low. However, when the gap was filled with water, water changed the wave impedance and functioned as a conduit of the stress wave, and much more stress wave energy traveled through the gap. The amplitude of the received signal dramatically increased in the frequency range 50–200 kHz. This test verified that the amplitude of the sensor signal corresponds to the crack and the water presence in the crack.

3.3.2. Water-quantity detection

In this test, SA-1 (actuator) was embedded into a small concrete cylinder before casting and SA-2 (sensor) was fixed by a cramp outside the concrete shown in Figures 7 and 8. There is a water layer between these two SAs. SA-1 generated a swept sine wave from 1 Hz to 1 MHz and SA-2 detected the signal response. The depths of the water between the concrete and SA-2 were set in three cases: (1) 1 mm, (2) 2 mm, and (3) 3 mm.

Figure 9 shows the amplitudes of the received signal corresponding to different water depths. The amplitudes of the sensor signal show a proportional trend of decreasing voltage vs. increasing water depth. This test verified that the amplitudes of the sensor signal correlated with the water quantity between SAs.

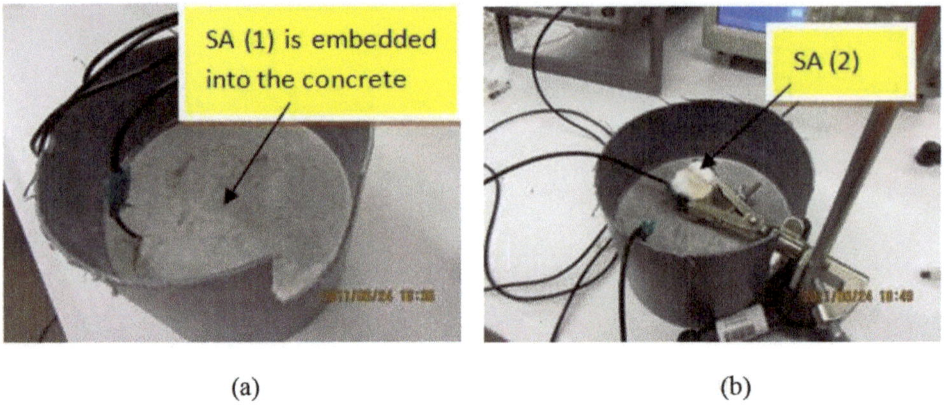

Figure 7. Test setup: (a) concrete specimen with an embedded SA; (b) concrete specimen with an external SA.

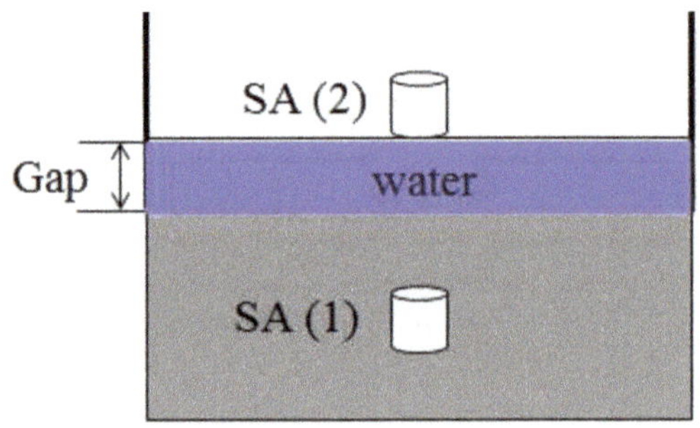

Figure 8. Diagram of quantitative water detection.

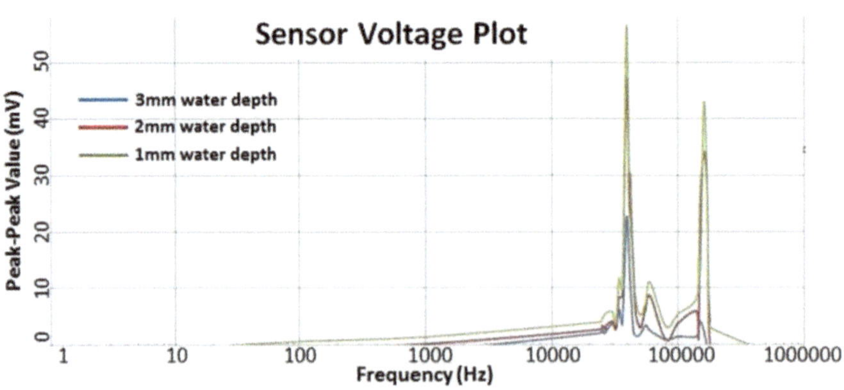

Figure 9. Amplitudes of the sensor signal with different water depths.

From the two simple experimental verifications, the active sensing approach successfully detected the case of water migrating in the crack. Water in the crack strengthened the stress wave propagation. Moreover, the verified tests provided potential feasibility to quantitatively monitor the water amount existing in the crack.

4. Experimental setup

4.1. Experimental procedures

Case 1: Concrete beam without cracks

In the experiment, a concrete beam (18″ × 6″ × 3″), as shown in Figure 10, was casted in the laboratory. Three SAs were embedded in predetermined locations before casting. SA-0, as the actuator, was placed in the middle of the concrete beam. SA-1 and SA-2, as sensors, were placed 2 inches from the ends of the concrete beam. The rebar on the bottom of the concrete beam was used to enhance the strength of the concrete beam. The location of the embedded SAs in the concrete beam is shown in Figure 11.

Case 2: Concrete beam with a large crack

A load test was conducted to artificially form a crack in the concrete beam, as shown in Figure 12. A continuously increasing load was applied to the concrete beam. After the loading process, an artificial crack was generated in the concrete beam, as shown in Figure 13.

Case 3: Crack filled with water

In case 3, the whole concrete beam was stored in a water environment for 24 hours. Therefore, the water completely filled in the concrete crack and the boundary condition of the concrete

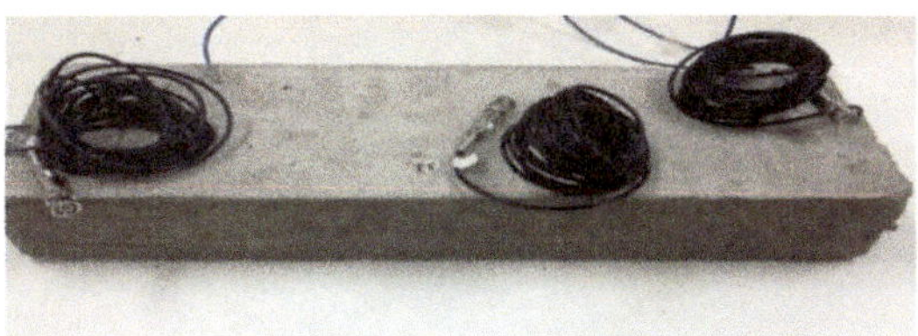

Figure 10. Casted concrete beam with three embedded SAs.

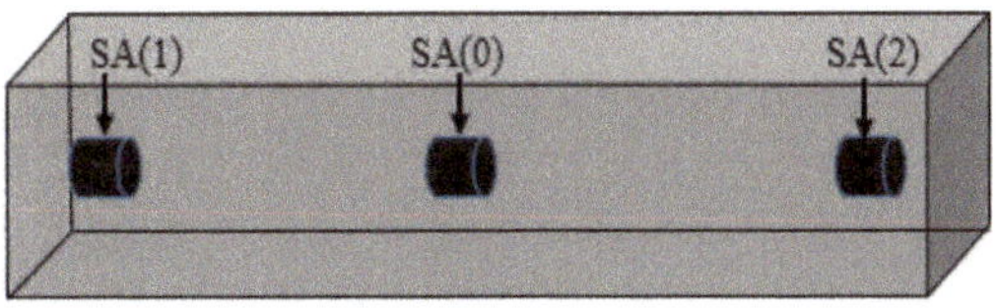

Figure 11. The schematic of SAs in the concrete beam.

Figure 12. Concrete beam loading setup.

Figure 13. The concrete beam with a large crack.

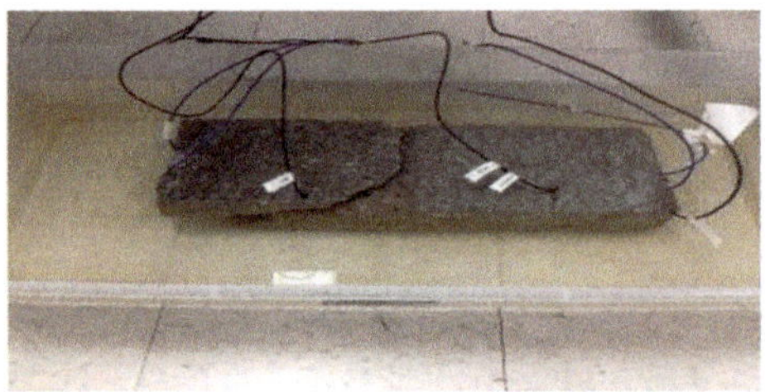

Figure 14. The concrete beam with water in the large crack.

beam was stable. Figure 14 is a photograph of the concrete beam stored in the water environment.

4.2. Excited swept frequency sine wave signal

In this experiment, SA-0 was used as the actuator to generate the desired swept sine wave using an Agilent 33120A function/arbitrary waveform generator. The detailed parameters of the guided wave are shown in Table 2, while the graph of the swept sine wave for 0.3 second of data is shown in Figure 15.

5. Experimental results

5.1. Time domain analysis

In each case mentioned above, SA-1 and SA-2 were both used as sensors to detect the stress wave generated by SA-0. The received sensor voltage plots, as shown below in Figure 16, demonstrated the sensor signal response. Figure 16(a), (c), and (e) corresponds to SA-1 while Figure 16(b), (d), and (f) corresponds to SA-2. Each of the signal graphics contained two periods of the sensor signal according to the swept excitation sine wave. Owing to the artificial crack between SA-0 and SA-1, the stress wave propagation was

Table 2. Detailed parameters of the swept sine wave signal.

Start frequency	100 Hz
Stop frequency	150 kHz
Sweep period	3 s
Amplitude	10 V

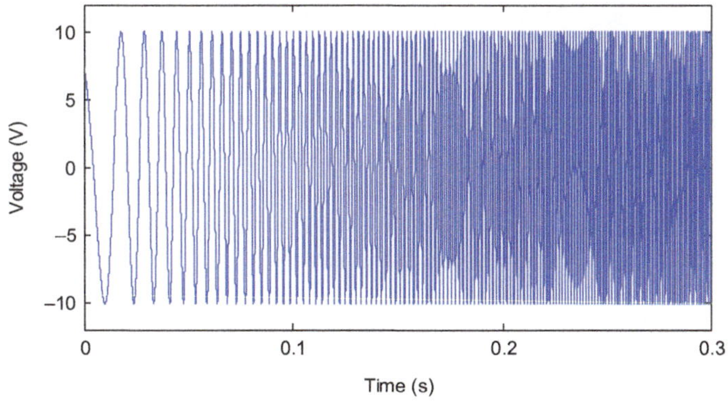

Figure 15. Graph of the swept sine wave for 0.3 second.

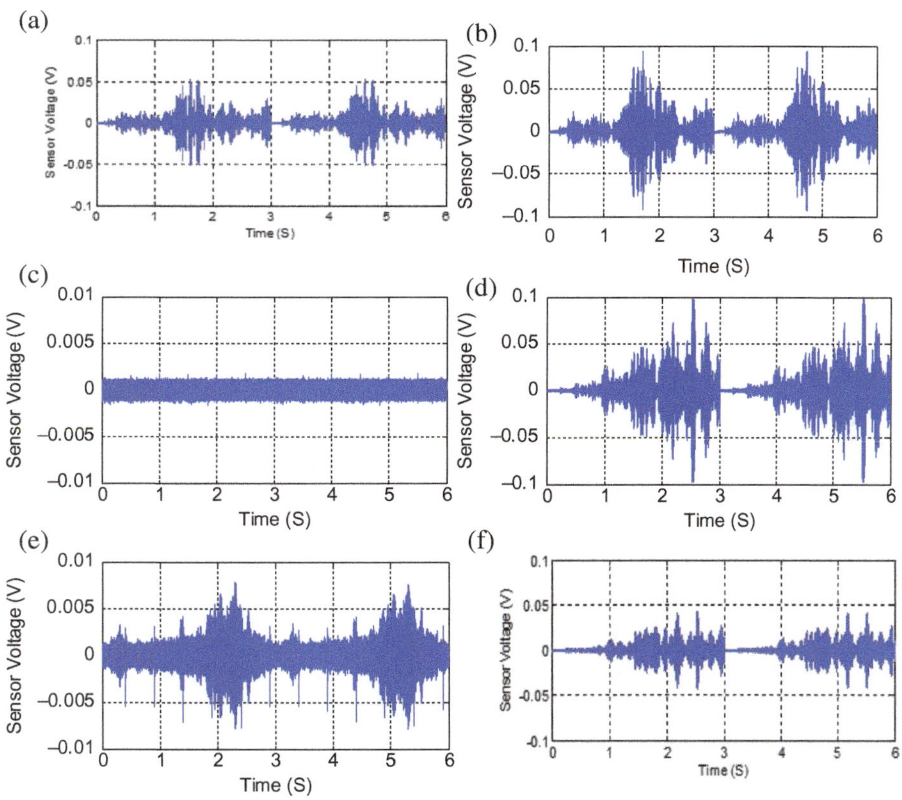

Figure 16. Sensor voltage graphic: (a) SA-1, no crack and no water; (b) SA-2, no crack and no water; (c) SA-1, with crack but no water; (d) SA-2, with crack but no water; (e) SA-1, with water in crack; and (f) SA-2, with water in crack.

extremely weakened and no signal response was detected. When the crack was filled with water, water changed the wave impedance and strengthened the stress wave propagation. As a result, the amplitude of the received signal increased, as shown in Figure 16(e). Figure 16(b), (d), and (e) presents the signal responses of SA-2. Compared with the crack existence condition between SA-0 and SA-1, no crack formed between SA-0 and SA-2. However, when comparing Figure 16(d) to Figure 16(b), Figure 16(d) demonstrates the same voltage level but a different signal response from the sensor voltage graphic. Figure 16(f) indicates a slight voltage value drop when the concrete beam is placed in the water environment. The reason is the boundary condition of the stress wave propagation change in both cases. Although the stress wave propagation between SA-0 and SA-2 is not directly related to the crack and water, the received stress wave response is still slightly influenced by the boundary condition change.

5.2. Wavelet packet analysis: sensor energy index

Based on the energy vector calculation using wavelet packet analysis, the received sensor energy values are presented in Figure 17. For SA-1, a sensor energy trend is demonstrated in each case. Without the crack and water, the received sensor energy level is close to 10^5, as shown with the left-most blue bar. The large crack in the concrete beam reduces the sensor energy to the degree of 10^2, as shown with the left-most green bar. When the crack is filled with water, the sensor energy rises to the degree of 10^3, as shown with the left-most brown bar. For SA-2, the boundary condition-induced energy change is negligible and the sensor energy maintains stable for all the cases, as seen in the right-hand portion of Figure 17.

6. Conclusions

In this paper, the SA-based active sensing approach is successfully developed to detect water presence in a concrete crack. The large crack in the stress wave path acts as a stress relief and attenuates the transition energy from the actuator to the sensor. Therefore, the amplitude of the received sensor voltage significantly reduces. However, in the case of water presence in the concrete crack, water changes the wave impedance and guides the excited stress wave to propagate through the crack. The water presence in the concrete crack can be confirmed by the increase of the received sensor signal amplitude. In addition, the wavelet packet-based energy index in this research quantitatively indicates the received sensor energy level corresponds to different concrete conditions such as health status, large crack damage and water presence in the concrete crack.

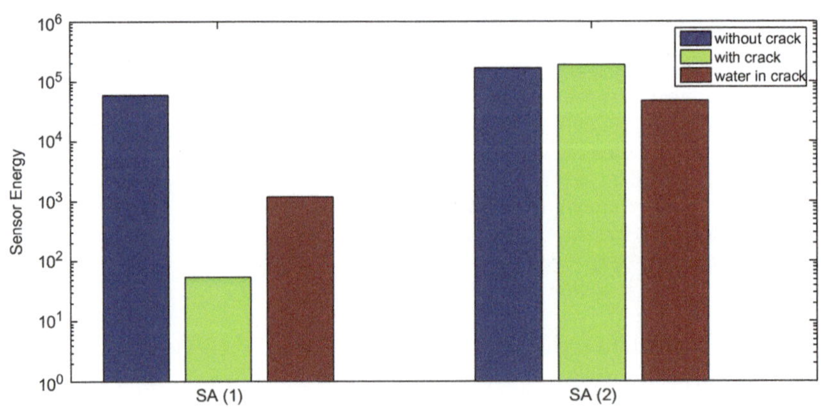

Figure 17. Sensor energy performance of SA-1 and SA-2 in three cases.

Disclosure statement

No potential conflict of interest was reported by the authors.

Funding

This work was partially supported by the Science Fund for Creative Research Groups under Grant No. [51121005] and a research project under Grant No. [51278084] from the National Science Foundation of China.

References

[1] A. Ito, Y. Aoki, and S. Hashimoto, *Accurate extraction and measurement of fine cracks from concrete block surface image*, Proceedings of IECON 2002, Seville, 2002, pp. 77–82.

[2] Y. Fujita, Y. Mitani, and Y. Hamamoto, *A method for crack detection on a concrete structure*, 18th International Conference on Pattern Recognition, ICPR 2006, Vol. 3, Hong Kong, 2006, pp. 901–904.

[3] T. Yamaguchi and S. Hashimoto, *Improved percolation-based method for crack detection in concrete surface images*, 19th International Conference on Pattern Recognition, ICPR 2008, Tampa, FL, December 2008, pp. 1–4.

[4] T. Yamaguchi and S. Hashimoto, *Fast crack detection method for large-size concrete surface images using percolation-based image processing*, MVA 21 (5) (2010), pp. 797–809.

[5] S. Tanaka, *Non-destructive inspection of concrete structures by ultrasonic sensor based on multi-reflected-wave and stationary wave models*, Trans. Soc. Instrum. Control Eng. 37 (6) (2001), pp. 484–492. doi:10.9746/sicetr1965.37.484.

[6] P. Antonaci, C.L.E. Bruno, P.G. Bocca, M. Scalerandi, and A.S. Gliozzi, *Nonlinear ultrasonic evaluation of load effects on discontinuities in concrete*, Cem. Concr. Res. 40 (2010), pp. 340–346. doi:10.1016/j.cemconres.2009.09.014.

[7] F. Hey, S. Bhalla, and K.C. Son, *Optimized parallel interrogation and protection of piezo-transducers in electromechanical impedance technique*, J. Intell. Mater. Syst. Struct. 17 (2006), pp. 457–468. doi:10.1177/1045389X06058792.

[8] K.K. Tseng and L. Wang, *Smart piezoelectric transducers for in situ health monitoring of concrete*, Smart Mater. Struct. 13 (2004), pp. 1017–1024. doi:10.1088/0964-1726/13/5/006.

[9] F.P. Sun, Z. Chaudhry, C.A. Rogers, and M. Majmundar, *Automated real-time structure health monitoring via signature pattern recognition*, Proc. SPIE 2443 (1995), pp. 236–247.

[10] C.K. Soh, K.K. Tseng, S. Bhalla, and A. Gupta, *Performance of smart piezoceramic patches in health monitoring of a RC bridge*, Smart Mater. Struct. 9 (2000), pp. 533–542. doi:10.1088/0964-1726/9/4/317.

[11] S. Bhalla and C.K. Soh, *High frequency piezoelectric signatures for diagnosis of seismic/blast induced structural damages*, NDT E. Int. 37 (2004), pp. 23–33. doi:10.1016/j.ndteint.2003.07.001.

[12] S. Park, S. Ahmad, C.-B. Yun, and Y. Roh, *Multiple crack detection of concrete structures using impedance-based structural health monitoring techniques*, Exp. Mech. 46 (2006), pp. 609–618. doi:10.1007/s11340-006-8734-0.

[13] X. Zhao and H. Li, *Health monitoring of reinforced concrete frame-shear wall using piezo-ceramic transducer*, J. Vib. Shock 25 (2006), pp. 82–84.

[14] V. Raju, G. Park, and H.H. Cudney, *Impedance-based health monitoring of composite reinforced structures*, 9th International Conference on Adaptive Structures and Technologies, Boston, MA, 1999, pp. 448–457.

[15] S. Multon and F. Toutlemonde, *Water distribution in concrete beams*, Mater. Struct. 37 (2004), pp. 378–386. doi:10.1617/14031.

[16] T.L. Yeo, D. Eckstein, B. McKinley, L.F. Boswell, T. Sun, and K.T.V. Grattan, *Demonstration of a fibre-optic sensing technique for the measurement of moisture absorption in concrete*, Smart Mater. Struct. 15 (2006), pp. N40–45. doi:10.1088/0964-1726/15/2/N03.

[17] S.C.M. Ho, L. Ren, H.N. Li, and G. Song, *A fiber Bragg grating sensor for detection of liquid water in concrete structures*, Smart Mater. Struct. 22 (5) (2013), pp. 055012. doi:10.1088/0964-1726/22/5/055012.

[18] J.B. Ong, Z. You, J. Mills-Beale, E.L. Tan, B.D. Pereles, and K.G. Ong, *A wireless, passive embedded sensor for real-time monitoring of water content in civil engineering materials*, IEEE Sens. J. 8 (12) (2008), pp. 2053–2058. doi:10.1109/JSEN.2008.2007681.
[19] S. Hou, H.B. Zhang, and J.P. Ou, *A PZT-based smart aggregate for compressive seismic stress monitoring*, Smart Mater. Struct. 21 (10) (2012), pp. 105035. doi:10.1088/0964-1726/21/10/105035.
[20] H. Gu, G. Song, H. Dhonde, Y.L. Mo, and S. Yan, *Concrete early-age strength monitoring using embedded piezoelectric transducers*, Smart Mater. Struct. 15 (2006), pp. 1837–1845. doi:10.1088/0964-1726/15/6/038.
[21] Q. Kong, S. Hou, Q. Ji, Y.L. Mo, and G. Song, *Very early age concrete hydration characterization monitoring using piezoceramic based smart aggregates*, Smart Mater. Struct. 22 (8) (2013), pp. 085025. doi:10.1088/0964-1726/22/8/085025.
[22] Q. Kong, R. Wang, G. Song, Z.J. Yang, and B. Still, *Monitoring the soil freeze-thaw process using piezoceramic-based smart aggregate*, J. Cold Reg. Eng. 28 (2014), pp. 2. doi:10.1061/(ASCE)CR.1943-5495.0000066.
[23] G. Song, C. Olmi, and H. Gu, *An overheight vehicle–bridge collision monitoring system using piezoelectric transducers*, Smart Mater. Struct. 16 (2007a), pp. 462–468. doi:10.1088/0964-1726/16/2/026.
[24] G. Song, H. Gu, Y.L. Mo, T.T.C. Hsu, and H. Dhonde, *Concrete structural health monitoring using embedded piezoceramic transducers*, Smart Mater. Struct. 16 (2007b), pp. 959–968. doi:10.1088/0964-1726/16/4/003.

10

Long-range attractive forces extending from the alumina's nanolayer surface in aqueous solutions

Leonid A. Kaledin*, Fred Tepper and Tatiana G. Kaledin

Argonide Corporation, Nanomaterial & Filtration Technologies, 291 Power Court, Sanford, FL 32771, USA

Aluminum oxide-hydroxide nanolayer with a thickness of approximately 1.2 nm is electroadhesively deposited onto silicious support material with large surface area of about 50 m^2/g, forming a highly electropositive composite of boehmite nanolayer in the form of monocrystalline oxide/hydroxide (α-$Al_2O_3 \cdot H_2O$) on the second electronegative solid. The composite can be viewed as a sphere with a rough surface and charge density of approximately 0.08 C/m^2. This creates a significant electric field with negligible screening ($ka \ll 1$) in the region close to the surface of the nanocomposite. This field attracts nano- and micron-sized particles from as far as 200 μm in a few seconds, many orders of magnitude greater than conventional Derjaguin–Landau–Verwey–Overbeek (DLVO) theory, which predicts only nanometer-scale effects arising from the presence of the surface. The strong electric field on the surface is then able to retain small particles such as viruses, atomically thin sheets of graphene oxide, RNA, DNA, proteins, dyes as well as heavy metals such as cobalt, arsenic, and lead.

Alumina's nanolayer surface can be further functionalized by adding other sub-micron or nano-sized particles to target a specific contaminant. An example is shown where alumina nanolayer is coated with nano-sized iron monohydrate to yield an arsenic sorbent that shows high sorption capacity.

Keywords: alumina nanolayer; electric double layer; electrostatic adsorption

1. Introduction

Recently, we described a filter media whose primary filtration mechanism for removing suspended solids and heavy metals is by electroadsorption [1]. The active component in the filter is boehmite in the form of crystalline oxide/hydroxide as (α-$Al_2O_3 \cdot H_2O$) nanofibers, only 2 nm in diameter and about 300 nm long. Nanoalumina is also known in the literature as 'boehmite whiskers'. In the filter media, the boehmite whiskers are bonded to a micro-glass non-woven scaffold. The pore size of the media is 2 μm, yet it can filter suspended solids as small as viruses and does so at high flow rates. A single layer can retain 6 LRV (log retention value) of *E. coli* bacteria and 4 LRV of MS2 viruses. A dual layer of media increases bacteria and virus retention to greater than 7 and 6 LRV, respectively. An alternative media, containing ultrafine (8 μm) powder activated carbon particles was also characterized for metal and particle retention. Its retention of colloidal and suspended solids,

---

*Corresponding author. Email: kaledin@argonide.com

including bacteria and virus, was found to be at least as good as that of the above media. Both are superior than ultraporous membrane filters for removing colloidal solids.

Electroadhesive adsorption is a mechanism that is less dependent upon pore size than is the case with mechanical filtration by either membranes or with typical depth filters. A non-woven filter format is best suited for separation via electroadhesion. Electroadhesion utilizes the difference in charge that may exist between a surface (or fiber) and a particle in an aqueous solution, where a distribution of charges has traditionally been called electrical double layer (EDL), although it is often more complex than just two layers. We showed that the standard expression for Coulombic force (charge-charge, $\sim 1/r^2$) quantitatively described electrokinetic motion in aqueous solution of MS2 viruses and *E. coli* bacteria in electric field of alumina nanofibers grafted onto glass microfibers, thereby forming a macroscopic assembly of alumina nanofibers on the second solid in highly organized matter [1]. The assembly can be viewed as straight cylinder with rough surface and charge density of approximately 0.08 C/m^2. This creates a significant electric field with negligible screening ($ka \ll 1$) in the region close to the surface of the assemblies. This field attracts nano- and micron-sized particles from as far as 300 μm in less than a few seconds, many orders of magnitude greater than the conventional Derjaguin–Landau–Verwey–Overbeek (DLVO) theory, which predicts only nanometer-scale effects arising from the presence of the surface.

In classical physics, it was well established more than a century ago that direct intermolecular forces are of long range, e.g., when a liquid is moving in turbulent flow, a zone exists next to a stationary solid wall in which flow is laminar. In a review article, Henniker [2] cites more than a hundred published studies confirming the long-range effect of varied surfaces on many liquids, including water. For example, in 1904, Nernst [3] and Brunner [4] estimated the thickness of the non-turbulent film in the range of 20 to 52 μm. Davis and Crandall [5] confirmed these data by examining the movement of particles suspended in water with the use of a microscope. They found that in quiescent water, the particles within 400 μm of the surface were stationary, those below being seen to move freely under the influence of random currents. Hardy suggested [6] that in close-packed structures such as fluids and solids, a strain transmitted from molecule to molecule may extend so deeply as to modify the molecular state of a skin water layer some hundreds of microns in depth from the surface.

However, the long-range forces were abandoned in favor of electrical forces effective for only a fraction of nanometer [7–9]. Electrokinetic phenomena are associated with the electric double layer (EDL) on charged surfaces of particles immersed in electrolyte solutions. Gouy [7] and Chapman [8] developed a diffuse layer (DL) theory. Stern extended that theory by proposing a model in which the interfacial layer is divided into two parts [9], the inner Helmholtz plane and the outer Helmholtz plane [10]. It has been shown [11] that the Stern model is applicable away from the surface, but cannot be taken literally near the surface. Static double DLs are electroneutral, meaning that the surface charge σ^{o} is exactly compensated by the countercharge. The latter consists of two parts, the stagnant layer, σ^{i} and the 'electrokinetically active' part, σ^{ek}. In most cases, σ^{i} closely coincides with the non-diffuse Stern layer, part of the double layer. The σ^{ek} can be determined from electrokinetics (e.g., from measuring zeta potential, ζ [11] and σ^{o} can be determined from titration of surface groups [12] and their difference can be used to obtain σ^{i}.

In the case of a flat DL, there is only one parameter, the Debye length, $1/k$, while in the case of a spherical or cylindrical DL, there is an additional geometric parameter, the radius of the particle, α. A vast majority of aqueous dispersions satisfy the $ka \gg 1$ condition (thin DL), where the DL depth is much thinner than the particle radius while the

$ka \ll 1$ condition (thick DL) where the DL depth is much larger than the particle radius. Theoretical treatments have been suggested for the thin DL ($ka > 10$) and for the thick DL ($ka < 1$) that include DL overlap [13]. At high electrolyte ionic strengths (>1 M), the Debye length becomes comparable with the size of the ions, implying that the DL collapses onto the particle surface. However, there are indications that in some cases, the DL still exists even at ionic strengths greater than 1 M.

Colloidal dispersion stability is often predicted by the DLVO theory [14,15], which was developed for smooth, homogeneous particles with ideal geometries and with no DL overlap. The electrostatic repulsive part of the potential between surface and solution is thought to drop off as e^{-rk}, where r is the distance from the surface and k is the inverse of the Debye length of the solution [16]. Despite the success of the DLVO theory, numerous investigators have modified the DLVO theory to factors not accounted for in the DLVO model. For example, haloes are formed when live biological cells (leucocytes, especially tumor cells) are suspended in concentrated solutions of inorganic electrolytes. In the medium made up of erythrocytes, the haloes form within 5–10 min as a result of the background cells drawing apart from the central halo-forming cell at a distance of 10–100 μm and more [17]. In the medium made of the Indian ink particles, the haloes form during 2–4 s and attain a thickness of about 10–20 μm [17].

Derjaguin and Churaev described the phenomenon involving the formation of condensates with anomalous properties from the vapors of water and other liquids on silicate surfaces [18], hypothesizing formation of stable associates of water molecules $(H_2O)_n$. This hypothesis was developed further by Lippincott *et al.* [19], who stated that the substance in question is *not* water. The substance is certainly built of oxygen and hydrogen, but their arrangement in a hexagonal lattice bears little resemblance to their arrangement in the water molecule. The authors considered it clear that this entity was chemically distinct from water. The most conclusive findings were the observations of void structures of the order of 25–50 μm in a highly purified dispersion of latex particles [20–23] and long-range exclusion zones of latex microspheres of the order of 100 μm from the gel surface [24,25]. Solute-free zones usually form next to hydrophilic surfaces in an aqueous solution. Anionic and cationic ion-exchange beads of 500 μm diameter form 200 μm deep zones with negative and positive net charges, respectively. Numerous details of this ordered phase of water have been recently documented in Ref [26].

Sogami suggested a different interpretation of colloidal forces [27]. The Sogami–Ise theory [28] claims that the attraction between colloidal particles comes from the influence of the intervening ions. The Sogami potential contains both short-range repulsion and long-range attraction. This gives the attraction a longer reach than in DLVO theory. Recently, Sogami argued that the colloidal dispersion is homogeneous with respect to variable T (Sogami–Ise theory [28]), and not with respect to volume, V (DLVO theory), which is not a suitable thermodynamic variable [29]. The screened Coulombic potential of pure repulsion obtained from the Helmholtz free energy in the DLVO theory was replaced by a new screened electric potential with a long-range attractive tail derived from the generalized Gibbs free energy, which is identified with the total sum of chemical potentials [29].

In this paper, we report on long-range attractive forces extending from the alumina's nanolayer surface in aqueous solutions and on associated mechanisms of filtering heavy metals such as cobalt, arsenic, lead as well as colloidal particles such as endotoxins, proteins, bacteria, and viruses using either a precoat filtration technique or dispersions of a new sorbent, aluminized diatomaceous earth (DE).

2. Materials and methods

2.1. Alumina nanolayer electropositive sorbent

2.1.1. Streaming and zeta potentials

The streaming potential apparatus is similar to that described in Ref [30]. The streaming potential is measured by means of a pair of Ag/AgCl electrodes located on both sides of two parts of a channel (ID-19 mm) separated by a septum (three layers of two-dimensional filter media having an irregular pore pattern and an irregular pore size of 0.1–0.2 mm. The aluminized DE powder (available from Argonide Corporation, Sanford, FL, USA) was packed inside the tube on the septum with depth of 2.5–3.5 cm). A solution of a known conductivity, as measured by Oakton ECTestr Pure, is placed in a pressurized, aluminized DE-packed column and pressure is applied to the vessel and therefore the pressure is transferred to the filter media. The streaming potential is found by applying at a given time an over-pressure (ΔP, bar) to the vessel and measuring the resulting potential difference ($\Delta\phi \equiv E_s, V$) on both sides of the filter bed. The E_s and ΔP values are measured at ±0.1 mV and ±0.02 bar, respectively. The bulk conductivity of the solutions ranged from 1.2 μS/cm for 0.01 mM NaCl to 284 μS/cm for 2 mM NaCl solutions in purified RO water with initial conductivity less than 0.1 μS/cm. The pH values were adjusted with 0.1 M NaOH or with 0.1 M HCl.

Zeta potential is an important and reliable indicator of the surface charge of filters. It can be deduced from experimental data, e.g., from measuring the streaming potential. It has been long recognized [31,32] that in the plug flow situation, the apparent zeta potential ($\zeta_{apparent}$) is a function of bulk conductivity of filtered liquid (λ_b) while the true zeta potential (ζ_{true}) is a function of bulk conductivity of filtered liquid (λ_b) as well as surface conductance (λ_s). The true zeta potential and specific surface conductance can be evaluated by plotting of reciprocal of the apparent zeta potential as a function of the resistivity of water. The apparent zeta potential ($\zeta_{apparent}$) for different grades of aluminized DE filter media was calculated using the weighted least-squared procedure [33] for at least three different conductivities of water adjusted with NaCl. The computed results for true zeta potential of aluminized DE18 filter media are shown in Figure 1.

Figure 1 indicates that aluminized DE18 powders are highly electropositive ($\zeta > +50$ mV) at aluminum oxide/hydroxide (α-$Al_2O_3 \cdot H_2O$) solids loading greater than ~20% as compared to the highly electronegative character of the bare DE18 support ($\zeta \sim -70$ mV).

2.1.2. X-ray diffraction (XRD), scanning electron microscopy (SEM) and transmission electron microscopy (TEM) analyses

XRD analysis shows that the coating is amorphous and most of the peaks seem to correspond to silica minerals underlying the coating. The aluminized DE material is transparent to XRD and seems to exist as a nanometer thick coating (Figures 2 and 3). The medium-grade aluminized DE has Brunauer, Emmet and Teller (BET)-specific surface area of 126.5 m^2/g (see Table 1). An estimation shows that the thickness of the coating is only 1.2 nm (or 12 Å), which is comparable to unit cell parameter b for boehmite (α-$Al_2O_3 \cdot H_2O$) a = 2.868 Å, b = 12.227 Å, and c = 3.700 Å.

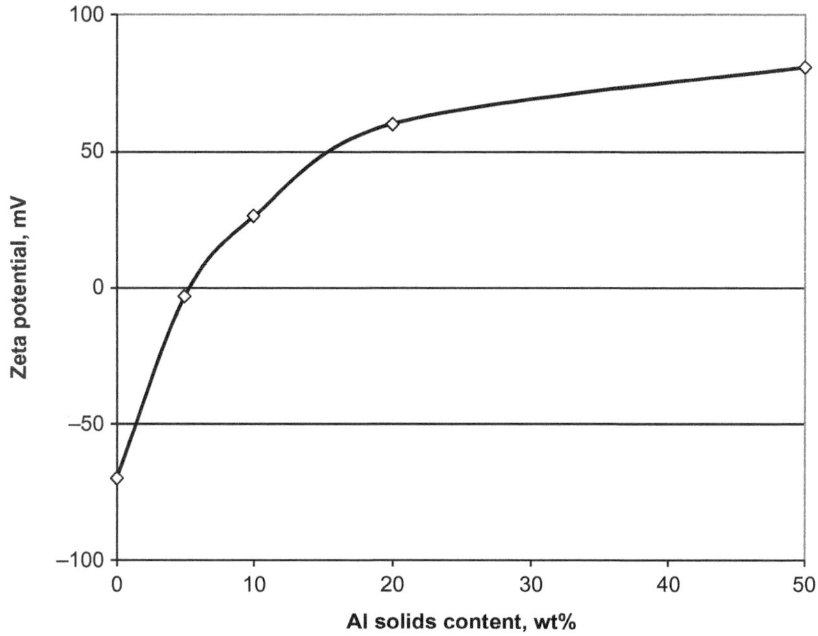

Figure 1. Zeta potential of DE18/aluminized DE18 as a function of aluminum solids.

Figure 2. SEM of aluminized DE60 particles.

2.1.3. Ironized alumina nanolayer on DE support for arsenic removal

Nano-sized iron hydroxide can be deposited on the aluminized DE nanolayer coating to form a sorbent capable of adsorbing arsenic in a precoat configuration. Figure 4 is an SEM of the iron composite. The features of the particles in Figure 4 are very similar to that of uncoated DE, with some agglomeration owing to the aluminized coating. The particles of iron hydroxide are too small to be featured in the photograph.

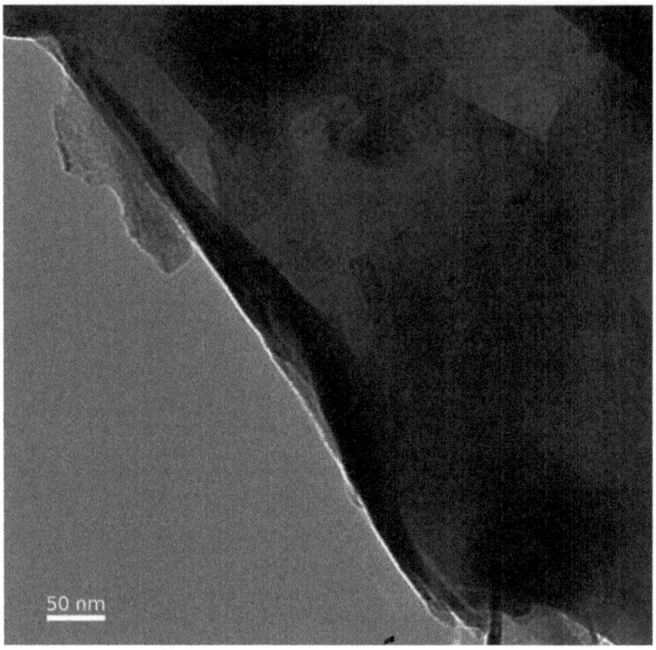

Figure 3. TEM of aluminized DE80 particles (Courtesy of Dr. A. Slesarev of Rice University).

Table 1. BET surface area of different grades of aluminized DE composites.

Grade	Mean particle size, μm	BET surface area, m^2/g	Mean pore volume, cm^3/g	Mean pore width, Å
Fine	5	140.3	0.0499	124.9
Medium	18	126.5	0.0499	133.0
Coarse	80	57.0	0.0203	125.4
Medium, coated with FeOOH	18	72.8	0.0449	133.0

2.1.4. Specific surface area analysis

Table 1 shows the BET surface area of different grades of aluminized DE composite.

Results of Table 1 show that the surface area of several grades of aluminized DE is on the same order of magnitude, i.e., the BET ratios of different grades are lesser than a factor of 3.

Table 2 compares the BET surface area of DE18, aluminized DE18, and ironized/ aluminized DE18-FeOOH materials.

Results of Table 2 show that the surface area of aluminized DE18 increased by 157% as well as the mean pore volume by 185% as compared to the uncoated DE. The mean pore width does not change within the accuracy of measurements. This suggests that surface properties of aluminized DE and aluminized DE-FeOOH significantly change upon the coating procedures.

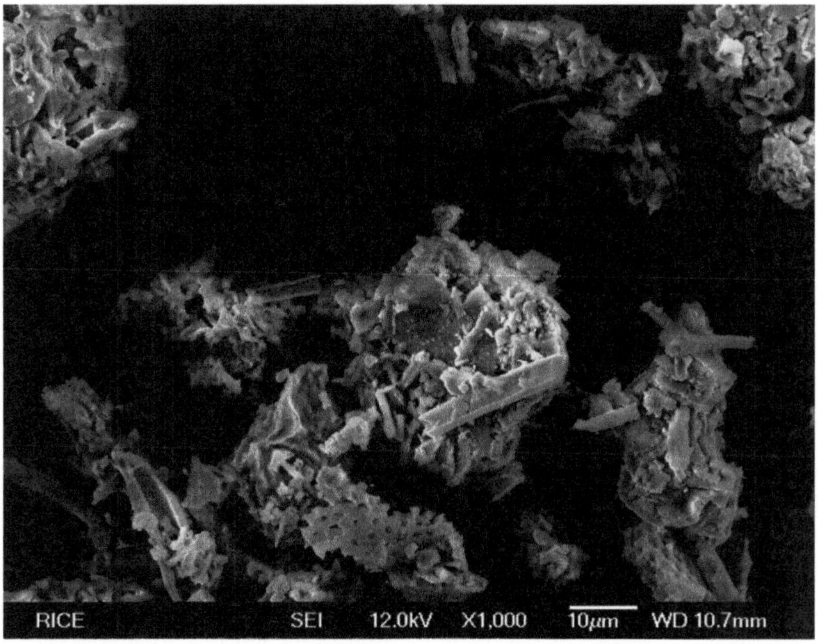

Figure 4. SEM of aluminized DE18/FeOOH.

Table 2. BET specific surface area of DE18, aluminized DE18, and aluminized DE18-FeOOH.

Aluminized DE grade	BET surface area of powders, A_g, m^2/g	Ratio of aluminized DE surface area to DE support	Mean pore volume, V_g, cm^3/g	Ratio of mean pore volume of coating to DE support	Mean pore width, Å
DE18	49.2	≡1	0.0175	≡1	117.8
DEAL18	126.5	2.57	0.0499	2.85	133.0
DEAL18-FeOOH	72.8	1.48	0.0449	2.57	133.0

2.2. Virus and bacteria isoelectric points

The pH value at which the net surface charge switches its sign is referred to as the isoelectric point (IEP) and is a characteristic parameter of a virus in equilibrium with its environmental water chemistry. The IEP value of MS2 bacteriophage in different environments, based on the averaging of 10 measurements, shows a mean IEP value of 3.5 ± 0.6 and with the discrepancy in IEP (ΔIEP) of 1.8 from 2.2 to 4.0 [34]. Recent studies by Duek et al. [35] and Chrysikopoulos et al. [36] determined the alteration of the IEP by a change in water chemistry (e.g., ionic strength or ionic composition) with IEP of the former case in the range from less than 2 at ionic strengths of 0.0 mM and 2.6 mM to IEP of 4.4 in the latter case that was measured in double distilled water. In the case where clean colloidal particles are suspended in absolute pure water, the point of zero charge (PZC), which is characteristic to the particle's surface, can be determined. The IEP and the

PZC do not necessarily differ in the presence of monovalent ions; however, the PZC is very difficult to measure experimentally [34].

MS2 (ATCC 15597-B1) and *fr* (ATCC 15767-B1) bacteriophages are icosahedral in shape with a diameter of 26–27 nm [35] and 19 nm [37], respectively. The isoelectric point of *fr* bacteriophage is 8.9–9.0 [37]. In this study, MS2 bacteriophage was chosen because it is often used as a surrogate for human enteric viruses and *fr* bacteriophage was chosen because it is electropositive in aqueous suspensions at pH range of 3–9.

E. coli bacterium has a cell isoelectric point of 3.5 and zeta potential of $\zeta \sim -35$ mV. Zeta potential is fairly insensitive within a pH range of pH6 to pH9 [38]. The zeta potential of MS2 ($\zeta_{MS2} \sim -(33–38)$ mV [36]) is similar to that of *E. coli*.

2.3. *Bacteriophages and bacteria preparation*

Suspensions of MS2 and *fr* bacteriophages with high titers of up to $\sim 10^{12}$ particles/mL were prepared as described in Ref [39], stored at 4°C, and then diluted in deionized reverse osmosis (DI RO) water with conductivity less than 0.1 μS/cm.

Stocks of *E. coli* (ATCC 15597) with titers up to $\sim 10^{9}$ particles/mL were prepared and stored at 4°C, and then diluted in DI RO water with conductivity less than 0.1 μS/cm.

3. Results and discussion

3.1. *Contaminant removal using the precoat process*

The Buchner funnel is a laboratory device that may be used to test efficiencies of filtering aid powders or slurries. The Buchner funnel test was modified to form a precoat DE or aluminized DE to simulate the filter septum. The removal efficiency of various contaminants was compared between samples exposed to either: (1) DE body fluid; (2) DE precoat + DE body fluid; (3) aluminized DE body fluid; (4) aluminized DE precoat + aluminized DE body fluid.

3.2. *Formation of the precoat*

The precoats were formed using Nalgene filter holders with receiver (available from Cole-Palmer, Cat # S-06730–53). The capacity of the upper chamber was 500 mL and that of the receiver was 1 L. Five layers of 47 mm diameter woven wire screen with an average pore size of 200 μm were used as a support to form the precoat. A vacuum pump was used to reduce pressure in the receiver chamber to provide a differential pressure with respect to the upper chamber, which was held at the ambient atmospheric pressure. The test was conducted in batches. Each individual batch was 1 L in volume. Four grams of either DE or aluminized DE powders were manually mixed in 1 L of DI RO water to form powder suspensions at concentrations of 4 g/L. An aliquot of 650 mL of the powder suspension of either DE or aluminized DE was poured into the upper chamber, and the suction valve was opened on the vacuum pump to control the flow rate. After filtering 650 mL of slurry, the suction valve on the vacuum pump was closed and the flow was terminated. Each precoat has a surface area of ~ 10 cm^2 and is ~ 3 mm thick.

3.3. *Formation of body fluid*

After the precoat was formed, the lower reservoir was emptied and flushed several times with reverse osmosis water or distilled water to reduce turbidity to below a detection limit of 0.05 NTU. A known amount of a contaminant was added to the remaining 350 mL of the powder suspension and was filtered through the precoat. The flow of the body fluid was initiated by opening the suction valve. After 350 mL of contaminant-seeded water was filtered through the precoat layer, the effluent was analyzed for the residuals.

3.4. *Contaminant removal using the precoat process*

3.4.1. *Bacteria removal*

Table 3 shows *E. coli* bacteria removal efficiency by aluminized DE18 and DE18 with the use of a ~3 mm precoat layer. The results indicate that aluminized DE has extremely high *E. coli* removal efficiency (>8.4 LRV), while untreated DE removes only about 50% (or 0.3 LRV) of the bacteria.

3.4.2. *Virus removal*

MS2 bacteriophage is extensively used by the US Environmental Protection Agency (USEPA) and others as a surrogate for viruses. While virtually all viruses are electronegative, there are a few that are electropositive at neutral pH. Virus *fr* is used in many water filter testing protocols to assure that the removal method encompasses all pathogenic viruses known to contaminate water. Table 4 shows the results of virus removal for both MS2 and *fr* viruses.

Results of Table 4 indicate that:

(1) Aluminized DE has extremely high MS2 removal efficiency.
(2) Aluminized DE functionalized with nano-sized iron monohydrate (FeO(OH)·H_2O) has high MS2 removal efficiency.
(3) Untreated DE has very low removal capability for MS2 (~50% or 0.3 LRV).
(4) Aluminized DE has high *fr* removal efficiency.
(5) Virus removal is very rapid in the mixing tank, suggesting a less expensive and less tedious process for solid separation as compared to the precoat process.

Table 3. Removal efficiencies of *E. coli* bacteria by aluminized DE and DE.

Powder	Mode of filtration	Contact time, min	Influent concentration, CFU[a]/mL	Removal efficiency, LRV[b]
Aluminized DE	Precoat + body fluid	30	$7.2 \cdot 10^6$	>8.4
	Precoat	30	$2.0 \cdot 10^{10}$	4.2
	Body fluid	1	$7.0 \cdot 10^7$	1.0
		3		0.8
		10		1.2
		40		1.1
DE	Precoat + body fluid	30	$2.1 \cdot 10^7$	0.3

Note: [a]Coliform forming units; [b]logarithm removal value.

Table 4. Removal efficiencies of MS2[a] and *fr*[b] by aluminized DE and DE at pH 7.0.

Powder	Mode of filtration	Bacteriophage	Contact time, min	Influent concentration PFU[c]/mL	Removal efficiency, LRV[d]
Aluminized	Precoat + body fluid	MS2	30	$1.0 \cdot 10^8$	>8.0
DE	Precoat		30	$1.3 \cdot 10^{10}$	9.0
	Body fluid		1	$1.0 \cdot 10^8$	6.7
			3		6.7
			10		6.8
			40		7.3
	Body fluid	*fr*	5	$1.1 \cdot 10^5$	4.0
			10		4.3
Aluminized DE/FeOOH	Body fluid	MS2	1	$5.9 \cdot 10^4$	>4.1
DE	Precoat + body fluid	MS2	30	$2.1 \cdot 10^7$	0.3

Note: [a]ATCC 15597-B1; [b]ATCC 15767-B1; [c]plaque forming units; [d]logarithm removal value.

3.4.3. Arsenic removal using an ironized, aluminized DE

Nano-sized iron hydroxide can be deposited on the aluminized DE coating to form a sorbent that is capable of adsorbing arsenic in a precoat configuration. We treated a high concentration of As V (As^{5+}) with 18 μm average particle size aluminized DE-FeOOH sorbent. Using the precoat process, we were able to reduce (Figure 5) arsenic V from 180 ppb to below 10 part-per-billion (ppb) for the most part of the experiment.

It should be noted that aluminized DE eliminates the traditional need to convert As^{3+} into As^{5+} for its removal from water. Conversion of As^{3+} to As^{5+} is a critical element in a great majority of arsenic treatment processes [40]. This conversion can be accomplished by adding an oxidizing agent such as chlorine or permanganate.

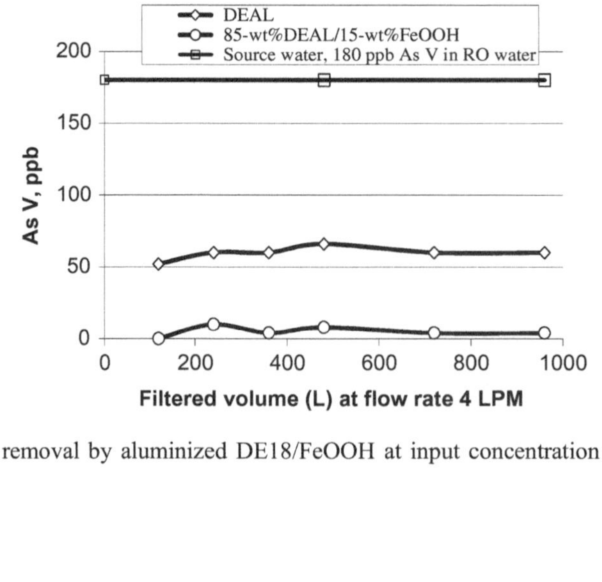

Figure 5. As V removal by aluminized DE18/FeOOH at input concentration of 180 ppb and at pH 6.5.

3.5. Sorption model

From the perspective of the design engineer, prediction of the breakthrough curve for a given filter, using only basic kinetic and equilibrium data is desirable. To extract dynamic information, the experimental adsorption isotherm(s) must be fitted to the theoretical response curve(s), calculated with a suitable dynamic model for the system.

A model was developed [41] that describes the adsorption of nanometer-sized particulates by the electropositively charged depth filter media and *the same model was applied to the case of the new aluminized DE filter media*. The surface area derived via BET adsorption for the aluminized DE powders indicates that the large surface area of alumina coating (up to ~140 m^2/g (see Table 1)) is external and available for rapid adsorption. This permitted us to reduce the complexity of the diffusion solutions. Although the expressions for the breakthrough curves derived from different models are algebraically different, the numerical difference is quite small and it became a common practice to use the simpler linear driving force models to further simplify the algebraic solutions [42].

We chose a general analytical solution [42] by assuming that: (1) the nature of equilibrium relationship is linear; (2) the nature of equilibrium is isothermal, i.e., heat transfer resistance can be neglected; (3) the concentration level of the adsorbable component is relatively low such that it does not influence basic properties of the carrier liquid, e.g., viscosity, conductivity, etc.; (4) the flow model is an ideal plug flow system when the axial dispersion term $-D_L\partial^2 c/\partial z^2$ can be neglected, thereby reducing the mass-balance equation to a first-order hyperbolic equation [42]; (5) the kinetic model assumes negligible mass transfer resistance due to the fact that instantaneous equilibrium can be assumed at all points because of the open character of the precoat layer. Under these assumptions, the analytical solution for a breakthrough curve is given by the following asymptotic equation (see, e.g., Ref [42]):

$$c_b{}^+ = \frac{1}{2}(1 - erf(\sqrt{z^+} - \sqrt{\theta^+})) \quad \text{for } z^+ \geq \theta^+ \quad \text{for } z^+ \gg 1 \tag{1}$$

where $c_b{}^+ = c_b/c_0$ is dimensionless concentration;

c_0 is challenge concentration;

c_b is bulk concentration;

θ^+ is a dimensionless value of corrected time θ defined as $t\text{-}z\cdot\varepsilon/v$;

z^+ is a dimensionless value of z;

z is depth of the precoat layer;

ε is porosity of the precoat layer;

v is flow velocity; and

erf is the error function.

Equation (1) therefore allows prediction of a breakthrough curve for large values of z^+ from constants derived from experimental measurements of the dynamic adsorption isotherms. To characterize the adsorption process of small particles by a filter media of given depth, a specific model can be constructed in the following way: (1) measure dynamic adsorption isotherms for a given thickness of precoat layer for at least three input particle concentrations at the same flow rate, pH, amount of total dissolved solids (TDSs), etc.; (2) fit dynamic adsorption isotherm to Equation (1), which allows determination of the value of z^+ and filter capacity; and (3) fit the capacity values to the Freundlich isotherm.

3.5.1. Adsorption and elution of proteins and separation processes

Aluminized DE is an effective sorbent for proteins. Concentrations of proteins (bovine serum albumin (BSA), hemoglobin, myoglobin and histone) were measured in the eluent after flowing through a 3 mm precoat layer of aluminized DE. Measurements were accomplished with the use of Bradford reagent protein assay procedure and a Genesis 10 UV spectrophotometer at a wavelength of 595 nm. Sorption capacity at a given input particle concentration was estimated from the breakthrough curves (see Figures 6 and 7

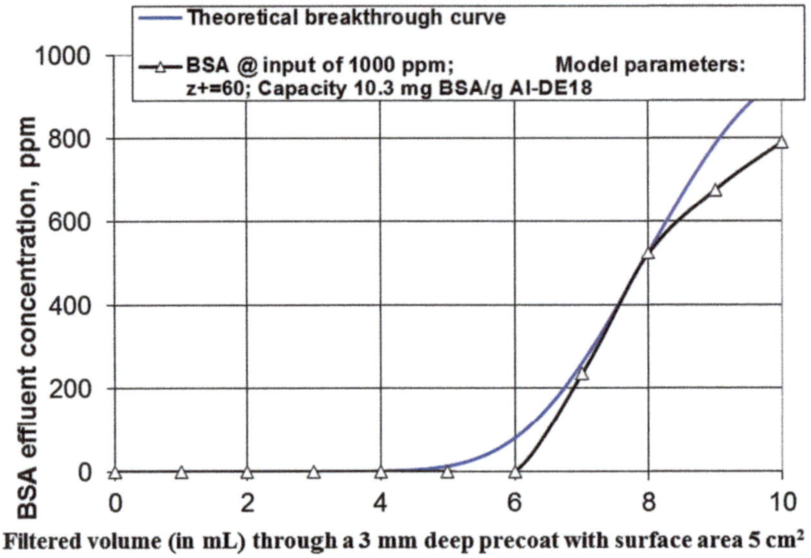

Figure 6. Removal efficiency of BSA particles by a 3 mm precoat layer of aluminized DE18.

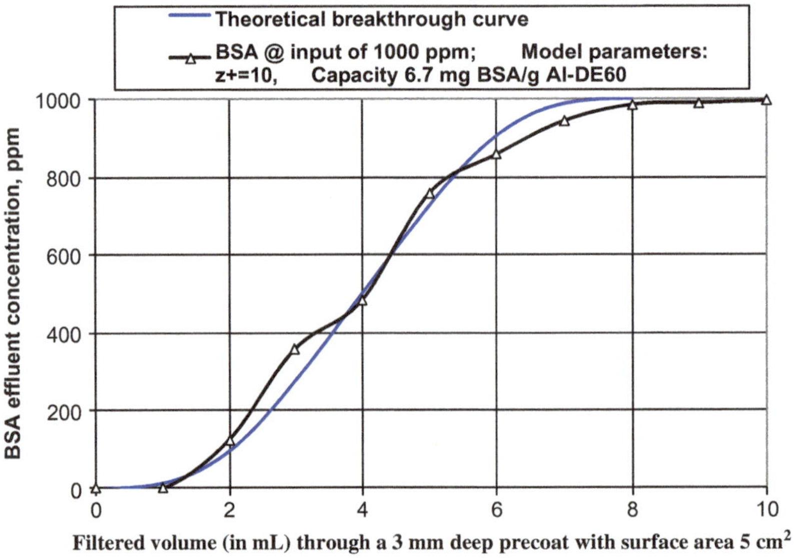

Figure 7. Removal efficiency of BSA by a 3 mm precoat layer of aluminized DE60.

Table 5. Protein removal by aluminized DE.

Sorbent	Protein	Protein input concentration, ppm	Flow velocity, cm/min	Sorption capacity, q, mg protein/g sorbent	Adsorption kinetics parameter, z^+
Aluminized DE18	Bovine serum albumin (BSA)	1000	1.5	10.3	60
		1000	0.15	16.5	20
		700	0.15	15.8	40
		400	0.38	15.0	20
		320	1.5	11.7	60
		200	0.75	13.1	20
		100	0.5	19.9	40
	Hemoglobin	1000	0.15	3.9	10
	Myoglobin	1000	0.15	3.2	20
	Histone	1000	0.15	2.6	30
Aluminized DE60	BSA	1000	1.5	6.7	10

for input concentration of 1000 ppm), assuming that the breakthrough curve is symmetric with respect to a 50% adsorption point on the curve.

The particle removal efficiency of the media sample can be calculated directly from the inlet and effluent values. These tests use relatively high contaminant challenge levels to reduce the time to breakthrough. We found that Equation (1) accurately describes breakthrough curves at a residence time greater than about 0.1 s through the filter media. Table 5 presents values of z^+ (Equation (1)) and filter capacity for the studied proteins obtained from the fits of the breakthrough curves for different input concentrations.

Results of Table 5 indicate that the model (Equation (1)) accurately describes breakthrough curves (see Figures 6 and 7).

Figure 8 shows elution efficiency of hemoglobin adsorbed on a 3 mm layer of aluminized-DE using 0.1 M NaOH and Na_2CO_3 eluents (0.15 cm/min for adsorption

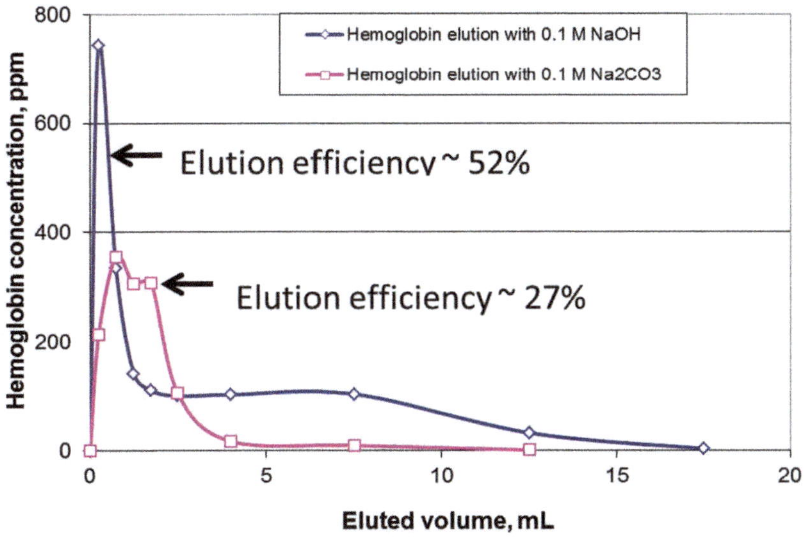

Figure 8. Elution efficiency of hemoglobin from an aluminized DE precoat layer.

and also for elution). The data of Figure 8 show that aluminized DE is a candidate for filtration of biopharmaceutical wastewater. Optimization of recovery efficiency as a function of sodium hydroxide and sodium carbonate concentrations for different proteins has not yet been determined. It appears that the sorbent can concentrate product even in the presence of saline. Data suggests that the smaller the particle, the more rapidly will aluminized DE adsorb it, and the adsorption capacity would also be greater. Moreover, reducing the size of the aluminized DE particle increases adsorption capacity on a weight/weight basis. We demonstrated a feasibility of chromatographic separation of two small viruses MS2 and $\alpha 3$ (ATCC 13706-B2) using similar electropositive media in the form of a 2 nm diameter boehmite oxide/hydroxide as (α-$Al_2O_3 \cdot H_2O$) fibers bonded to a microglass non-woven scaffold (see Figures 5–7 of Ref. [41]). These two viruses have nearly identical particle sizes (27 nm for MS2 and 25 nm for $\alpha 3$).

3.5.2. Adsorption of dyes

Adsorption is an effective method for removing dissolved organic compounds from waste streams. Activated carbon has been evaluated extensively for waste treatment of different classes of dyes. Adsorbability of activated carbon is lowest for metanillic isomers in the aniline molecule (=21 mg (metanillic acid)/g Calgon Filtersorb F400) [43].

The dimensions of metanil yellow (MY) are approximately 1.8 nm long and 0.9 nm wide [44]. Residual concentrations of MY dye molecule in 100 mL filtered aliquots by a 3 mm precoat layer of aluminized DE18 particles in gravity filtering mode were measured at 25°C with the use of Genesis-10 UV spectrophotometer at a wavelength of 674 nm. Sorption capacity at a given input dye concentration was estimated from the breakthrough curves (see Figure 9 for input concentration of 1000 ppm), assuming that the breakthrough curve is symmetric with respect to a 50% adsorption point on the curve.

The particle removal efficiency of the media sample can be calculated directly from the inlet and effluent values. These tests use relatively high contaminant challenge levels to reduce the time to breakthrough. We found that Equation (1) accurately describes

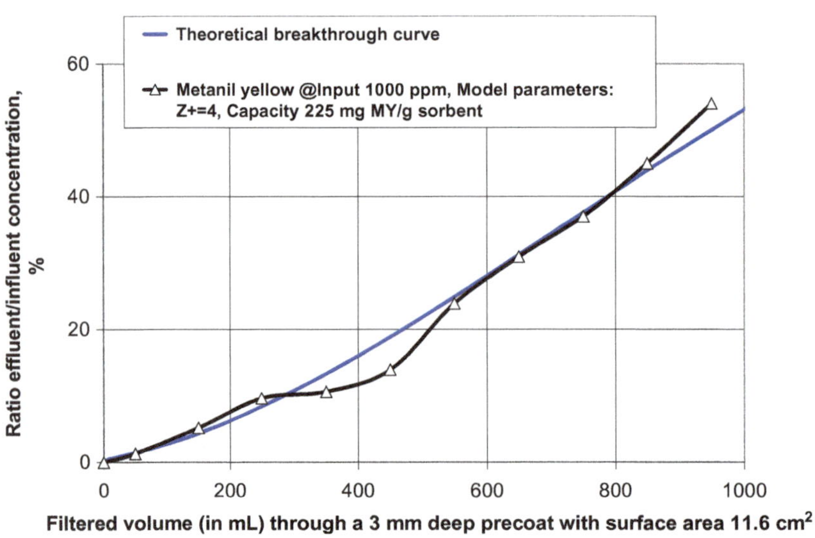

Figure 9. Removal efficiency of MY by a 3 mm precoat layer of aluminized DE18.

Table 6. Dissolved organic compounds (MY) reduction by aluminized DE18.

Sorbent	Input concentration, c, ppm	Capacity, q, mg (dye)/g sorbent	Parameter z^+ (see Eq. (1))
Aluminized DE18	400	70 ± 5	4 ± 1
	700	140 ± 5	4 ± 1
	1000	225 ± 5	4 ± 1

breakthrough curves at a residence time greater than about 0.1 s through the filter media. Table 6 presents values of z^+ (Equation (1) and filter capacity for MY obtained from the fits of the breakthrough curves for input concentrations of 400, 700 and 1000 ppm.

Results of Table 6 indicate that aluminized DE has high sorption capacity for MY at pH 8.2 as compared to activated carbon [43]. In addition to soluble dyes, aluminized DE would also retain pigments. Pigments are particles that are only nanometers in size, but with low water solubility.

The fit of experimental data of Table 6 to the Freundlich empirical isotherm:

$$q = K \cdot c^{1/n} \tag{2}$$

where K and n are constants that give the following parameter values: $K = 0.040 \pm 0.003$, $n = 0.80 \pm 0.05$. The Freundlich relationship is useful in that it develops a parameter K (Equation (2)), which is a measure of sorption capacity, and a parameter $1/n$, which is a measure of adsorption strength [45]. The closer parameter n is to unity, the better the adsorption strength of a sorbent [45], thereby reducing a possibility of shedding of adsorbed particles.

3.5.3. Adsorption of inks

The model was also tested via Equation (1) for removal efficiency of ink particles at a flow rate of 1 GPM/ft^2 and at pH 7. Model parameters q and z^+ were determined by fitting procedure for each individual breakthrough curve. Figure 10 shows the removal efficiency of ink particles from RO water by a 3 mm precoat of aluminized DE60 particles as measured by Genesis-10 UV spectrophotometer in a 1 cm long cuvette at a wavelength of 560 nm. Absorbance of stock solution was 0.402 (arbitrary units).

The fits of experimental data with the use of the model described by Equation (1) suggest that adsorption has near-ideal conformance with mass transfer equations at residence times through the filter greater than 0.1 s. This suggests that a mass transfer equation could be developed, which describes the chromatographic process, and that such a model would be applicable to particles other than dyes and inks.

3.5.4. Depyrogenation

Depth filtration using asbestos-containing filters has long been recognized as an effective method of pyrogen control [46]. Moreover, non-asbestos charge-modified depth filter media [47] as well as nanoceramic filter media [1] were used for this purpose. At neutral pH, most naturally occurring particles are electronegatively charged materials in aqueous solution. DE material has negative surface charge (see Figure 1) and therefore can serve only as a mechanical filter.

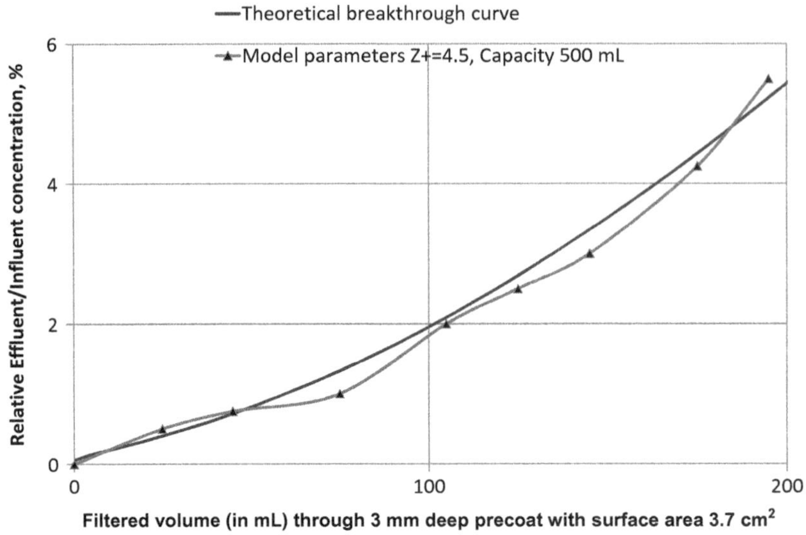

Figure 10. Removal efficiency of ink particles by a 3 mm precoat of aluminized DE60 particles.

3.5.5. Removal efficiencies of endotoxins at a high input concentration of 1 · 10^5 EU/ml by aluminized DE at pH 7

Table 7 shows initial removal efficiencies of *E. coli* (strain 055:B5) endotoxins at input concentration of 1 · 10^5 Endotoxin Units (EU)/mL by aluminized DE at neutral pH. Removal efficiency by aluminized DE media in a precoat configuration was measured using the lymulus amebocyte lysate (LAL) gel technique. Small volumes of endotoxins at *high* input concentrations of 1 · 10^5 EU/mL in milli-Q water with conductivity less than 0.1 μS/cm were passed through a precoat layer and output endotoxin levels were significantly reduced (see Table 7).

3.5.6. Protecting sea water reverse osmosis (SWRO) membrane from polysaccharides

SWRO is growing as much of the world's fresh water resources are depleted. One of the limitations of SWRO membranes is that they are readily fouled by nano-size particles such as polysaccharides and sub-micron biological debris. Experiments were done to determine how efficient aluminized DE would be in removing a nano-size particle in seawater. We used MS2 virus as a surrogate. Note that under all conditions of salinity, aluminized DE removed greater than 99% (> 2 LRV, (Table 8)) of all particles in less than 1 min contact time in the fluid. It is very likely that a coarser aluminized DE (e.g., 25 μm) would be very efficient at removal of RO foulants.

Table 7. Initial removal efficiencies of *E. coli* (strain 055:B5) endotoxins at input concentration of 1 · 10^5 EU/mL by Al-DE at neutral pH.

Al-DE	Precoat thickness, mm	Flow velocity, cm/min	Endotoxin load, EU/cm^2	Initial removal efficiency, %
Al-DE-18 μm	4.5	0.15	100,000	99.9999
Al-DE-60 μm	6	0.15	100,000	>99.9999

Table 8. Removal efficiencies[a] of MS2 by aluminized DE5[b] at pH 7.0 and 8.5 and at high salinity.

Challenge water (body fluid)		Removal efficiency[d], LRV
pH	TDS[c], g/L	
7	0	3.4 ± 0.5
8.5	0	3.1 ± 0.5
7	30 (typical sea water)	2.8 ± 0.5
8.5	30	3.2 ± 0.5
7	200 (near saturation)	4.9 ± 0.5
8.5	200	4.2 ± 0.5

Note: [a]Contact time, 1 min; [b]mean particle size of 5 μm; [c]adjusted with NaCl; [d]at input concentration of $2.2 \cdot 10^8$ PFU/mL.

Table 9. Heavy metal removal by different grades of aluminized DE with a precoat[a].

Ion	pH	Precoat thickness, mm	Flow rate, GPM/ft^2	Input, ppb	Output, ppb	Initial removal efficiency, %
Co^{2+}	8	50	15	656	<5	>99.2[a,c]
	10	50	15	635	<5	>99.2[a,c]
	9.1	25	15	2010	164	92[a,c]
Pb^{2+}	6.5	3	0.1	150	7	95[b]
	6.5	3	0.25	150	6	96[c]
As^{3+}	6.5	3	0.15	330	<2	>99.4[d]
	6.5	3	0.23	330	14	96[e]

Note: [a]A 250 mL sample was collected after passing 10 bed volumes; [b] median particle size of 18 μm; [c]median particle size – 60 μm; [d]aluminized DE/FeOOH with median particle size of 18 μm; [e]aluminized DE/FeOOH with a median particle size of 32 μm.

3.5.7. Heavy metal removal

Table 9 shows that aluminized DE adsorbs heavy metals in a precoat configuration.

Soluble transition metals such as iron, cobalt, and manganese as well as a many other transition metals can be converted to colloidal oxides/hydroxides by adjusting the pH to that of the bulk solution. The electropositive layer closest to the surface has lower pH than the bulk fluid. The fluid farther from the electropositive surface is alkaline, that is, it has higher pH than the bulk fluid, thereby maintaining the overall electroneutrality. The formation of oxides/hydroxides can be further promoted by adding an oxidant such as dilute hydrogen peroxide or sodium hypochlorite.

3.5.8. Testing with radionuclides in low level liquid waste

The first test of aluminized DE with radioisotopes was done at the Brunswick Nuclear Plant in North Carolina [48]. The test was performed on off-standard water that could not be reprocessed by plant systems and needed to meet specific guidelines for discharge. Analysis noted that there were traces of cesium and cobalt in the water (Table 10). In a multiple test series, several ion exchange resins, along with aluminized DE were tested in order to evaluate methods for cleaning the water. The aluminized DE

Table 10. Raw water chemistry.

Nuclide	Activity (nCi/mL)	Mass of 1 Ci,[a] mg	Mass fraction, ppb
Mn-54	0.1171	0.129	0.015
Co-58	0.03508	0.0314	0.0011
Co-60	4.591	0.884	4.06
Zn-65	0.1599	0.121	0.0193
Cs-134	0.03513	0.773	0.0276
Cs-137	1.030	11.565	11.91
Total Gamma Activity	5.968		16.033

Note: [a]1 Ci = $3.7 \cdot 10^{10}$ Bq/ln(2)$N_A \cdot t_{1/2}$ (s), where N_A is Avogadro constant and $t_{1/2}$ is half-life in seconds.

Table 11. Removal efficiency of radionuclides from raw water.

Nuclide	Influent activity (nCi/mL)	Effluent activity (nCi/mL)	Removal efficiency, %
Mn-54	0.1171	0.00703	94%
Co-58	0.03508	Not done	
Co-60	4.591	0.04994	98.9%
Zn-65	0.1599	Not done	
Cs-134	0.03513	0.01451	59.4%
Cs-137	1.030	0.3774	63.4%
Total Gamma Activity	5.968	0.4489	92%

experiment involved adding sodium hydroxide solution to bring the pH from 3.1 to 7.8, thereby favoring the conversion of soluble radioactive metals to form colloidal hydroxides [49–51]. The neutralized solution was passed through a column containing 50 mL of packed aluminized DE60. The glass column was approximately 300 mm tall and 15 mm in diameter. The flow rate was approximately 10 mL/min and the detection time duration was 15 min.

Table 11 shows the analytical and radiochemistry measurements of the effluent. The results of radiochemistry measurements could be converted to dimensionless quantities (that is, to the dilution factors) of ppb level for input concentrations and part-per-trillion (ppt) for filtered water. For example, Table 11 shows the activity of Co-60 as 4.591 nCi/mL that is equivalent to 4.054×10^{-12} g/mL or 4.054 ppb, while the effluent is equivalent to 4.59 ppt.

3.6. Ultra-far electric field penetration from a precoat into surrounding water

3.6.1. MS2 bacteriophage adsorption by alumina's nanolayer surface

The precoats were formed over a septum consisting of a stack of five wire screens with an average pore size of 200 μm. The filter stack was held within Nalgene filter holders. A vacuum pump was used to reduce pressure in the receiver chamber to provide a differential pressure with respect to the upper chamber that was held at the ambient atmospheric pressure. The test was conducted in batches. Each individual batch was 0.5 L in volume. Four grams of aluminized DE powders with average particle size of 18 μm were manually mixed in 0.5 L of DI RO water to form powder suspensions at concentrations of 8 g of aluminized DE/L. The powder suspension of aluminized DE was poured into the upper chamber and the suction valve was opened on the vacuum pump to control the flow rate. After filtering all slurry, the suction valve on the vacuum pump was closed and the flow was terminated. Each precoat has a surface area of ~10 cm^2 and is ~3 mm thick.

Several aliquots of DI RO water were slowly added to the precoat to saturate the precoat with water and create a thin visible sheen layer on the top of the precoat. Air from an air compressor was used to increase pressure in the receiver chamber to provide a differential pressure of ~100 Pa with respect to the upper chamber, which was held at the ambient atmospheric pressure. This prevents gravity filtration of the liquid phase to the lower chamber from the water- saturated precoat layer. Then a 400 μL aliquot of MS2 suspension in DI RO water (low conductivity, in the range from 15 to 20 μS/cm) and an input concentration of ~(1–3)×10^8 PFU/mL was added to the precoat, which creates a ~0.4 mm deep supernatant layer of MS2 suspension above the aluminized DE precoat surface. The content was allowed to dwell either for 5, 30 or 300 s. Then the Buchner funnel was tilted at 45°, facilitating collection of the supernatant into a small area on the precoat. A 10 μL aliquot out of original 400 μL was collected and rapidly injected into 15 mL centrifuge tubes prefilled with 10 mL of DI RO water. After each experiment, the entire volume of the supernatant was collected and its volume was measured. The collected volumes were found to be comparable within ~25% with the initial 400 μL aliquot. The 15 mL tubes were then centrifuged for 10 min at 1300 *g*. The supernatant was serially diluted by a factor of 10 and analyzed for MS2 virus as plaque forming units with the use of the double layer method [39,52]. Briefly, dilutions of each sample were mixed with host cells, plated on nutrient agar, and incubated overnight. The resulting plaque counts were converted to PFU/mL.

To assure that there was no interference from trace particulate of aluminized DE, several control tests were conducted to make sure that the above handling procedure, after collection of the 10 μL sample from the precoat, does not interfere with the following analysis. In this case, a 400 μL aliquot of DI RO water instead of the MS2 suspension was added above the precoat. A 10 μL aliquot of the supernatant water was collected, injected into 15 mL centrifuge tubes prefilled with 10 mL of DI RO water. Then a 10 μL aliquot of MS2 suspension in DI RO water at an input concentration of ~(1–3)×10^8 PFU/mL was added to the centrifuge tube. The tubes were manually shaken and centrifuged for 10 min at 1300 *g*. No measurable difference in the MS2 counts at the level of 0.0 ± 0.2 LRV was found as compared to the stock solution. This control demonstrates that there was no detectable aluminized DE in the effluent and that the sample handling procedure has no interference with the results presented in Table 12.

Table 12. Dependence of MS2 removal efficiency from a 400 μm deep supernatant layer of water by aluminized DE18 at pH 6.5 vs. contact time.

Contact time, s	Number of replicates	MS2 removal efficiency, LRV	Depth of electric field penetration, r_0 (Eq. 3), μm	Brownian displacement, x, μm
5	23	1.0 ± 0.2	200	23.5
30	22	2.0 ± 0.2	300	57.6
300	30	4.0 ± 0.2	380	182

3.6.2. Electric field extension

We define the depth of the electric field penetration from a precoat into surrounding water to be equal to an average distance from the surface where water has the same charge as the surface, i.e., it is electropositively charged in the case of aluminized DE. A small particle travels from the farthest point P from the precoat surface with $r = r_0$, to the surface of the precoat during a fixed time. The removal efficiency of the particle (*e.g.*, MS2) from the suspension could be used to estimate r_0,

$$r_0 = A\left(1 - \sqrt[3]{\frac{C_{\text{final}}}{C_{\text{initial}}}}\right) \tag{3}$$

where A is the average depth of the supernatant layer, C_{initial} and C_{final} are the concentrations of MS2 viruses before and after a certain dwelling time when the sample was taken and analyzed. Table 12 shows the dependence of MS2 removal efficiency vs. contact time and compares with Brownian displacement x estimated from the Stokes–Einstein equation $x^2/2\ t = \mathrm{k}T/6\pi\mu a$) during the same period of time under experimental condition with $\mu = 900$ μPa·s at 25°C and the MS2 radius $a \sim 13.75$ nm.

Results of Table 12 indicate that electric field penetrates as far as 380 μm and it attracts and adsorbs approximately 99.99% of MS2 viruses during 5 min and the contribution from the Brownian motion is insignificant at short contact times of 5 and 30 s. *The depth of the penetration is far greater by many orders of magnitude than that predicted by the DLVO theory* [14,15], which was developed for smooth, homogeneous particles with ideal geometries. This creates a significant electric field in the region far from the surface of the aluminized DE particle. As a result, the local water composition in the suspension becomes highly electropositive, which differs significantly from the DI RO water at neutral pH. It manifests itself in a highly unusual adsorption behavior of metal ions, e.g., initial adsorption of cobalt, which does not depend on the flow rates (see Table 9).

3.7. Mechanism of adsorption

To elucidate the mechanism of adsorption, we measured the pH change in water above the precoat layer. A pH microelectrode (available from Lazar Research Laboratories, Inc., model PHR-146) was inserted into the bulk water 1 mm above the precoat layer of aluminized DE-18. The pH values dropped by 0.5 units as the electrode was manually driven with an increment of 25 μm toward the precoat surface. The increase of acidity level of water (lower pH) close to the solid interface is an indicative of high level of hydronium ions, the only available positive charge in water. The pH went back to its original level of the bulk water at distances greater than about 1000 μm from the precoat when the electrode was manually driven back away from the precoat.

This work demonstrates that a thin (~12 Å) monocrystalline α-$Al_2O_3 \cdot H_2O$ coating deposited on the second electronegative solid such as DE presented in this work possesses a combination of several unique features:

(1) the screening around the alumina nanolayer becomes negligible ($ka \ll 1$) at electrolyte strengths less than ~10 mM;
(2) as a result, the entire volume of aqueous suspension becomes electropositively charged in close vicinity of the alumina nanolayer;

(3) the resulting polarization field created by the alumina nanolayer extends far beyond the DL of the adsorbate particle. The *near field* has a range of $O(1/k)$ and the *far field* of $O(\alpha)$ [53], where α is the equivalent surface spherical diameter of the adsorbate particle (~0.882 μm for *E. coli* bacteria [54] and ~20–30 nm for the MS2 and *fr* viruses);

(4) the expressions for the potential $\psi(r)$ and for the strength of electric field $\vec{E}(r)$ acquire the forms for charged flat surface without countercharge in the range close to the alumina nanolayer, roughly estimated to be ~1/20 of the aluminized DE average particle diameter or up to 1 and 4 μm for aluminized DE18 and DE80, respectively, that is, the Coulomb case created by a flat surface with negligible electric field $\vec{E}(r)$ decay at distances r of order of a few microns from the surface (see Section 3.6 and Table 12).

The above mechanism is similar to that described in Ref [1], which showed that the standard expression for the Coulombic force (charge-charge, $\sim 1/r^2$) quantitatively described electrokinetic motion in aqueous solution of MS2 viruses and *E. coli* bacteria in electric field of alumina nanofibers grafted onto glass microfibers.

3.8. Electrokinetic motion of MS2 in electric field of the aluminized DE composite

3.8.1. Electrophoretic mobility of MS2

The electrophoretic mobility, μ, is defined as a ratio of field-induced particle velocity, v, to the homogeneous field strength, E_{ext}, in the liquid at a large distance from the particles. In the present case, we can define an average static electrophoretic mobility $<\mu>$ as:

$$\langle\mu\rangle = \frac{\langle v\rangle}{\langle E\rangle} \tag{4}$$

From the data of Table 12, for average velocity of MS2 virus of 40 μm/s at 5 s contact time, the electrophoretic velocity can be calculated by the Helmholtz–Smoluchowski equation for electrophoresis [55]:

$$v_e = \frac{\varepsilon_m \varepsilon_0 \zeta}{\mu} E \tag{5}$$

where ζ is the zeta potential, μ is the dynamic viscosity (for water at 25°C is equals to ~900 μPa·s). The average strength of electric field in aqueous solution at 25°C with v_e of 40 μm/s (Table 12), and ζ(MS2) = − 40.4 ± 3.7 mV [34] at pH7 can be estimated at 12.8 V/cm, which is in good agreement with the value of 15.2 V/cm created by nanoalumina fibers grafted onto microglass fibers [1].

This good agreement in values of generated electric field by alumina surfaces in form of 2 nm diameter nanofiber and alumina nanolayer grafted onto the DE particle provides strong evidence that *both materials have the same crystallographic structure of boehmite oxide/ hydroxide as (α-$Al_2O_3 \cdot H_2O$)* as it was determined for a macroscopic assembly of alumina nanofibers on the second solid [1].

Moreover, the characteristic time scale for attracting and adsorbing of MS2 particles in aqueous suspensions by alumina nanolayer of 5 s (Table 12) is comparable with about 2 s by nanoalumina/microglass assembly [1] and with formation times of haloes (2–4 s) by the Indian ink particles [17].

4. Conclusions

This paper demonstrates how the strong electropositive charge properties of alumina's nanolayer composite dispersed in polar liquids and the precoat configuration can be used as a means of collecting and separating submicron and micron particles such as viruses, bacteriophages, and bacteria. The composite in a precoat configuration creates a strong electric field that extends as far as 380 μm from alumina's nanolayer surface into an aqueous dispersive medium. This is many orders of magnitude (>4) greater than that predicted by the DLVO theory.

The composite media adsorbs submicron particles predominately on the basis of a combination of their inherent charge and acquired induced polarization charge, rather than by physical entrapment based on particle size. The media has a fast dynamic response for adsorption, allowing purification within a very shallow bed (~3 mm thick) and at flow rates an order of magnitude or greater than can be achieved with an ultraporous membrane. A given particle can be eluted by replacement with one that is more adherent. The media is suggested for purification of virus and proteins, where a high dynamic binding capacity is combined with high flow velocities and at low differential pressure. Chromatographic separations of a range of submicron particles of biological relevance, such as viruses, chromosomes, DNA, RNA, and macromolecules, at pressure drops less than 1 bar over ambient, in thin beds and with high flow rates, appear feasible.

After more than a century of research in the field of colloid science pioneered by Helmholtz in 1853, interactions between charged colloidal particles are still a very puzzling matter. A high-level quantum mechanical research is needed to resolve the controversy between the DLVO and the Sogami theories. Whatever the mechanism of attraction and adsorption of nano- and micron-sized particles by alumina's nanolayer composite presented in this work, the long-range phenomena are novel and striking and may have important implications not only for surface science but also for physics and life sciences.

Acknowledgments

We thank Mr. Kirk Scammon of Materials Characterization Facility, University of Central Florida for providing us with XRD and SEM images of aluminized DE powders and Dr. A. Slesarev of Rice University for providing us with XRD, SEM, and TEM images of aluminized DE powders.

Disclosure statement

No potential conflict of interest was reported by the authors.

References

[1] L.A. Kaledin, F. Tepper, and T.G. Kaledin, *Long-range attractive forces extending from alumina nanofiber surface*, Int. J. Smart Nano Mater. 5 (2014), pp. 133–151. doi:10.1080/19475411.2014.919970.

[2] J.C. Henniker, *The depth of the surface zone of a liquid*, Rev. Mod. Phys. 21 (1949), pp. 322–341. doi:10.1103/RevModPhys.21.322.

[3] W. Nernst, *Theorie der reaktionsgeschwindigkeit in heterogenen systemen*, Zeits. f. physik. Chemie. 47 (1904), pp. 52–55.

[4] E. Brunner, *Die bestimmung der borsäure*, Zeits. f. physik. Chemie. 47 (1904), pp. 56–102.

[5] H.S. Davis and G.S. Crandall, *The role of the liquid stationary film in batch absorptions of gases. I. Absorptions involving no irreversible chemical reactions*, J. Am. Chem. Soc. 52 (1930), pp. 3757–3768.
[6] W.B. Hardy, *The tension of composite fluid surfaces and the mechanical stability of films of fluid*, Proc. Roy. Soc. Lond. A86 (1912), pp. 610–635. doi:10.1098/rspa.1912.0053.
[7] G. Gouy, *Sur la constitution de la charge electrique a la surface d'un electrolyte*, Compt. Rend. 149 (1910), pp. 654–657.
[8] D.L. Chapman, *A contribution to the theory of electrocapillarity*, Philos. Mag. 25 (1913), pp. 475–481. doi:10.1080/14786440408634187.
[9] O. Stern, *The theory of electrolytic double layer*, Z. Electrochem. 30 (1924), pp. 508–516.
[10] A.V. Delgado, F. González-Caballero, R.J. Hunter, L.K. Koopal, and J. Lyklema, *Measurement and interpretation of electrokinetic phenomena (IUPAC technical report)*, Pure Appl. Chem. 77 (2005), pp. 1753–1805. doi:10.1351/pac200577101753.
[11] H. Zhang, A.A. Hassanali, Y.K. Shin, C. Knight, and S.J. Singer, *The water-amorphous silica interface: Analysis of the stern layer and surface conduction*, J. Chem. Phys. 134 (2011), pp. 024705. doi:10.1063/1.3510536.
[12] D.J. Bonthuis and R.R. Netz, *Unraveling the combined effects of dielectric and viscosity profiles on surface capacitance, electro-osmotic mobility, and electric surface conductivity*, Langmuir 28 (2012), pp. 16049–16059. doi:10.1021/la3020089.
[13] A.S. Dukhin and P.J. Goetz, *Characterization of Liquids, Nano- and Microparticulates, and Porous Bodies Using Ultrasound*, 2nd ed., Elsevier, Amsterdam, 2010.
[14] B.V. Derjaguin and L. Landau, *Theory of stability of strongly charged lyophobic sols of the adhesion of strongly charge particles in solution electrolytes (in Russian)*, Acta Physicochim. USSR. 14 (1941), pp. 633–662.
[15] E.J. Verwey and J.T.G. Overbeek, *Theory of the Stability of Lyophobic Colloids*, Elsevier, Amsterdam, 1948.
[16] J. Israelachvili, *Intermolecular and Surface Forces*, Academic Press, San Diego, CA, 1992.
[17] B.V. Derjaguin and M.V. Golovanov, *On long-range forces of repulsion between biological cells*, Coll. Surf. 10 (1984), pp. 77–84. doi:10.1016/0166-6622(84)80009-8.
[18] B.V. Derjaguin and N.V. Churaev, *Nature of "anomalous water"*, Nature 244 (1973), pp. 430–431. doi:10.1038/244430a0.
[19] E.R. Lippincott, R.R. Stromberg, W.H. Grant, and G.L. Cessac, *Polywater*, Science 164 (1969), pp. 1482–1487. doi:10.1126/science.164.3887.1482.
[20] N. Ise and H. Yoshida, *Paradoxes of the repulsion-only assumption*, Acc. Chem. Res. 29 (1996), pp. 3–5. doi:10.1021/ar950019k.
[21] B.V.R. Tata, E. Yamahara, P.V. Rajamani, and N. Ise, *Amorphous clustering in highly charged dilute poly(chlorostyrene-styrene sulfonate) colloids*, Phys. Rev. Lett. 78 (1997), pp. 2660–2663. doi:10.1103/PhysRevLett.78.2660.
[22] N. Ise, *Recent study on counterion-mediated attraction between colloidal particles*, Coll. Surf. 146 (1999), pp. 347–357. doi:10.1016/S0927-7757(98)00775-4.
[23] N. Ise, *Ordering of ionic solutes in dilute solutions through attraction of similarly charged solutes – A change of paradigm in colloid and polymer chemistry*, Angew. Chem. Int. Ed. Engl. 25 (1986), pp. 323–334. doi:10.1002/(ISSN)1521-3773.
[24] J.-M. Zheng and G.H. Pollack, *Long-range forces extending from polymer-gel surfaces*, Phys. Rev. E 68 (2003), pp. 031408. doi:10.1103/PhysRevE.68.031408.
[25] J.-M. Zheng and G.H. Pollack, *Solute exclusion and potential distribution near hydrophilic surfaces*, in *Water and the Cell*, G.H. Pollack, I.L. Cameron, and D.N. Wheatley, eds., Springer, Dordrecht, 2006, pp. 165–174.
[26] G.H. Pollack, *The Fourth Phase of Water: Beyond Solid, Liquid, and Vapor*, Ebner and Sons Publishers, Seattle, WA, 2013.
[27] I. Sogami, *Effective potential between charged spherical particles in dilute suspension*, Phys. Lett. A96 (1983), pp. 199–203. doi:10.1016/0375-9601(83)90704-1.
[28] I. Sogami and N. Ise, *On the electrostatic interaction in macroionic solutions*, J. Chem. Phys. 81 (1984), pp. 6320–6332. doi:10.1063/1.447541.
[29] I. Sogami, Long-range attractive tail of colloidal interaction, Chem. Lett. 41 (2012), pp. 1331–1333. doi:10.1246/cl.2012.1331.
[30] C.S. Oulman and E.R. Baumann, *Streaming potentials in diatomite filtration of water.* J. AWWA. 56 (July 1964), pp. 915–930.

[31] D.R. Briggs, The determination of the ζ potential on cellulose-a method, J. Phys. Chem. 32 (1928), pp. 641. doi:10.1021/j150287a001.

[32] C.E. Mossman and S.G. Mason, *Surface electrical conductance and electrokinetic potentials in networks of fibrous materials*, Can. J. Chem. 37 (1959), pp. 1153–1164. doi:10.1139/v59-170.

[33] D.W. Marquardt, *An algorithm for least-squares estimation of nonlinear parameters*, J. Soc. Ind. App. Math. 11 (1963), pp. 431–441. doi:10.1137/0111030.

[34] B. Michen and T. Graule, *Isoelectric points of viruses*, J. Appl. Microbiol. 109 (2010), pp. 388–397.

[35] A. Duek, E. Arkhangelsky, R. Krush, A. Brenner, and V. Gitis, *New and conventional pore size tests in virus-removing membranes*, Water Res. 46 (2012), pp. 2505–2514. doi:10.1016/j.watres.2011.12.058.

[36] C.V. Chrysikopoulos and V.I. Syngouna, *Attachment of bacteriophages MS2 and ΦX174 onto kaolinite and montmorillonite: Extended-DLVO interactions*, Coll Surf. B 92 (2012), pp. 74–83. doi:10.1016/j.colsurfb.2011.11.028.

[37] G. Herath, K. Yamamoto, and T. Urase, *Removal of viruses by microfiltration membranes at different solution environments*, Water Sci. Technol. 40 (1999), pp. 331–338. doi:10.1016/S0273-1223(99)00515-6.

[38] D.-Q. Lin, P.J. Brixius, J.J. Hubbuch, J. Thommes, and M.-R. Kula, *Biomass/adsorbent electrostatic interactions in expanded bed adsorption: A zeta potential study*, Biotechnol. Bioeng. 83 (2003), pp. 149–157. doi:10.1002/bit.10654.

[39] International Standard Organization (ISO 10705-1), *Part 1, Enumeration of F-specific RNA bacteriophages*, 1st ed., ISO, Geneva, 1995.

[40] B. Petrusevski, S. Sharma, J.C. Schippers, and K. Shordt, *Arsenic in drinking water*. Thematic Overview Paper 17, IRC International Water and Sanitation Centre, Delft, March 2007

[41] F. Tepper, L. Kaledin, and T. Kaledin, *Non-woven electrostatic media for chromatographic separation of biological particles*, J. Liq. Chromatogr. Relat. Technol. 32 (2009), pp. 607–627. doi:10.1080/10826070802711006.

[42] D.M. Ruthven, *Principles of Adsorption and Adsorption Processes*, Wiley-Interscience, New York, 1987, 464 pp.

[43] A. Reife and H.S. Freeman, *Environmental Chemistry of Dyes and Pigments*, John Wiley & Sons, Inc, New York, 1996.

[44] D. Graham, *Characterization of physical adsorption systems. III. The separate effects of pore size and surface acidity upon the adsorbent capacities of activated carbons*, J. Phys. Chem. 59 (1955), pp. 896–900. doi:10.1021/j150531a022.

[45] P.W. Atkins, *Physical Chemistry*, W. H. Freeman and Co, San Francisco, 1978.

[46] H. Kaden, *The use of asbestos filter beds in the production of sterile and pyrogen-free solutions*, Pharmazie 30 (1975), pp. 752.

[47] C.P. Gerba, K.C. Hou, R.A. Babineau, and J.V. Fiore, *Pyrogen control by depth filtration*, Pharm. Technol. 4 (1980), pp. 83.

[48] F. Tepper and L. Kaledin, *Scavenging cobalt from radwaste*, Nucl. Eng. Int., October (2014), pp. 34–35.

[49] N. Takeno, *Atlas of Eh-pH diagrams– Intercomparison of thermodynamic databases*, Geological Survey of Japan Open File Report No. 419, National Institute of Advanced Industrial Science and Technology, Research Center for Deep Geological Environments, Tsukuba, 2005.

[50] M. Pourbaix, *Atlas of Electrochemical Equilibria in Aqueous Solutions*, National Association of Corrosion Engineers, Houston, TX, 1966.

[51] D.G. Brookins, *Eh-pH Diagrams for Geochemistry*, Springer-Verlag, New York, 1988.

[52] I.L. Pepper and C.P. Gerba, *Environmental Microbiology. A Laboratory Manual*, 2nd ed., Elsevier AP, Amsterdam, 2004.

[53] J. Lyklema, *Fundamentals of Interface and Colloid Science: Solid-Liquid Interfaces* Vol. 2: Chapter 3, Academic Press, London, 1995.

[54] S.E. Truesdail, J. Lukasik, S.R. Farrah, D.O. Shah, and R.B. Dickinson, *Analysis of bacterial deposition on metal (hydr)oxide-coated sand filter media*, J. Coll. Interface Sci. 203 (1998), pp. 369–378. doi:10.1006/jcis.1998.5541.

[55] M. Smoluchowski, *Zur theorie der elektrischen kataphorese und der ober-henleitung*, Phys. Z. 6 (1905), pp. 529–531.

11

Electromechanical behavior of fiber-reinforced dielectric elastomer membrane

Chi Li, Yuhan Xie, Guorui Li, Xuxu Yang, Yongbin Jin and Tiefeng Li*

Institute of Applied Mechanics, Zhejiang University, 38 Zheda Road, Hangzhou 310027, China

Based on its large deformation, light weight, and high energy density, dielectric elastomer (DE) has been used as driven muscle in many areas. We design the fiber-reinforced DE membrane by adding fibers in the membrane. The deformation and driven force direction of the membrane can be tuned by changing the fiber arrangements. The actuation in the perpendicular direction of the DE membrane with long fibers first increases and then decreases by the increasing of the fiber spacing in the perpendicular direction. The horizontal actuation of the membrane decreases by decreasing the spacing of short fibers. In the membrane-inflating structure, the radially arranged fibers will break the axisymmetric behavior of the structure. The top area of the inflated balloon without fiber will buckle up when the voltage reaches a certain level. Finite element simulations based on nonlinear field theory are conducted to investigate the effects of fiber arrangement and verify the experimental results. This work can guide the design of fiber-reinforced DE.

Keywords: Dielectric elastomer; fiber reinforced; electromechanical

1. Introduction

As an electro-active polymer (EAP), dielectric elastomer (DE) possesses the attributes of large deformation, light weight, and noise-free operation [1–10]. The electromechanical coupling behavior of DEs has been studied extensively in numerous applications, including actuators [11–15], soft robotics [16–22], artificial muscles [23–25], and energy harvesters [26,27]. Recently, the electromechanical behaviors of DE have been widely investigated. Specially, based on the giant voltage–induced actuation of the DE, various methods have been used to enhance the electromechanical performances of the DEs such as synthesizing composite DE material [24], harnessing the electromechanical instabilities of DE [4], and improving the properties of electrodes [23]. Those approaches mainly focus on tuning the stiffness of the DE membrane or the effective stiffness of the DE structure to improve electromechanical properties of the DE under the voltage loads [28,29]. A simple way to make DE composite material is to embed fibers into the DE membrane [13]. This approach can restrain the actuation of the DE structure in some directions, and then enhance the actuation in other directions. We can design the arrangement of fibers to tune the actuation of DE structure in specific directions. This paper presents a fiber-reinforced DE structure to improve the driven forces and the

*Corresponding author. Email: blacksmith86@163.com

actuation in specific directions. We investigate the effects of fiber arrangements on the performance of DE structure. FEM analysis have been conducted to verify the experimental results and to illustrate the stress and strain distribution of the fiber-reinforced DE membrane under electromechanical loads [30]. The paper is organized as follows. In Section 2, we present the fabricating procedures and the experimental results of DE in-plane actuator. Section 3 shows the procedures and the experimental results of DE membrane inflating structure. Section 4 presents the working principle and FEM simulation results of the DE structure. Finally, conclusions are given in Section 5.

2. Fiber-reinforced dielectric elastomer in-plane actuator

In-plane actuator with DE membranes and rigid frames is a typical DE structure [31–33]. In this structure, the DE membranes are pre-stretched and then fixed by the rigid frames. We investigate the electromechanical behaviors of DE membrane with two types of arrangement of fibers, which are parallel arranged long fibers and parallel arranged short fibers.

In our experiment, we use 3 M-VHB 4910 as the DE and nylon fishing lines (~0.3 mm in diameter) as the fibers. In order to improve the strength of bonding between fibers and DE membranes, we sandwich the fibers between two layers of DE membranes. We use laser cutter to cut ABS plastic plates as the rigid frame. The length of the outer side of the frame is 80 mm, the length of the inner side of the frame is 60 mm, the width of the frame is 10 mm, and the thickness of the frame is 1 mm; the frame will be used to fix the pre-stretched membrane [11]. Second, we use the same way to cut A4 paper into exactly same shape with the ABS plastic plates frame, this paper frame will be used to ensure the membrane cannot detach from the plastic frame. Then we manufacture the fiber-reinforced DE membranes. As shown in Figure 1, we use 3 M-VHB 4910 as the DE, which has been bi-axially pre-stretched three

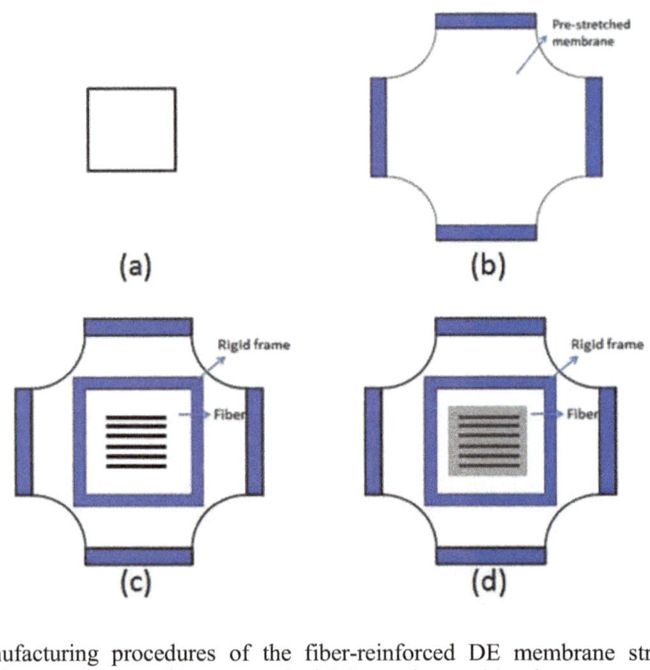

Figure 1. Manufacturing procedures of the fiber-reinforced DE membrane structure. (a) The reference state of DE membrane; (b) pre-stretched DE membrane; (c) enhancement of the membrane by aligned fibers; (d) electrode coating.

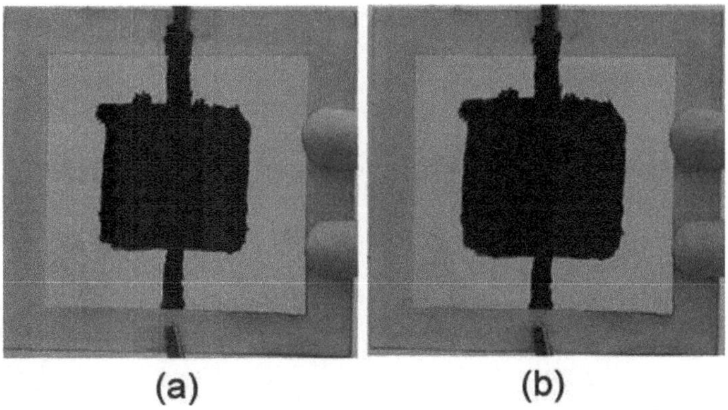

Figure 2. DE membrane actuator in (a) the pre-stretched state, and (b) the actuated state by voltage.

times by the biaxial stretching machine before attaching on the ABS plastic frame [10]. After pre-stretch, we fix the elastomer on plastic frame and then attach the paper frame on the other side of the membrane, make the membrane sandwiched by the plastic frame and the paper frame. Compliant electrodes (MG Chemical 846–80 G carbon grease) are coated on both sides of the DE membrane. Figure 2 shows (a) the DE membrane actuator in the pre-stretched state and (b) the actuated state by voltage.

2.1. *Parallel arranged long fibers*

As shown in Figure 1(d), the long fibers are continuously embedded in the DE membrane through the horizontal direction. The long fibers are parallel distributed with fixed spacing in the perpendicular direction. We use nylon fishing lines as the fibers with 80 mm in length and 0.3 mm in diameter. We attach several fibers on one side of the pre-stretched membrane and then put another pre-stretched membrane on the fiber and make the fiber sandwiched between two membranes, and then exclude the air between the two membranes. In order to investigate the effects of fiber spacing on the actuation performances of the membrane, we vary the fiber spacing in the perpendicular direction and the number of fibers in the DE actuator. As shown in Figure 3(a)–(c), the fiber spacing in the perpendicular direction are 2, 4, and 6 mm.

We coat square-shaped compliant electrodes (carbon grease MG Chemical 846–80 G) in the central area and on both sides of the DE membrane. The coated compliant electrodes are not beyond the area of fibers in both horizontal and perpendicular directions. The length of the square electrode is 25 mm. We wire the electrode out and connect it with high-voltage power supply (TREK-610E) to apply high voltage on the DE membrane. Figure 3 shows the behavior of DE membranes with different fiber spacing under the same high voltage load (0 V and 8 kV). The results show that once the high voltage is applied on the membrane, electrode-coated areas expand in plane. Because the membranes have been bi-axially pre-stretched three times, if there is no fiber to restrain the DE membrane, the electrode-coated area will expand in both horizontal and perpendicular directions. Figure 3(a)–(c) shows the horizontal actuation of the DE membrane is restrained by the long fibers, while the DE membranes expand along the perpendicular direction. Because the horizontal deformation of the membrane is restrained, the

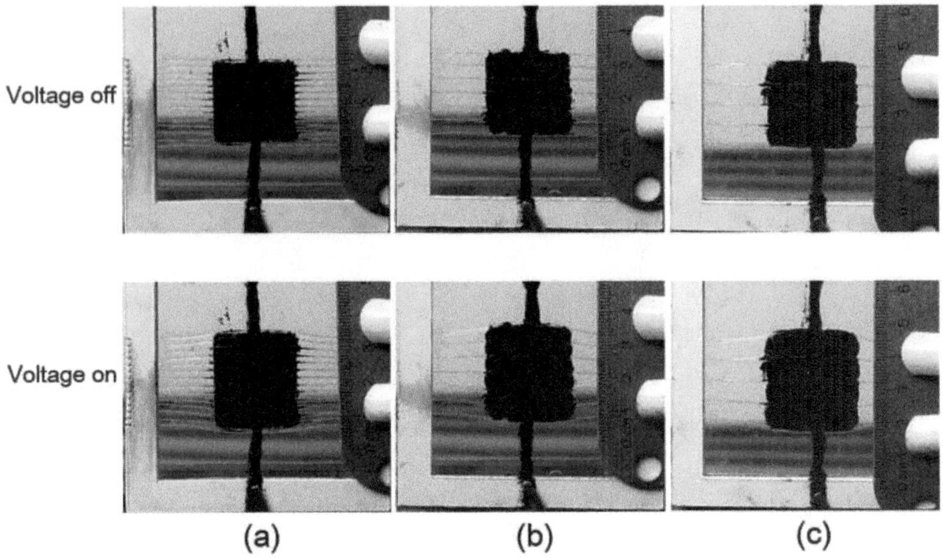

Figure 3. DE membrane in-plane actuator with long fibers in its pre-stretched state and actuated state under 8 kV. The fiber spacing in the perpendicular direction are (a) 2 mm, (b) 4 mm, and (c) 6 mm.

perpendicular deformation increases. We further compare the behaviors of DE membrane with different fiber spacing. In Figure 3, we find when the fiber spacing increases, the perpendicular actuation will increase. The displacements of central point of the lateral fiber in Figure 3(a)–(c) are 2.3, 2.5, and 2.8 mm. The limit case of the fiber spacing is zero and infinite, which corresponds the DE membrane with fully occupied fibers and no fiber. From above we know that the long fiber will transfer part of the horizontal deformation into perpendicular deformation, but at the same time, since the perpendicular deformation will make fibers bending, will also block perpendicular deformation to a certain extent. That means there must be a most reasonable spacing that will induce the maximum perpendicular deformation. Parallel-arranged long fibers can help us eliminate deformation in one direction and increase the deformation in the other direction, and we can change the deformation value along the other direction through changing the fiber spacing.

2.2. Parallel-arranged short fibers

Short fiber can also be used to reinforce the DE membrane. The short fibers are not continuously embedded through the whole membrane. The short fibers are aligned in straight lines with gaps between each other in the horizontal direction. We use the spacing in the horizontal direction to define the short fiber arrangements (Figure 4). The manufacturing method of the parallel-arranged short fibers is the same as that of long fibers.

After we apply high voltage through the thickness direction of the DE membrane, we find that the short fiber will partly restrain the horizontal deformation, but not restrain the deformation along the perpendicular direction (Figure 4). Figure 4(a)–(c) also shows that when the fiber spacing in the horizontal direction decreases, the limitation to the

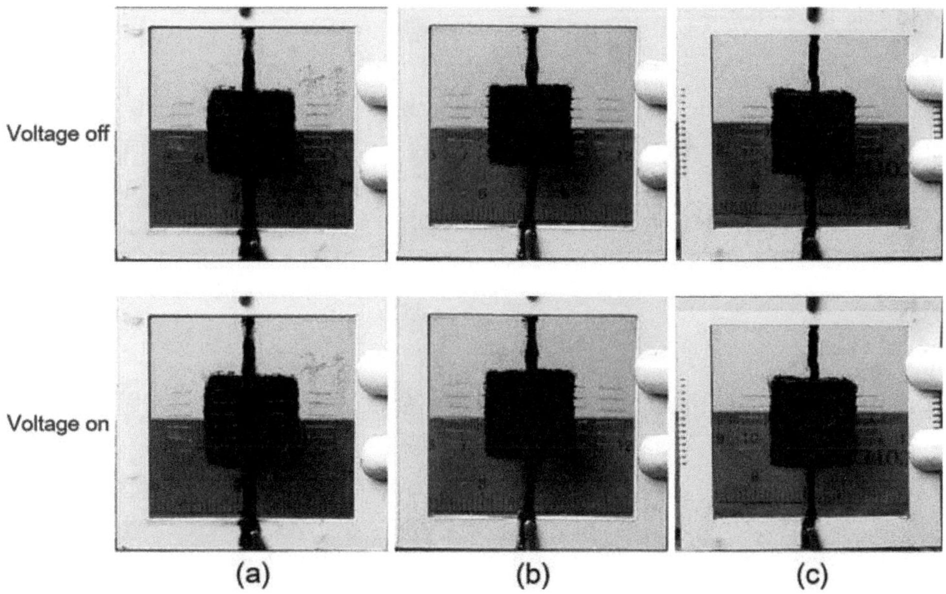

Figure 4. DE membrane in-plane actuator with short fiber in its pre-stretched state and actuated state under 8 kV. The fiber spacing in the perpendicular direction is 4 mm. The fiber spacing in the horizontal direction are (a) 8 mm, (b) 4 mm, and (c) 2 mm.

horizontal deformation increases. From these results, we find that parallel-arranged short fibers can help us change the deformation ratio along both the horizontal and perpendicular directions.

3. Fiber-reinforced dielectric elastomer membrane inflating structure

In this part, we experimentally investigate the electromechanical behavior of DE membrane inflating structure with radially arranged fibers. Figure 5 shows the fabricating procedure and the operation states of the DE membrane. The DE membrane is first pre-stretched and fixed by the rigid ring with the outer diameter of 80 mm and the inner diameter of 60 mm. We arrange eight fibers on the pre-stretched membrane along the radial direction. The fibers locate from the edge of the membrane and have the length of 26 mm. After we finish manufacturing the radially arranged fiber DE membrane, we fix it at one end of a cylindrical chamber of volume of 800 ml. After sealing the ring frame, DE membrane, and the chamber, we inflate air into the chamber using air pump through a one-way air valve. The membrane will be inflated to a spherical shape without applying voltage (Figure 5(d)). The amount of air in the chamber is fixed during the actuation of DE membrane by voltage loads. Figure 6 shows the experimental results of the DE membrane. Because the fiber is not connected at the center of the circular membrane, when the membrane expands to a semi-sphere [8], the part without fiber at the center of the membrane will have a larger deformation. The top area of the DE membrane buckles up (the buckled area is above the dashed line). Therefore, we observe that when we apply high voltage through the thickness direction of the DE membrane [23], the main deformation concentrates in the middle part of the membrane and produces great sonic characteristics.

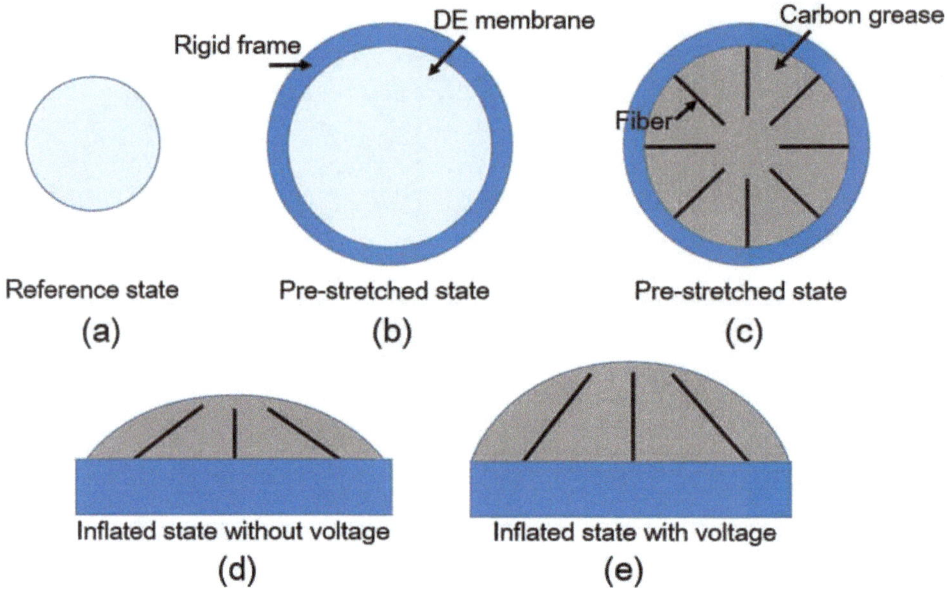

Figure 5. DE membrane inflating actuator with radially arranged fibers in various states: (a) the reference state of a DE membrane; (b) DE membrane is pre-stretched and fixed by a rigid frame; (c) fibers and carbon grease are integrated on the pre-stretched membrane. (d) the structure is inflated with only pressure but no voltage; (e) the structure is further actuated by voltage.

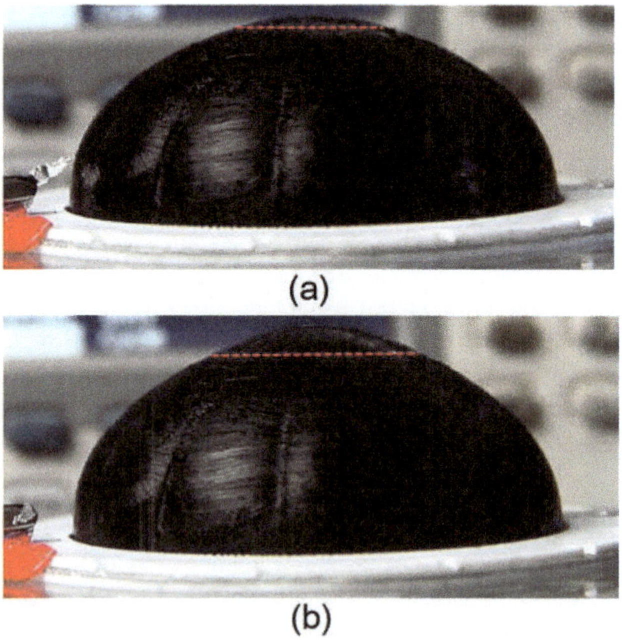

Figure 6. Experimental results of DE membrane inflating actuator: (a) the structure is inflated with only pressure but no voltage; (b) the structure is further actuated by 8 kV.

4. FEM simulation of DE in-plane actuator

Based on the experimental results in Section 2, we investigate the fiber-reinforced DE membrane with finite element simulations. The stress and strain distributions of parallel-arranged long fibers and parallel-arranged short fibers are illustrated by finite element simulations using ABAQUS 6.10. In the FEM simulation, the DE structure is consisted of two parts, the DE part and the fiber part. The actuation of the DE under the high voltage is simulated by using UMAT [28,33]. C3D8RH meshes are used in the simulation. The dimensionless shear modulus is 1, and the dimensionless permittivity is 0.3. The fiber part is set as linear elastic material, with the elastic modulus 15 times the shear modulus of the DE part [10]. The voltage loading is applied on the structure as temperature in the simulation. In order to compare the different structures, the pre-stretch of the DE part and the elastic modulus of the fiber part are the same, the load (equivalent to voltage in experiment) on the DE also equals. In the experiment, we sandwich the fiber between two pre-stretched membranes, but in the FEM simulation, we implant the fiber into the membrane. That means we first model the DE membrane as one structure, and then define part of its section as the fiber material. We change the number and the spacing of the fibers to show the comparison. The number of the fibers are 3, 5, and 7 with the corresponding spacing of 10, 6, and 4 mm. Figure 7 shows the comparison of long fiber–reinforced membrane of different numbers and spacing of

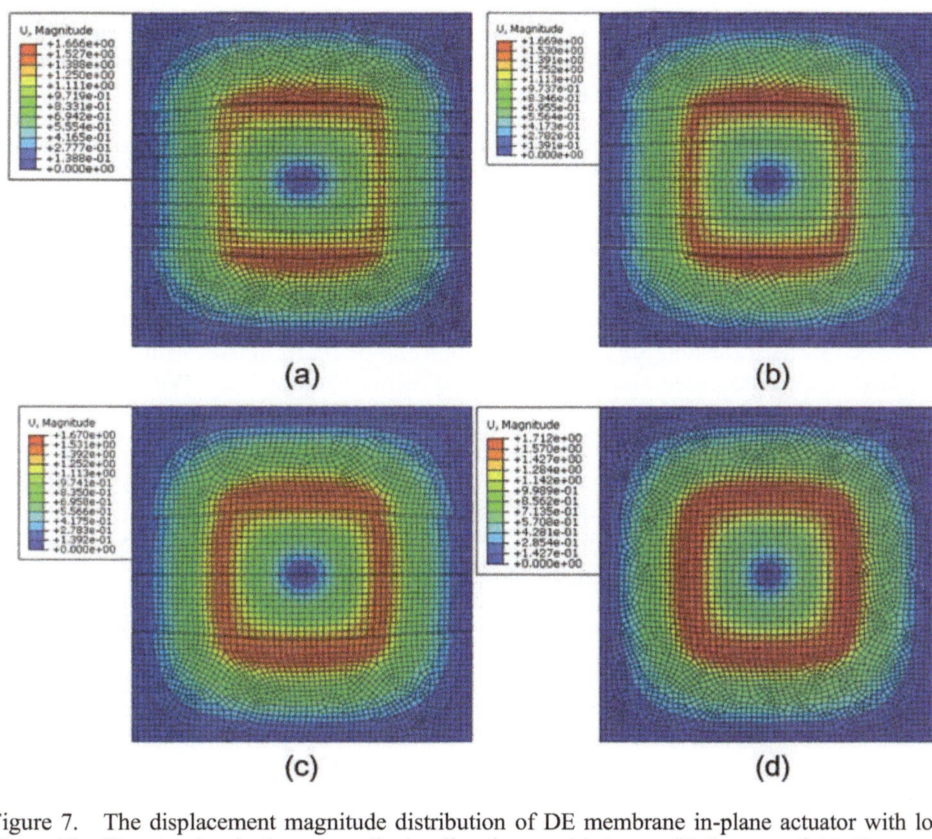

Figure 7. The displacement magnitude distribution of DE membrane in-plane actuator with long fibers. The fiber spacing in the perpendicular direction are (a) 4 mm, (b) 6 mm, and (c) 10 mm. (d) No fiber is used in the DE membrane.

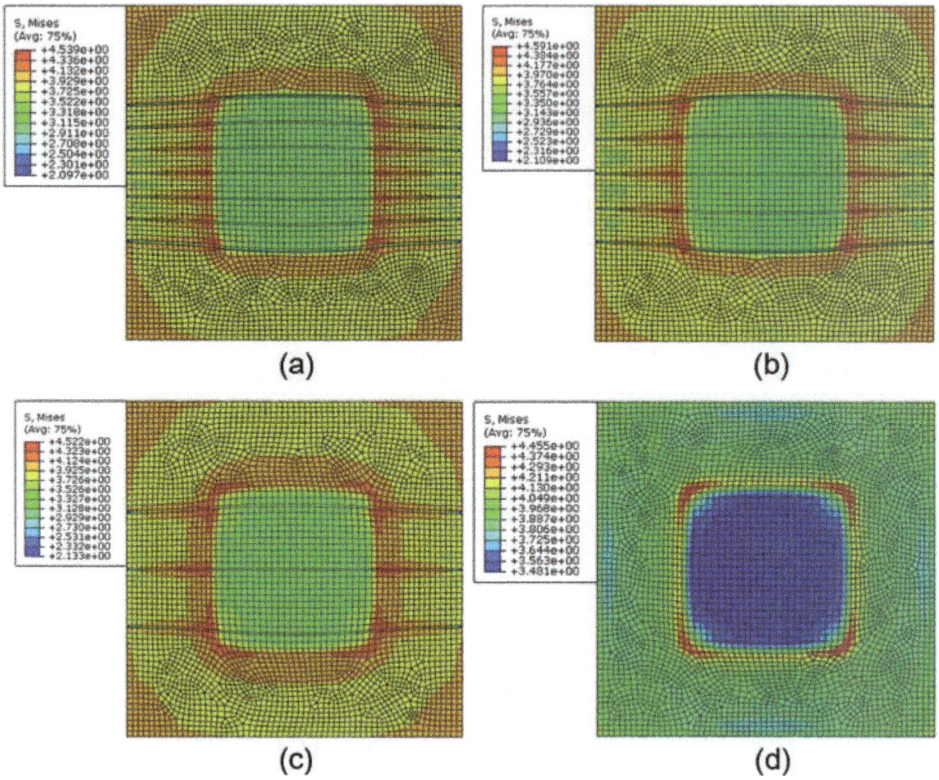

Figure 8. The Mises stress magnitude distribution of DE membrane in-plane actuator with long fibers. The fiber spacing in the perpendicular direction are (a) 4 mm, (b) 6 mm, and (c) 10 mm. (d) No fiber is used in the DE membrane.

fibers. The FEM results show the distribution of displacement on the DE membrane. We find if there is no fiber, the deformations of the membrane along horizontal direction and perpendicular direction are nearly the same (Figure 7(d)). In the DE membrane with long fibers, the deformation along horizontal direction is restrained. When we increase the number of fibers and decrease the spacing of fibers, the limitation to the horizontal deformation increases.

We also study the effects of driven force of the DE membrane with parallel long fibers. The distributions of Mises stress magnitudes are investigated to illustrate the driven force of the DE membrane. From Figure 8, we find if there is no fiber the stress of the membrane along horizontal direction and perpendicular direction are basically the same. But once we add long fibers into the membrane, the stress along horizontal direction is limited. From the analysis, we know that the displacement along horizontal direction is also limited. We can obtain the driven force by multiplying the stress and displacement. Therefore, we get the results that parallel long fibers can limit the driven force in one direction and only output force from the other direction.

Figure 9 shows the FEM simulation of DE membrane with parallel-arranged short fibers. We vary the fiber spacing in the horizontal direction to show the comparison. When the fiber spacing in the horizontal direction is small (Figure 9(a)), the limitation to horizontal

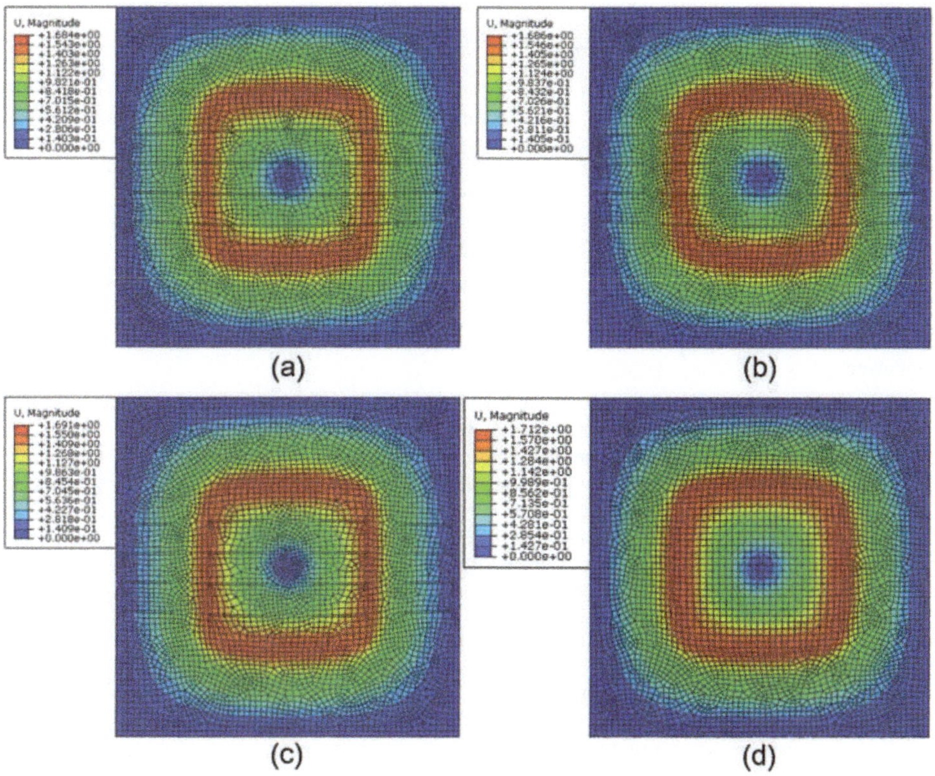

Figure 9. The displacement magnitude distribution of DE membrane in-plane actuator with short fibers. The fiber spacing in the perpendicular direction is 4 mm. The fiber spacing in the horizontal direction are (a) 8 mm, (b) 4 mm, and (c) 2 mm. (d) No fiber is in the membrane.

deformation is almost the same with parallel-arranged long fibers. And when the fiber spacing in the horizontal direction increases, the limitation to the horizontal deformation decreases.

5. Conclusion

In this paper, we investigate the electromechanical behaviors of fiber-reinforced DE membrane structure. The actuation of the DE structures is investigated with both experiments and FEM simulations. Fiber arrangements largely affect the voltage-induced actuation of the DE membrane. The actuation in the perpendicular direction of the DE membrane with long fibers first increases and then decreases by increasing the fiber spacing in the perpendicular direction. The actuation in the perpendicular direction of the DE membrane with short fibers only depends on fiber spacing in the perpendicular direction, whereas the actuation in the horizontal direction depends on fiber spacing in the horizontal direction. In the membrane inflating structure, the radially arranged fibers will break the axisymmetric behavior of the structure. The top area of the inflated balloon without fiber will buckle up when the voltage reaches a certain level.

Disclosure statement

No potential conflict of interest was reported by the authors.

Funding

This work is supported by the National Natural Science Foundation of China [grant number 11302190, 11321202], Zhejiang Provincial Natural Science Foundation of China [grant number Q13A020002], Doctoral Fund of Ministry of Education of China, Qianjiang Talent Program D of Zhejiang Province, Research Foundation of Educational Commission of Zhejiang Province, CAST Research Foundation, and the Fundamental Research Funds for the Central Universities.

References

[1] T. Li, S. Qu, Z. Li, W. Tao, and M. Wang, Investigation of the mechanical properties of the Ni–P–CNTs coated copper composite materials: experiments and modeling, Mater. Sci. Engineering. 500 (25) (2009), pp. 182–187. doi:10.1016/j.msea.2008.09.035

[2] T. Li, S. Qu, and W. Yang, Electromechanical and dynamic analyses of tunable dielectric elastomer resonator. Int. J. Solids Struct. 49 (2012), pp. 3754–3761.

[3] T. Li, S. Qu, and W. Yang, Energy harvesting of dielectric elastomer generators concerning inhomogeneous fields and viscoelastic deformation, J. Appl. Phys. 112 (2012), p. 034119.

[4] T. Li, C. Keplinger, R. Baumgartner, S. Bauer, W. Yang, and Z. Suo, Giant voltage-induced deformation in dielectric elastomers near the verge of snap-through instability, J. Mech. Phys. Solids 61 (2013), pp. 611–628. doi:10.1016/j.jmps.2012.09.006.

[5] T. Li, Z. Zou, G. Mao, S. Qu, and M. Shi, Electromechanical bistable behavior of a novel dielectric elastomer actuator, J. Appl. Mech. 81 (2014), p. 0410191.

[6] T. Li, C. Li, G. Mao, Z. Zou,Y. Xie, and S. Qu, Novel dielectric elastomer structures with electromechanical instability, SPIE Smart Structures and Materials+ Nondestructive Evaluation and Health, San Diego, CA, 2014.

[7] T. Li, W. Li, Z. Zou, Z. Hong, and S. Qu, Effects of stretching rate and size on the rupture of acrylic dielectric elastomer, Int. J. Appl. Mech. (2014).

[8] G. Mao, T. Li, Z. Zou, and S. Qu, Pre-stretch effect on snap-through instability of short-length tubular elastomeric balloons under inflation, Int. J. Solids Struct. 51 (2014), pp. 2109–2115. doi:10.1016/j.ijsolstr.2014.02.013.

[9] Z. Zou, T. Li, S. Qu, and Y. Honghui, Active shape control and phase coexistence of dielectric elastomer membrane with patterned electrodes, J. Appl. Mech. 81 (2014), p. 031016.

[10] C. Li, Y. Xie, X. Huang, J. Liu, Y. Jin, and T. Li, *Novel dielectric elastomer structure of soft robot*, Proceedings of SPIE 9430, Electroactive Polymer Actuators and Devices (EAPAD), 1 April 2015. doi:10.1117/12.2084015.

[11] F. Carpi, A. Migliore, and D. De Rossi, *A new contractile linear actuator made of dielectric elastomers*, Proceedings of SPIE 5759, Smart Structures and Materials 2005: Electroactive Polymer Actuators and Devices (EAPAD), 64, 23 May 2005. doi:10.1117/12.593918.

[12] H. Kim, O. Sinjong, K. Hwang, H. Choi, J. Jeon, and J. Nam, *Actuator model of electrostricive polymers (EPs) for microactuators*, Proceedings of SPIE 4329, Smart Structures and Materials 2001: Electroactive Polymer Actuators and Devices, 482, 16 July 2001. doi:10.1117/12.432681.

[13] S. Arora, T. Ghosh, and J. Muth, Dielectric elastomer based prototype fiber actuators, 25th Anniversary Sens. Actuators A: Phy. 136 (1 May 2007), pp. 321–328. doi:10.1016/j.sna.2006.10.044

[14] F. Carpi and D.R. De Rossi, Dielectric elastomer cylindrical actuators: Electromechanical modelling and experimental evaluation, Mater. Sci. Engineering: C. 24 (1 June 2004), pp. 555–562. doi:10.1016/j.msec.2004.02.005

[15] E. Biddiss and T. Chau, Dielectric elastomers as actuators for upper limb prosthetics: Challenges and opportunities, Med. Eng. Phys. 30 (May 2008), pp. 403–418. doi:10.1016/j.medengphy.2007.05.011.

[16] R.F. Shepherd, F. Ilievski, W. Choi, S.A. Morin, A.A. Stokes, A.D. Mazzeo, X. Chen, M. Wang, and G.M. Whitesides, Multigait soft robot, Proc. Natl. Acad. Sci. 108 (2011), pp. 20400–20403. doi:10.1073/pnas.1116564108.

[17] S.A. Morin, R.F. Shepherd, S.W. Kwok, A.A. Stokes, A. Nemiroski, and G.M. Whitesides, Camouflage and display for soft machines, Science 337 (2012), pp. 828–832. doi:10.1126/science.1222149.
[18] J.-S. Plante, *Dielectric elastomer actuators for binary robotics and mechatronics*. Massachusetts Institute of Technology (2006).
[19] A.A. Stokes, R.F. Shepherd, S.A. Morin, F. Ilievski, and G.M. Whitesides, A hybrid combining hard and soft robots, Soft Robotics 1 (2014), pp. 70–74. doi:10.1089/soro.2013.0002.
[20] R.V. Martinez, J.L. Branch, C.R. Fish, L. Jin, R.F. Shepherd, R. Nunes, Z. Suo, and G.M. Whitesides, Robotic tentacles with three-dimensional mobility based on flexible elastomers, Adv. Mater. 25 (2013), pp. 205–212. doi:10.1002/adma.201203002.
[21] H.-T. Lin, G.G. Leisk, and B. Trimmer, Goqbot: A caterpillar-inspired soft-bodied rolling robot, Bioinspir. Biomim. 6 (2011), p. 026007.
[22] K.Y. Ma, P. Chirarattananon, S.B. Fuller, and R.J. Wood, Controlled flight of a biologically inspired, insect-scale robot, Science 340 (2013), pp. 603–607. doi:10.1126/science.1231806.
[23] P. Brochu and Q. Pei, Advances in dielectric elastomers for actuators and artificial muscles, Macromol. Rapid. Commun. 31 (2010), pp. 10–36. doi:10.1002/marc.v31:1.
[24] T. Mirfakhrai, J.D.W. Madden, and R.H. Baughman Polymer artificial muscles, Mater. today. 10 (April, 2007), pp. 30–38.
[25] I.A. Anderson, T.A. Gisby, T.G. Mckay, B.M. Obrien, and E.P. Calius, Multi-functional dielectric elastomer artificial muscles for soft and smart machines, J. Appl. Phys. 112 (2012), pp. 041101-041101-20.
[26] L. Liu, X. Luo, Y. Liu, and J. Leng, Dielectric elastomer energy harvesting undergoing polarization saturation, Proc. SPIE 8340. Electroactive Polymer Actuators and Devices (EAPAD) 83401V (2012).
[27] Y. Liu, L. Liu, and Z. Zhang, Y. Jiao, S. Sun, and J. Leng, Analysis and manufacture of an energy harvester based on a Mooney-Rivlin–type dielectric elastomer, Europhys. Lett. 90 (2010), p. 36004.
[28] Y. Liu, L. Liu, Z. Zhang, L. Shi, and J. Leng, Comment on "method to analyze electromechanical stability of dielectric elastomers, Appl. Phys. Lett. 93 (2008), p. 106101.
[29] Y. Liu, L. Liu, J. Leng, K. Yu, and S. Sun, Electromechanical stability of dielectric elastomer, Appl. Phys. Lett. 95 (2009), p. 211901.
[30] Y. Liu, L. Liu, Z. Zhang, and J. Leng, Dielectric elastomer film actuators: Characterization, experiment and analysis, Smart Mater. Struct. 18 (2009), p. 095024.
[31] L. Liu, Y. Liu, Y. Kai, and J. Leng, Thermoelectromechanical stability of dielectric elastomers undergoing temperature variation, Mechanics Mater. 72 (2014), pp. 33–45. doi:10.1016/j.mechmat.2013.05.013.
[32] L. Liu, Y. Liu, X. Luo, B. Li, and J. Leng, Electromechanical instability and snap-through instability of dielectric elastomers undergoing polarization saturation, Mechanics Mater. 55 (2012), pp. 60–72. doi:10.1016/j.mechmat.2012.07.009.
[33] X. Zhao and Z. Suo, Method to analyze programmable deformation of dielectric elastomer layers, Appl. Phys. Lett. 93 (2008), p. 251902.

12

XRD study of intercalation in statically annealed composites of ethylene copolymers and organically modified montmorillonites. 2. One-tailed organoclays

Sara Filippi* and Giovanni Polacco

Department of Civil and Industrial Engineering, University of Pisa, Largo Lucio Lazzarino, 1, 56126 Pisa, Italy

Ethylene copolymers with different polar comonomers, such as vinyl acetate, methyl acrylate, glycidyl methacrylate, and maleic anhydride, were used for the preparation of polymer/clay nanocomposites by statically annealing their mechanical mixtures with different commercial or home-made organically modified montmorillonites containing only one long alkyl tail. The nanostructure of the products was monitored by X-ray diffraction, and the dispersion of the silicate particles within the polymer matrix was qualitatively evaluated through microscopic analyses. The effect of the preparation conditions on the structure and the morphology of the composites was also addressed through the characterization of selected samples with similar composition prepared by melt compounding. In agreement with the findings reported in a previous paper for the composites filled with two-tailed organoclays, intercalation of the copolymer chains within the tighter galleries of the one-tailed clays occurs easily, independent of the application of a mechanical stress. However, the shear-driven break-up of the intercalated clay particles into smaller platelets (exfoliation) seems more hindered. A collapse of the organoclay interlayer spacing was only observed clearly for a commercial one-tailed organoclay – Cloisite® 30B – whereas the same effect was almost negligible for a home-made organoclay with similar structure.

Keywords: ethylene copolymers; one-tailed organoclays; annealing; X-ray diffraction

1. Introduction

The stability of polymer/clay nanocomposites (PCNs) depends on the characteristics of polymer, silicate platelets, and organic clay modifier [1]. From the thermodynamic point of view, the equilibrium structures (intercalated/exfoliated) achievable when blending a polymer/organoclay couple can be reliably forecast [1]. In contrast, the kinetics and the mechanism of intercalation of the surfactant or polymer into the clay stacks are less understood even if studied since long time ago [2]. The intercalation process is influenced by the transport of polymer into the primary particles of the silicate [3], as well as by the degree of polymerization and of the length of the surfactant tails and/or the presence of functional groups along the macromolecules [4]. An important aspect is that the mobility of the polymer chains in the organoclay galleries is comparable to that in the bulk polymer melt. Therefore, PCN forms rapidly if the temperature is sufficiently higher than the glass

*Corresponding author. Email: s.filippi@diccism.unipi.it

transition, the average molecular weight is moderate, and the coverage of the silicate surface by organic surfactants is appropriate. This is confirmed by the fact that intercalation may be obtained also under static conditions [5–8].

Ethylene copolymers have been extensively employed in the preparation of PCNs, and reviews dedicated to polyolefin-based PCNs are available in the scientific literature [9–14]. The level of silicate layers dispersion in PCNs based on ethylene copolymers depends on the type, the concentration and the location of the functional groups on the polymer chains, the structure of the surfactant used for clay modification, the organoclay loading, etc. Usually, the composites prepared by solution blending contain unintercalated clay tactoids that underwent fast intercalation upon thermal treatment carried out at temperatures higher than the polymer melting point [15,16]. Recent studies suggested that organoclays with two long alkyl tails undergo better dispersion than those with a single tail [17–19]. However, opposite results were also described by other researchers [20–22]. Martins *et al.* suggested that these conflicting reports are possibly due to the different processing conditions [23]. This point is discussed in detail in this paper. In the previous paper of this series, the intercalation in statically annealed dry blends of different functionalized polyethylenes with organoclays modified with ammonium ion surfactants containing two long alkyl groups was investigated [24]. The structure and the morphology of the PCNs prepared by this procedure were compared with those of analogous materials produced with different methods, such as melt compounding and solution blending. The intercalation of the copolymer chains within the galleries of the organoclays occurs rapidly, independent of the application of shear stress. On the other hand, solution blending, proved unable, in the absence of subsequent thermal treatments, to yield intercalation. Moreover, the preliminary dispersion of micron-sized clay particles achievable by solution blending leads to even faster intercalation when the mixture is melted. For most of the polymer/organoclay couples studied in part 1 of this work [24], mixed intercalated/exfoliated structures were thermodynamically stable. However, the highest levels of exfoliation were obtained for the melt compounded PCNs prepared from polymer/clay couples with optimal match of polar characteristics.

In this work, the above study was extended to analogous composites loaded with modified montmorillonites (MMTs) organically modified with surfactants containing only one long alkyl tail, and the effect of a lowered interlayer distance and a higher polarity of the inorganic filler was investigated.

Commercially available organoclays differ for the structure of the organic modifier, the lateral dimensions, the charge density, and the cation exchange capacity (CEC). The latter plays an important role in the process of polymer chains intercalation. For this reason, most of the organoclays used in this work were prepared by modification of only one commercial Na-MMT (Dellite® LVF, with a CEC of 1.05 meq/g). The effect of a difference in the CEC values was also addressed using a couple of organoclays based on Cloisite-Na^+ (with a CEC of 0.926 meq/g), that is, the commercially available Cloisite® 30B and the home-synthesized C-ArqHT.

The ethylene copolymers used for the PCNs preparation were the same already used in the previous work [24]. Except for HDMA, which is a linear HDPE grafted with maleic anhydride (MA), all the other materials possess the branched structure typical for LDPE. HDMA is certainly the less polar copolymer, as the MA content is only 1 wt%. On the other hand, the presence of MA and glycidyl methacrylate (GMA) groups in the chains does probably enhance the interaction of these copolymers with the hydroxyl groups of the silicate layers.

2. Experimental part

2.1. Materials

The organoclays used in this work are indicated in Table 1, with their sources, the CEC, the milliequivalent exchange ratio (MER), the loss on ignition, and the d_{001} spacing determined by XRD. The surfactants used for MMT modification are the quaternary ammonium chlorides indicated by symbols, where H is hydrogen, C18 is *n*-octadecyl, M is methyl, T is 'tallow', that is, a partly unsaturated alkyl chain with approximate composition: 65% C18; 30% C16; 5% C14, (HT) is 'hydrogenated tallow', (HE) is 2-hydroxyethyl, and O is oleyl.

Cloisite® 30B (from Southern Clay Products Inc., USA), and C-ArqHT were prepared from Cloisite-Na^+, having a CEC of 0.926 meq/g. D-ODA, D-ArqHT, and D-Eth were synthesized from Dellite® LVF (CEC = 1.05 meq/g; Laviosa Chimica Mineraria S.p.A., Italy) by cation exchange with, respectively, octadecyl ammonium (ODA) chloride and the commercial $M_3(HT)_1$ and $M_1(HE)_2O$ surfactants (Arquad HT-50 and Ethoquad O/12, by Akzo Nobel).

The polymers used as matrixes are listed in Table 2, together with their sources and some of their characteristics. EVA14, EVA28, EMA, EGMA, and EAGMA are random

Table 1. Organoclays[a].

Clay code	Commercial designation	Surfactant[b]	CEC[c] (meq/g)	MER[d] (meq/g)	Loss on ignition (wt%)	d_{001}[e] (nm)
D-ODA	Experimental	H_3C_{18}	1.05	1.05[f]	28	2.03
D-ArqHT	Experimental	$M_3(HT)_1$	1.05	1.05[f]	29	2.22
D-Eth	Experimental	$M(HE)_2O$	1.05	1.05[f]	29	2.06
30B	Cloisite® 30B	$M(HE)_2T$	0.926	0.90	30	1.85
C-ArqHT	Experimental	$M_3(HT)_1$	0.926	0.926[f]	27	1.92

Notes: [a]The data for Cloisite® 30B, produced by Southern Clay Products Inc., are from the supplier's product bulletin. [b]cf. 'Experimental' for an explanation of the symbols. [c]Cation exchange capacity. [d]Milliequivalent exchange ratio (modifier used per g of native clay). [e]Interlayer spacing. [f]cf. 'Experimental'.

Table 2. Ethylene copolymers (supplier's data).

Sample code	Functional comonomer(s) (wt%)	Commercial designation (supplier)	Structure	MFI (dg/min)	Density (kg/m^3)	T_m (°C)
EVA14	VA[a] 14	Greenflex® FC45 (Polimeri Europa)	E-*ran*-VA	0.3	—	93
EVA28	VA 28	Greenflex® HN70 (Polimeri Europa)	E-*ran*-VA	6	—	73
EMA	MeA[b] 20	Elvaloy® 1820AC (DuPont)	E-*ran*-MeA	8	942	92
EGMA	GMA[c] 8	Lotader® AX8840 (Arkema)	E-*ran*-GMA	5	940	109
EAGMA	GMA – MeA 3–24	Lotader® AX8930 (Arkema)	E-*ran*-GMA-*ran*-MeA	6	940	66
HDMA	MA[d] 1	Polybond® 3009 (Chemtura)	HDPE-*g*-MA	3–6	950	127

Notes: [a]Vinyl acetate. [b]Methyl acrylate. [c]Glycidyl methacrylate. [d]Maleic anhydride.

copolymers with branched structure. The polymers were dried in a vacuum oven at 65°C for 2 days before use.

2.2. *Procedures*

D-ODA, D-ArqHT, D-Eth, and C-ArqHT were synthesized as follows: 15 g of Dellite® LVF (or Cloisite-Na) were stirred for 2 h in 1000 mL of deionized water at 80°C; 200 mL of a water solution of the appropriate surfactant (20% excess, with respect to stoichiometry) was added, and the mixture was mechanically stirred for 24 h at 80°C. The precipitate was collected on a sintered glass filter, washed with warm water and, successively, with EtOH in order to remove the excess surfactant. Thus, the MER of organoclays is nominally indicated in Table 1 as equal to the CEC of the starting MMT. The organoclays were dried in a vacuum oven at 40°C for 24 h, milled in a mortar and sieved (200 mesh).

The statically annealed (s.a.) composites were prepared as follows: mechanical blending the polymers and the clays, as powders of micrometric dimensions; shaping the mixtures into tablets of 20 mm diameter and 2 mm thickness in a Carver press; heating the tablets at temperatures slightly higher than the polymer melting point for about 10 min; slow cooling to room temperature.

The melt compounded (m.c.) composites were prepared in a Brabender Plasticorder mixer of 50 mL capacity, preheated to 120°C (for EVA14, EVA28, EMA, EGMA, and EAGMA) or 150°C (for HDMA). The rotor speed was maintained at 30 rpm for about 3 min after addition of the polymer/clay mixtures, and was then increased gradually (in 30 s) to 60 rpm. During the kneading time (10 min), the temperature increased by 10–15 K due to stress heating. The molten composites were extracted from the mixer and cooled naturally in air.

The loss on ignition of the experimental organoclays was determined by thermogravimetric analysis with a Q500 TA Instruments balance using a sample of about 10 mg. The test was made in air (60 mL/min), in the 30–900°C temperature range, with a scanning rate of 10°C/min.

X-ray diffraction measurements (CuK_{α}) were made in reflection mode with a Siemens D500 Krystalloflex 810 apparatus at a scan rate of 1.0° min^{-1}. The clay was analyzed as a powder, whereas the composites were in the form of disks of 20 mm diameter and 2 mm thickness.

Polarized optical microscopy (POM) observations were made at 150°C on a Leitz Ortholux microscope equipped with a Linkam TMS 93 hot stage and a digital camera JVC TK-1085E. A small fragment of the composite samples was placed on a glass slide, heated at 150°C, and pressed with a cover slip to obtain a film of 80–100 μm thickness. Observations were made under crossed polarizers, whereby the micron-sized clay agglomerates, if any, were easily revealed as highly birefringent particles.

3. Results and discussion

The XRD patterns of D-ODA and those of its 5 wt% m.c. and s.a. composites with EVA14, EVA28, and EMA are reported in Figure 1. Figure 2 shows similar patterns for the composites with HDMA, EGMA, and EAGMA.

The first-sight conclusion one can draw from the two figures is that there is a qualitative similarity between the structures of the corresponding m.c. and s.a. composites. In fact, the position of the low-angle peaks on the 2θ scale is independent of the

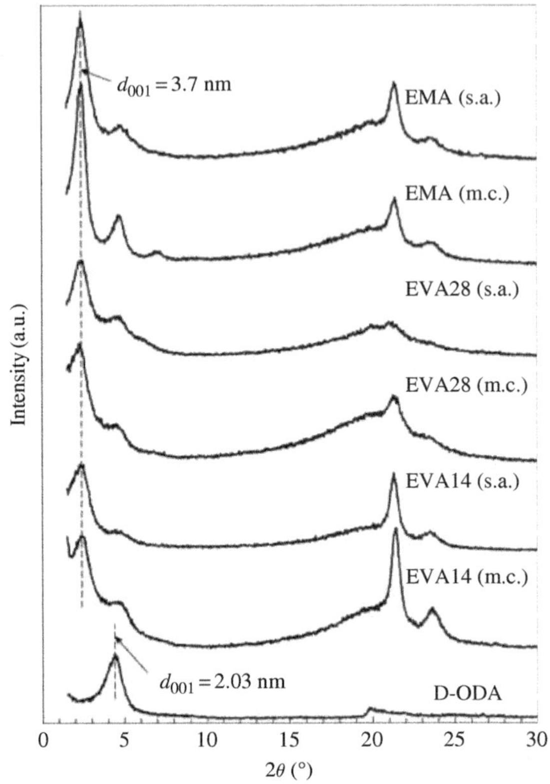

Figure 1. XRD patterns of D-ODA and its 5 wt% m.c. and s.a. composites with EVA14, EVA28, and EMA.

preparation protocol. This means that the shear stress acting on the polymer/clay mixtures during the melt compounding has little influence, if any, on nanostructure formation.

All the composites of Figure 1 display an intense reflection at 2θ€ ≈ 2.38°, with higher orders at 4.75° and 7.1°, corresponding to a regular d_{001} spacing of 3.7 nm, in fair agreement with the results found by Tian *et al.* [25] for an EVA/organoclay nanocomposite. This demonstrates that the D-ODA was intercalated by the EVA and EMA chains, and the penetration of the copolymer chains caused a considerable increase of the 2.03 nm interlayer distance of the organoclay. Moreover, the similarity of the XRD profiles of the m.c. and s.a. samples suggests that the resulting structure is thermodynamically stable for these systems. It should be pointed out that melt compounding EVA with an ODA-modified MMT was found by Zanetti *et al.* [26] to cause a shift to wider angles of the basal reflection of the nanofiller. This effect, attributed by the authors to the possible presence of unmodified MMT left in the organoclay, or to partial degradation of the latter [26], was apparently absent in our system as well as in that of Tian *et al.* [25].

The behavior of the composites shown in Figure 2 is somehow different. Let us deal with the EGMA and EAGMA composites, first. Here also, intercalation occurred, independent of the application of a shear stress during the preparation. However, the extent of gallery height expansion was slightly larger than that observed for the EVA and EMA composites. The capability of the GMA groups of the copolymers to efficiently interact with the silicate layer surface may probably explain this difference. It can also be observed that,

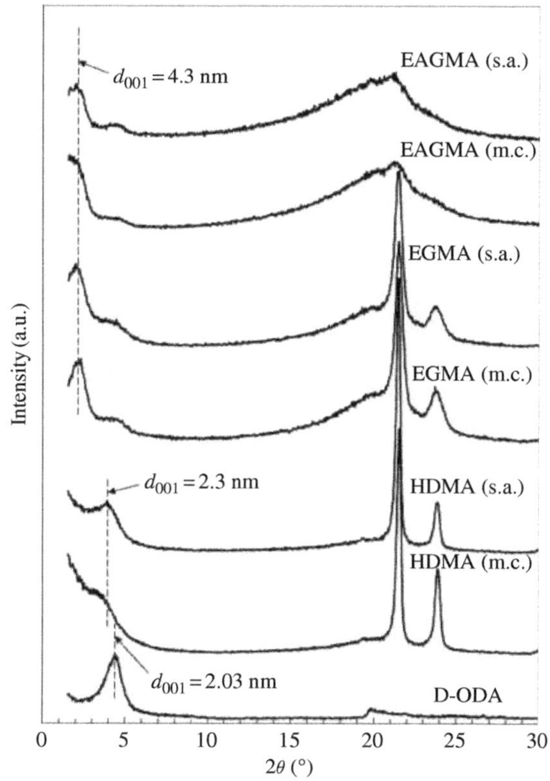

Figure 2. XRD patterns of D-ODA and its 5 wt% m.c. and s.a. composites with HDMA, EGMA, and EAGMA.

especially for the composite based on the more polar EAGMA, and notably for the m.c. sample, the basal reflection is very weak and broad and is accompanied by a continuous increase of reflected intensity at low angles, thus suggesting a level of clay exfoliation.

Finally, in the XRD patterns of the D-ODA composites based on EGMA and EAGMA, the weaker peak seen at 2θ angles of 4.35° probably comprise a contribution of unintercalated organoclay stacks rather than being just the second order of the basal reflection, which should otherwise appear at $2\theta \approx 4.1°$. By the way, a second-order peak would not be expected to be visible in these XRD patterns, given the low intensity and broadness of the basal reflections. Thus, for these copolymer/clay mixtures, the intercalation process is not so fast so that a small amount of unintercalated stacks is still present at the end of the sample preparation. However, despite the slower penetration of the copolymer chains within the D-ODA galleries, once intercalation has occurred, it tends to proceed further, particularly in the melt kneaded composite, toward exfoliation (notice that the basal reflection did almost fade in the m.c. sample).

The XRD patterns of the HDMA/D-ODA composites shown in Figure 2 suggest that the behavior of these materials differs even more from that of the systems of Figure 1. As already pointed out, HDMA comprises predominantly linear chains with a very limited number of grafted succinic anhydride groups. The overall polarity of these polymers is rather low, but the anhydride functionalities can probably lead to hydrogen bonding with the silicate layers. As shown in Figure 2, intercalation of the HDMA chains within the

D-ODA galleries seems to take place rapidly, though leading to a modest expansion of the interlayer space. However, a comparison of the diffractograms of the s.a. and m.c. samples suggests that the mechanical stress acting on the system during melt compounding helps disrupt progressively the intercalated stacks so that the level of exfoliation increases. Actually, the XRD profile of the m.c. sample retains just a broad shoulder at $2\theta \approx 3.8°$, ascribable to intercalation, on a background of continuously increasing reflected intensity pointing to exfoliation. This conclusion is also supported by the finding that the I_{110}/I_{200} intensity ratio of the reflections at $2\theta \approx 21.6°$ and 24.0° of the crystalline structure of HDMA is considerably lower for the m.c. sample (~2.5 vs. ~4 for the s.a. composite). As it was shown elsewhere [16], an intensity ratio lower than ~3 is indicative of a preferred orientation of both polymer crystallites and silicate platelets parallel to the specimen surface and the effect is particularly evident for highly exfoliated, m.c., and compression molded PCNs. Thus, the XRD patterns of the HDMA/D-ODA nanocomposites shown in Figure 2 suggest that the intercalated clay stacks, formed upon melting the polymer/clay mixture, slowly decompose into smaller particles (exfoliation [27]), and the process is favored by mechanical treatment.

The behavior of the composites loaded with D-ArqHT was found to resemble that discussed before for the analogous materials containing the D-ODA clay. For the sake of brevity, only the XRD traces recorded for the s.a. composites based on EVA14, EVA28, EMA, EGMA, and EAGMA are given in Figure 3. The XRD patterns of the analogous

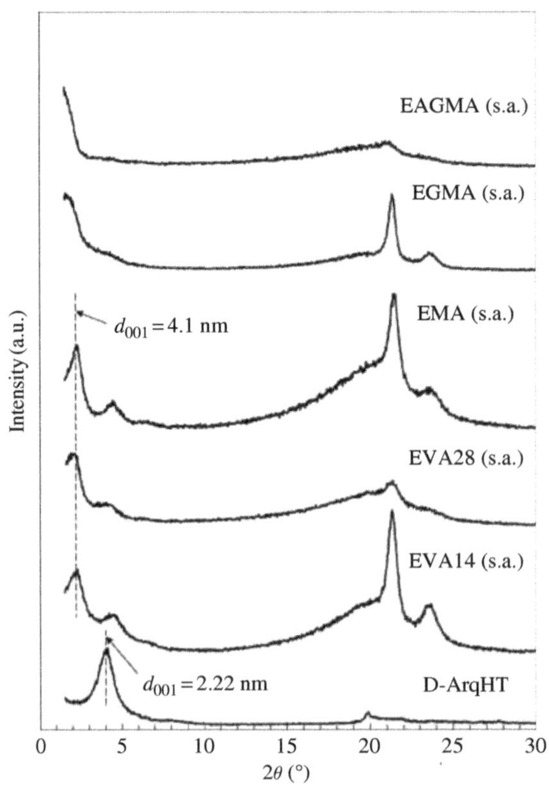

Figure 3. XRD patterns of D-ArqHT and its 5 wt% s.a. composites with EVA14, EVA28 and EMA, EGMA, and EAGMA.

m.c. materials (not shown) were very similar. Thus, the conclusions drawn before with reference to the effect played on intercalation by the stress acting on the system in the melt compounding preparations are also valid for the present PCNs.

The substitution of the three hydrogen atoms of the D-ODA modifier with three methyl groups leads to an increase of the organoclay spacing to d_{001} = 2.22 nm. The interlayer distance measured by XRD for the intercalated EVA and EMA nanocomposites filled with D-ArqHT is slightly larger than that of the analogous nanocomposites loaded with the D-ODA clay, shown in Figure 1 (4.1 vs. 3.7 nm).

The increased hydrophobicity of D-ArqHT compared with that of D-ODA has probably improved the affinity of the organoclay toward the EGMA and EAGMA copolymers. The XRD patterns of the composites based on the latter copolymers are almost free of the low-angle reflections of regularly stacked silicate layers, thus suggesting that the interaction of the D-ArqHT organoclay with these molten copolymers (in particular, EAGMA) led to extensive decomposition of the large silicate booklets into smaller and/or disorganized particles.

In agreement with the results found by others for the EVA composites with similar clays [28,29], no clear indication of D-ArqHT degradation was found in this study.

The XRD profiles recorded for the m.c. and s.a. composites of all the copolymers with the D-Eth nanofiller are reported in Figures 4 and 5. Once more, the qualitative close

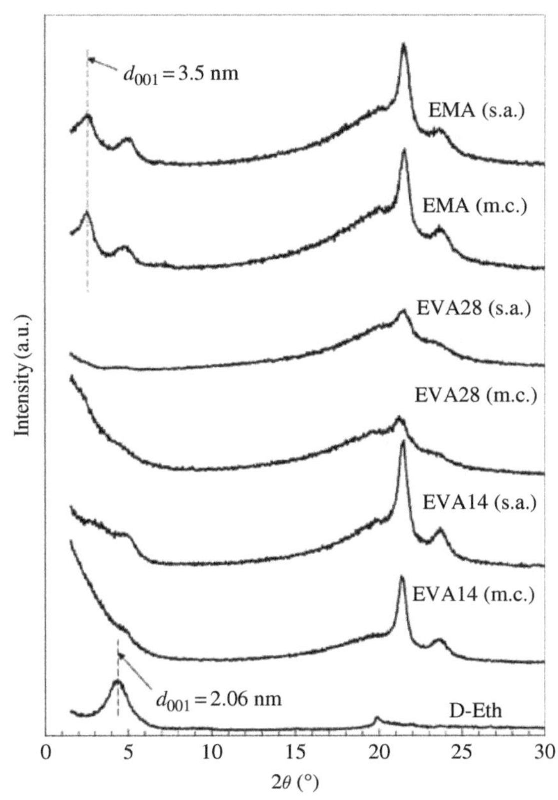

Figure 4. XRD patterns of D-Eth and its 5 wt% m.c. and s.a. composites with EVA14, EVA28, and EMA.

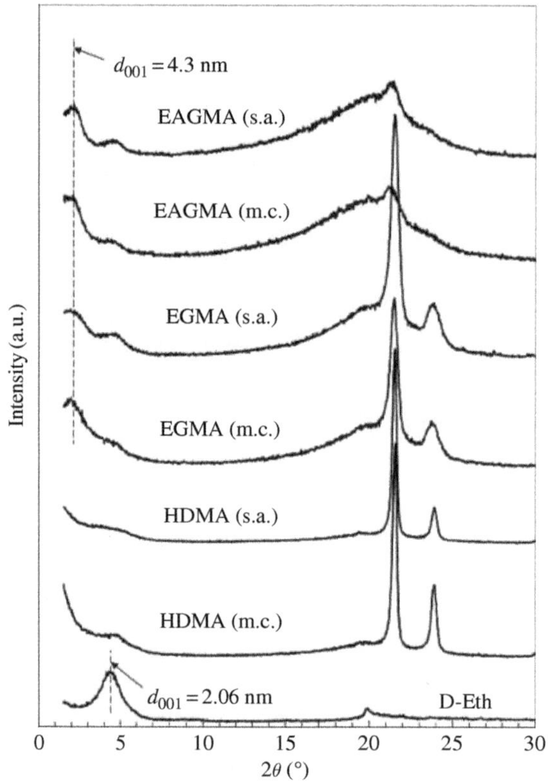

Figure 5. XRD patterns of D-Eth and its 5 wt% m.c. and s.a. composites with HDMA, EGMA, and EAGMA.

similarity between the structures of the m.c. and s.a. composites is confirmed for the majority of the materials considered in the figures.

The EVA composites, particularly that based on EVA28, display almost silent XRD patterns in the low-angle region. This indicates a high level of exfoliation and/or disordered intercalation. It should be remembered that the structure of the D-Eth organoclay is very similar to that of the commercial Cloisite® 30B (30B) that was extensively used for the preparation of EVA/organoclay nanocomposites [18,20–22,29–33]. The latter composites were also found to possess an appreciable level of exfoliation.

The XRD spectra of the EVA/D-Eth composites, especially that based on EVA14, comprise a weak reflection at wider angles compared with that of D-Eth ($2\theta \approx 5.0°$ vs. 4.33°), suggesting that a small amount of the organoclay may undergo a modest interlayer spacing collapse, similar to that often observed for the EVA/30B composites [18,29–33]. This point is discussed in more detail below.

The XRD patterns of the EMA/D-Eth composites shown in Figure 4 are typical for intercalated nanocomposites characterized by a d_{001} interlayer distance of 3.7 nm. Given that the functional groups of EMA and those of the EVA copolymers are isomeric and their contents are comparable, it may be surprising that the structures of the relevant nanocomposites are different. We cannot offer any definite explanation for this. Another apparent peculiarity of the EMA/D-Eth spectra is that the reflection at $2\theta \approx 5.0°$ is perhaps too intense to be ascribed to the second order of the basal reflection of the intercalated clay stacks at $2\theta \approx 2.51°$. This peculiarity might be rationalized assuming that a collapse of the d-spacing of the D-Eth organoclay, such as that previously envisaged for the EVA composites, has also taken place during the preparation of the EMA composites. This would lead to a superposition of the two peaks. However, we do not avail of any clear evidence supporting this hypothesis.

The XRD patterns of the EGMA and EAGMA composites shown in Figure 5 comprise low-angle reflections ($2\theta \approx 2.05°$) ascribable to intercalation. In place of the second-order peaks that, if present, would be expected to appear at $2\theta \approx 4.1°$, broad reflections around $2\theta \approx 4.6°$ are seen in the spectra. Once more, the overlapping of the second-order reflection and a peak due to collapsed tactoids of the D-Eth clay might explain this observation. Although we have no direct proof in favor of such hypothesis, a circumstantial supporting evidence is provided by the XRD patterns of the HDMA/D-Eth composites shown in the same figure. Indeed, neither of the HDMA patterns displays reflections at 2θ angles lower than that of the organoclay, thus demonstrating the absence of regularly intercalated silicate booklets. Just a continuous increase of reflected intensity suggesting exfoliation, plus a reflection close to $2\theta \approx 5°$, is visible in these spectra. The latter reflection must therefore be ascribed to silicate layers stacks with decreased spacing. Thus, the hypotheses illustrated above, that a collapse of the D-Eth organoclay may have taken place during the preparation and processing of the other nanocomposites, too, may appear reasonable.

Collapse of the interlayer spacing of organically modified clays, more notably those containing one-tailed quaternary ammonium ions, was observed in many studies, and generally described as due to thermal degradation of the organic component, with expulsion of the degradation products. In particular, several nanocomposites produced by melt compounding from EVA [18,29–33] and other polymers [5,34–38], with 30B as nanofiller, were found to display XRD patterns comprising reflections at 2θ angles wider than that of the organoclay. Only in a few reports on the preparation and characterization of EVA/30B nanocomposites [20–22] a shift of the basal reflection toward wider angles was apparently not found.

In a recent investigation, Benali *et al.* [39] compared the behavior of 30B with that of other organoclays modified with surfactants containing hydrogenated alkyl tails in place of the partly unsaturated [40] 'tallow' (T) groups of the 30B. The authors concluded that the decrease of the d-spacing is probably due to oxidation and cross-linking of the unsaturated alkyl tails.

As shown in Figure 6, the m.c. EVA14/30B composite prepared by us does show a well-resolved reflection at $2\theta \approx 6°$, corresponding to a collapsed d-spacing of ~1.47 nm. The XRD patterns of 30B and its EVA14 composite are compared in Figure 6 with those, already discussed above, of D-Eth and the analogous EVA14 composite.

From Figure 6 we conclude that, even if the shoulder seen at $2\theta \approx 5°$ ($d_{001} \approx 1.77$ nm) in the XRD pattern of the EVA14/D-Eth is actually due to tactoids with collapsed interlayer spacing, the amount of such tightened tactoids is considerably lower than that in the analogous 30B composite. In this respect, it may be interesting to point out that D-Eth was prepared by cation exchange with a surfactant (Ethoquad O12, by Akzo Nobel) comprising oleyl groups. Therefore, most of the long tails of this organoclay contain a double bond, whereas the modifier employed to synthesize the 30B has only 50% of unsaturated alkyl groups [40]. Thus, unsaturation of the organic component cannot be the (only) parameter responsible for the behavior of 30B. On the other hand, this organoclay was found to undergo a d-spacing collapse even when processed at temperatures lower than that (~180°C) of the onset of thermal degradation. Moreover, the presence of excess

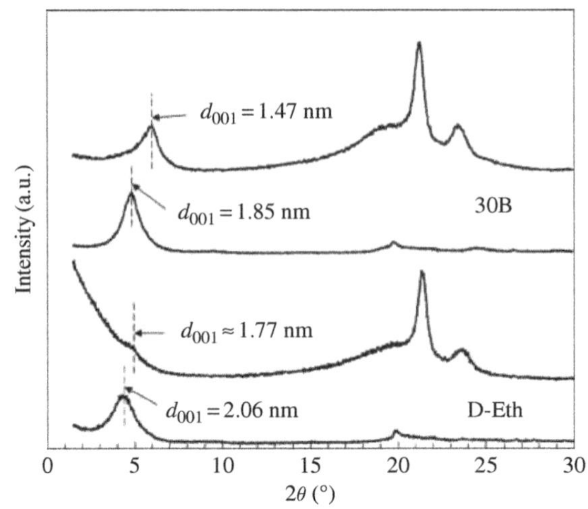

Figure 6. XRD patterns of D-Eth and 30B, together with their 5 wt% m.c. composites with EVA14.

surfactant (and/or other impurities) in the commercial product was shown to lower the thermal stability of this organoclay [41] and to enhance the phenomenon of gallery height shrinkage [42]. All this matter is being studied further in our laboratories, and the results will be published in a forthcoming paper.

In Figure 7, the effect of the organoclay CEC on the intercalation process is addressed. The two organoclays C-ArqHT and D-ArqHT differ from each other for the charge density of the starting MMT's: Cloisite Na^+ and Dellite® LVF (CEC = 0.926 vs. 1.05 meq/g). Due to the higher content of organic component, the interlayer spacing of D-ArqHT (2.22 nm) is larger than that of C-ArqHT (1.92 nm). As shown in the figure, however, intercalation of the EMA chains within the galleries of these organoclays leads, for both composites, to the same d-spacing of 4.1 nm.

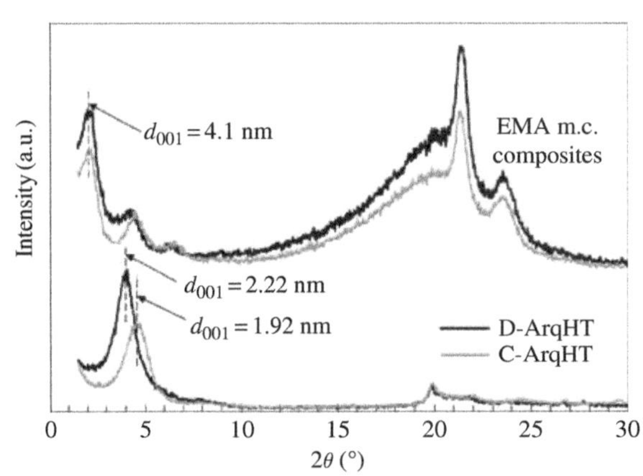

Figure 7. XRD patterns of C-ArqHT and D-ArqHT, and those of their 5 wt% m.c. composites with EMA.

The same behavior was already observed in the previous paper [24] for a couple of two-tailed clays with different CEC.

The Cloisite® 20A and D-ArqHC organoclays, modified with the same $M_2(HT)_2$ surfactant, displayed d-spacings of 2.42 and 3.05 nm, respectively, due to the different CEC of their precursor MMTs. However, the EVA14 PCNs loaded with the two clays possess an intercalated nanostructure with the same interlayer spacing of 3.91 nm (cf. Figures 1 and 2 of [24]).

The finding that CEC has almost no effect on the d-spacing of the intercalated nanocomposites prepared with different organoclays (modified with the same organic cations) may appear surprising, at first sight. However, this behavior can probably be rationalized considering that the thermodynamics of the intercalation process taking place in highly compatible polymer/organoclay mixtures depends on the equilibrium interlayer distance, that is, on the volume of organic material filling the galleries, rather than on the relative content of the two components of this material (ionically bonded clay modifier and polymer chains).

The morphology of the nanocomposites prepared from functionalized polyethylenes and one-tailed organoclays differs from that of the analogous composites loaded with two-tailed organoclays described in part 1 of this work [24]. For the latter PCNs, the morphology depends not only on the affinity between polymer and clay but also on the preparation method. Although intercalation occurs rapidly, independent of the application of mechanical stress, an excellent dispersion of submicrometric silicate particles within the polymer matrix was only observed when the molten polymer/clay mixture was kneaded in a mixer. The composites prepared by static annealing displayed morphologies characterized by the presence of poorly dispersed particles with dimensions of microns or even tens of microns. For the PCNs described in this paper, on the contrary, the effect of the preparation protocol seems to be less important, if any. In Figure 8, a few examples of the optical micrographs taken under crossed polarizers on molten films of some s.a. and m.c. composites are shown and compared with those of analogous m.c. samples loaded with a two-tailed organoclay (Cloisite® 20A).

The POM micrographs (a) and (e) of the EMA/D-Eth andEVA14/30B s.a. composites, respectively, demonstrate that the morphology of these materials is very similar to that of the analogous s.a. composites with two-tailed organoclays described before.

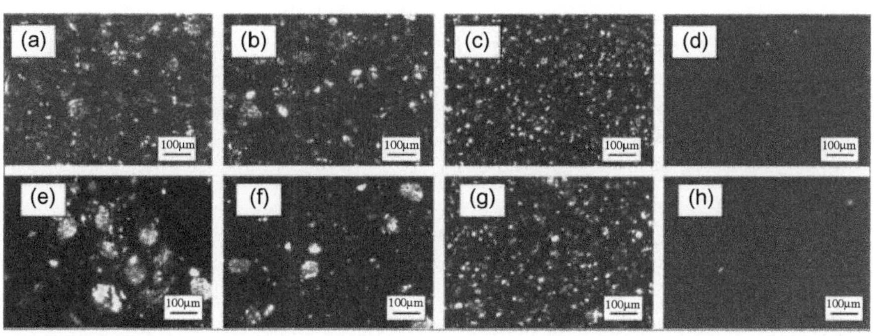

Figure 8. POM images of the EMA (above) and EVA14 (below) nanocomposites with 5 wt% of (a) D-Eth (s.a.); (b) D-Eth (m.c.); (c) 30B (m.c.); (d) Cloisite® 20A (m.c.); (e) 30B (s.a.); (f) D-Eth (m.c.); (g) 30B (m.c.); (h) Cloisite® 20A (m.c.).

In contrast, it was not expected that, as demonstrated by the micrographs (b), (c), (f), and (g), even after melt compounding, the micron-sized aggregates of the one-tailed organoclays failed to completely crumble into submicrometric particles as it was observed for the two-tailed organoclays (cf. the micrographs (d) and (h) in Figure 8). Results very similar to those illustrated in Figure 8 were also found from the morphological analysis carried out either by POM and by scanning electron microscopy on all the other polymer/clay couples described in this work.

It may be interesting to compare the POM micrographs (f) and (g) of Figure 8, taken on the m.c. composites EVA14/D-Eth and EVA14/30B, in the light of the XRD patterns of the same samples, shown in Figure 6. The second composite is known from the literature [18,20–22,29–33] to possess a high level of exfoliation. On the other hand, the XRD pattern displays a fairly intense reflection at $2\theta \approx 6°$ associated with unintercalated tactoids with collapsed d-spacing, and the bright micrometric particles seen in the POM image of this sample (Figure 8g) might be ascribed to their presence. Therefore, the EVA14/30B composite comprise a dispersed phase of unintercalated clay particles and a matrix, appearing as a black background in the micrograph, made up of a fine dispersion of exfoliated silicate platelets. The XRD pattern of the EVA14/D-Eth sample, on the contrary, shows just a hint of a reflection ascribable to collapsed tactoids and, consequently, the latter would be expected to eventually appear in the relevant POM image as very few and small bright spots. The micrograph of Figure 8f depicts a different situation: the black area is larger and the number of bright particles is lower, but the size of some of them is much larger. Considering that the small-angle region of the XRD spectrum of this sample displays a strong progressive increase of diffracted intensity, one might speculate that this is due to a population of silicate booklets or aggregates, characterized by an increasing level of intercalation and low packing regularity. At any rate, whatever is the true origin of the different behavior of 30B and the structurally similar D-Eth, it is no doubt the result of the different CEC. This point also needs further investigation.

4. Conclusions

The results of the present study indicate that the investigated ethylene copolymers do easily glide within the galleries of one-tailed organoclays, independent of the application of mechanical stress. The speed of the process, probably because of the smaller d-spacing, is a bit lower than with two-tailed clays. For both types of organoclays, the intercalated structures revealed by XRD are generally similar for the s.a. and m.c. composites, suggesting that they correspond to thermodynamic equilibrium. Exfoliation of the intercalated silicate booklets or aggregates, that is, their shear-driven decomposition into smaller platelets [27], is more difficult and slower for the MMTs modified with one-tailed surfactants. As a result, the morphology of the m.c. composites filled with the latter organoclays was characterized by the presence of many micron-sized, intercalated (according to XRD), silicate particles.

Probably due to the moderate temperatures employed in this work for the preparation and processing of the nanocomposites, the *d*-spacing collapse often reported for the clays with only one long alkyl group was shown to be remarkable only for the commercial 30B organoclay. For the structurally similar D-Eth clay, synthesized in our laboratory, the above effect was almost negligible despite the higher level of unsaturation of the alkyl tails.

References

[1] Q.H. Zeng, A.B. Yu, and G.Q. Lu, *Multiscale modeling and simulation of polymer nanocomposites*, Prog. Polym. Sci. 33 (2) (2008), pp. 191–269. doi:10.1016/j.progpolymsci.2007.09.002.

[2] S.A. Solin, *Kinetics and diffusion in graphite intercalation compounds*, in *Intercalation in Layered Materials*, Vol. 148, NATO ASI Ser. B, Phys., M.S. Dresselhaus, ed., Plenum Press, New York, 1986, pp. 173–183. doi:10.1007/978-1-4757-5556-5_14.

[3] R.A. Vaia, K.D. Jandt, E.J. Kramer, and E.P. Giannelis, *Kinetics of polymer melt intercalation*, Macromolecules 28 (24) (1995), pp. 8080–8085. doi:10.1021/ma00128a016.

[4] E. Manias, H. Chen, R. Krishnamoorti, J. Genzer, E.J. Kramer, and E.P. Giannelis, *Intercalation kinetics of long polymers in 2 nm confinements*, Macromolecules 33 (2000), pp. 7955–7966. doi:10.1021/ma0009552.

[5] J.T. Yoon, W.H. Jo, M.S. Lee, and M.B. Ko, *Effects of comonomers and shear on the melt intercalation of styrenics/clay nanocomposites*, Polymer 42 (1) (2001), pp. 329–336. doi:10.1016/S0032-3861(00)00333-5.

[6] S.W. Kim, W.H. Jo, M.S. Lee, M.B. Ko, and J.Y. Jho, *Effects of shear on melt exfoliation of clay in preparation of nylon 6/organoclay nanocomposites*, Polym. J. 34 (3) (2002), pp. 103–111. doi:10.1295/polymj.34.103.

[7] D. Homminga, B. Goderis, S. Hoffman, H. Reynaers, and G. Groeninckx, *Influence of shear flow on the preparation of polymer layered silicate nanocomposites*, Polymer 46 (23) (2005), pp. 9941–9954. doi:10.1016/j.polymer.2005.07.059.

[8] M. Bousmina, *Study of intercalation and exfoliation processes in polymer nanocomposites*, Macromolecules 39 (12) (2006), pp. 4259–4263. doi:10.1021/ma052647f.

[9] M. Alexandre and P. Dubois, *Polymer-layered silicate nanocomposites: Preparation, properties and uses of a new class of materials*, Mater. Sci. Eng. Reports 28 (1–2) (2000), pp. 1–63.

[10] S.S. Ray and M. Okamoto, *Polymer/layered silicate nanocomposites: A review from preparation to processing*, Prog. Polym. Sci. 28 (11) (2003), pp. 1539–1641. doi:10.1016/j.progpolymsci.2003.08.002.

[11] S. Pavlidou and C.D. Papaspyrides, *A review on polymer-layered silicate nanocomposites*, Prog. Polym. Sci. 33 (12) (2008), pp. 1119–1198. doi:10.1016/j.progpolymsci.2008.07.008.

[12] S. Filippi and P. Magagnini, *Polyolefin nanocomposites by solution blending method*, in *Advances in Polyolefin Nanocomposites*, V. Mittal, ed., Taylor & Francis, Boca Raton, FL, 2011.

[13] E. Passaglia and S. Coiai, *Functional polyolefins for polyolefin/clay nanocomposites*, in *Advances in Polyolefin Nanocomposites*, V. Mittal, ed., Taylor & Francis, Boca Raton, FL, 2011.

[14] K. Chrissopoulou and S.H. Anastasiadis, *Polyolefin/layered silicate nanocomposites with functional compatibilizers*, Eur. Polym. J. 47 (4) (2011), pp. 600–613. doi:10.1016/j.eurpolymj.2010.09.028.

[15] S. Filippi, E. Mameli, C. Marazzato, and P. Magagnini, *Comparison of solution-blending and melt-intercalation for the preparation of poly(ethylene-co-acrylic acid)/organoclay nanocomposites*, Eur. Polym. J. 43 (5) (2007), pp. 1645–1659. doi:10.1016/j.eurpolymj.2007.02.015.

[16] S. Filippi, C. Marazzato, P. Magagnini, A. Famulari, P. Arosio, and S.V. Meille, *Structure and morphology of HDPE-g-MA/organoclay nanocomposites: Effects of the preparation procedures*, Eur. Polym. J. 44 (4) (2008), pp. 987–1002. doi:10.1016/j.eurpolymj.2008.01.011.

[17] Q. Zhang, X. Ma, Y. Wang, and K. Kou, *Morphology and interfacial action of nanocomposites formed from ethylene–vinyl acetate copolymers and organoclays*, J. Phys. Chem. B. 113 (35) (2009), pp. 11898–11905. doi:10.1021/jp903448t.

[18] L. Cui, X. Ma, and D.R. Paul, *Morphology and properties of nanocomposites formed from ethylene–vinyl acetate copolymers and organoclays*, Polymer 48 (21) (2007), pp. 6325–6339. doi:10.1016/j.polymer.2007.08.040.

[19] K.M. Lee and C.D. Han, *Rheology of organoclay nanocomposites: Effects of polymer matrix/organoclay compatibility and the gallery distance of organoclay*, Macromolecules 36 (19) (2003), pp. 7165–7178. doi:10.1021/ma030302w.

[20] X. Li and C.S. Ha, *Nanostructure of EVA/organoclay nanocomposites: Effects of kinds of organoclays and grafting of maleic anhydride onto EVA*, J. Appl. Polym. Sci. 87 (12) (2003), pp. 1901–1909. doi:10.1002/app.11922.

[21] D.S. Chaudhary, R. Prasad, R.K. Gupta, and S.N. Bhattacharya, *Clay intercalation and influence on crystallinity of EVA-based clay nanocomposites*, Thermochim. Acta 433 (1–2) (2005), pp. 187–195.

[22] S. Peeterbroeck, M. Alexandre, and R. Jérôme, P.H. Dubois, *Poly(ethylene-co-vinyl acetate)/clay nanocomposites: Effect of clay nature and organic modifiers on morphology, mechanical and thermal properties*, Polym. Degrad. Stab. 90 (2) (2005), pp. 288–294. doi:10.1016/j.polymdegradstab.2005.03.023.

[23] C.G. Martins, N.M. Larocca, D.R. Paul, and L.A. Pessan, *Nanocomposites formed from polypropylene/EVA blends*, Polymer 50 (7) (2009), pp. 1743–1754. doi:10.1016/j.polymer.2009.01.059.

[24] S. Filippi, M. Paci, F. Baldanzi, D. Rossi, and G. Polacco, *XRD study of intercalation in statically annealed composites of ethylene copolymers and organically modified montmorillonites. 1. Two-tailed organoclays*, J. Taiwan Inst. Chem. Eng. 44 (2013), pp. 123–130. doi:10.1016/j.jtice.2012.09.002.

[25] Y. Tian, H. Yu, S. Wu, G. Ji, and J. Shen, *Study on the structure and properties of EVA/clay nanocomposites*, J. Mater. Sci. 39 (13) (2004), pp. 4301–4303. doi:10.1023/B:JMSC.0000033412.92494.ee.

[26] M. Zanetti, G. Camino, R. Thomann, and R. Mülhaupt, *Synthesis and thermal behaviour of layered silicate–EVA nanocomposites*, Polymer 42 (10) (2001), pp. 4501–4507. doi:10.1016/S0032-3861(00)00775-8.

[27] J.E.F.C. Gardolinski and G. Lagaly, *Grafted organic derivatives of kaolinite: II. Intercalation of primary n-alkylamines and delamination*, Clay Miner. 40 (4) (2005), pp. 547–556. doi:10.1180/0009855054040191.

[28] W. Zhang, D. Chen, Q. Zhao, and Y.Y. Fang, *Effects of different kinds of clay and different vinyl acetate content on the morphology and properties of EVA/clay nanocomposites*, Polymer 44 (26) (2003), pp. 7953–7961. doi:10.1016/j.polymer.2003.10.046.

[29] F. Cser and S.N. Bhattacharya, *Study of the orientation and the degree of exfoliation of nanoparticles in poly(ethylene–vinyl acetate) nanocomposites*, J. Appl. Polym. Sci. 90 (11) (2003), pp. 3026–3031. doi:10.1002/app.13026.

[30] M. Zanetti, T. Kashiwagi, L. Falqui, and G. Camino, *Cone calorimeter combustion and gasification studies of polymer layered silicate nanocomposites*, Chem. Mater. 14 (2) (2002), pp. 881–887. doi:10.1021/cm011236k.

[31] C.H. Jeon, S.H. Ryu, and Y.W. Chang, *Preparation and characterization of ethylene vinyl acetate copolymer/montmorillonite nanocomposite*, Polym. Int. 52 (2003), pp. 153–157. doi:10.1002/pi.1066.

[32] S. Duquesne, C. Jama, M. Le Bras, R. Delobel, P. Recourt, and J.M. Gloaguen, *Elaboration of EVA–nanoclay systems – characterization, thermal behaviour and fire performance*, Comp. Sci. Technol. 63 (2003), pp. 1141–1148. doi:10.1016/S0266-3538(03)00035-6.

[33] M.C. Costache, D.D. Jiang, and C.A. Wilkie, *Thermal degradation of ethylene–vinyl acetate coplymer nanocomposites*, Polymer 46 (2005), pp. 6947–6958. doi:10.1016/j.polymer.2005.05.084.

[34] K.-N. Kim, H. Kim, and J.-W. Lee, *Effect of interlayer structure, matrix viscosity and composition of a functionaiized polymer on the phase structure of polypropylene–montmorillonite nanocomposites*, Polym. Eng. Sci. 41 (2001), pp. 1963–1969. doi:10.1002/pen.10892.

[35] C.I. Park, M.H. Kim, and O.O. Park, *Effect of heat treatment on the microstructural change of syndiotactic polystyrene/poly(styrene-co-vinyloxazolin)/clay nanocomposite*, Polymer 45 (2004), pp. 1267–1273. doi:10.1016/j.polymer.2003.12.011.

[36] B.N. Jang, D. Wang, and C.A. Wilkie, *Relationship between the solubility parameter of polymers and the clay dispersion in polymer/clay nanocomposites and the role of the surfactant*, Macromolecules 38 (2005), pp. 6533–6543. doi:10.1021/ma0508909.

[37] F. Perrin-Sarazin, M.-T. Ton-That, M.N. Bureau, and J. Denault, *Micro- and nano-structure in polypropylene/clay nanocomposites*, Polymer 46 (2005), pp. 11624–11634. doi:10.1016/j.polymer.2005.09.076.

[38] C.O. Rohlmann, M.F. Horst, L.M. Quinzani, and M.D. Failla, *Comparative analysis of nanocomposites based on polypropylene and different montmorillonites*, Eur. Polym. J. 44 (2008), pp. 2749–2760. doi:10.1016/j.eurpolymj.2008.07.006.

[39] S. Benali, S. Peeterbroeck, J. Larrieu, F. Laffineur, J.-J. Pireaux, M. Alexandre, and P. Dubois, *Study of interlayer spacing collapse during polymer/clay nanocomposite melt intercalation*, J. Nanosci. Nanotech. 8 (2008), pp. 1707–1713. doi:10.1166/jnn.2008.020.
[40] T.D. Fornes, P.J. Yoon, and D.R. Paul, *Polymer matrix degradation and color formation in melt-processed nylon 6/clay nanocomposites*, Polymer 44 (2003), pp. 7545–7556. doi:10.1016/j.polymer.2003.09.034.
[41] G. Edwards, P. Halley, G. Kerven, and D. Martin, *Thermal stability analysis of organo-silicates, using solid phase microextraction techniques*, Thermochim. Acta 429 (2005), pp. 13–18. doi:10.1016/j.tca.2004.11.020.
[42] A. Montagnani, *Preparazione e caratterizzazione morfologica di nanocompositi a base di polietileni funzionalizzati*, Thesis, University of Pisa, 2010.

13

Mechanical and thermo-mechanical properties of carbon nanotube reinforced composites

Chensong Dong*

Department of Mechanical Engineering, Curtin University, GPO Box 1987, Perth, WA 6845, Australia

A study on the mechanical and thermo-mechanical properties of carbon nanotube (CNT) reinforced nanocomposites is presented in this article. Mori–Tanaka method is used for modeling the effective stiffness and coefficient of thermal expansion. Regression formulas were developed to describe the effects of CNT orientation, aspect ratio, and CNT volume fraction. Given the statistical distributions of CNT orientations and aspect ratios, the effective properties can be conveniently derived by numerical integration using these formulas.

Keywords: carbon nanotube (CNT); stiffness; coefficient of thermal expansion (CTE)

Nomenclature

a	Aspect ratio
$\mathbf{A}$	Concentration factor relating to the average strain in the effective reinforcement to that of the unknown effective material in which it is embedded
$\mathbf{C}$	Stiffness tensor of composites in the local coordinate system
$\overline{\mathbf{C}}$	Stiffness tensor of composites in the global coordinate system
C_{2D}	Stiffness of 2D random composites
C_{3D}	Stiffness of 3D random composites
$\mathbf{C_f}$	Stiffness tensor of fibers
$\mathbf{C_m}$	Stiffness tensor of matrix
$l_1, l_2, l_3, m_1, m_2, m_3, n_1, n_2, n_3$	Directional cosines
$\mathbf{S}$	Eshelby's tensor
$\mathbf{T}$	Transformation matrix for stress
$\mathbf{T}_\varepsilon$	Transformation matrix for strain
V_f	Fiber volume fraction
$\boldsymbol{\alpha}$	CTE in the local coordinate system
$\overline{\boldsymbol{\alpha}}$	CTE in the global coordinate systems
α_{2D}	CTE of 2D random composites
α_{3D}	CTE of 3D random composites
$\boldsymbol{\alpha}_\mathbf{f}$	CTE of fibers
$\boldsymbol{\alpha}_\mathbf{m}$	CTE of matrix
$\boldsymbol{\varepsilon}$	Strain tensors in the local coordinate system

*Email: c.dong@curtin.edu.au

$\bar{\varepsilon}$	Strain tensors in the global coordinate system
Φ, θ, β	Euler angles
$\boldsymbol{\sigma}$	Stress tensors in the local coordinate system
$\bar{\boldsymbol{\sigma}}$	Stress tensors in the global coordinate system

1. Introduction

Since the discovery of multi-walled carbon nanotubes in 1991 by Iijima [1] and subsequent synthesis of single-walled carbon nanotubes (SWCNTs) [2,3], numerous experimental and theoretical studies have been carried out to investigate the electronic, chemical, and mechanical properties of carbon nanotubes (CNTs). SWCNT–polymer composites are theoretically predicted to have both exceptional mechanical and special functional properties that carbon fiber–polymer composites cannot offer [4].

The magnitude of the coefficient of thermal expansion (CTE) depends on the structure of the materials. For single-phase materials, CTE is determined from atomic bonding, molecular structure, and molecular assembly. An elevated temperature would increase thermal energy and lead to increasing atomic movement. Weak atomic bonding with a low bonding energy would show a large CTE owing to an increasing interatomic distance. For multi-phase materials, such as composites, the CTE is dependent on each component phase and also the interactions between each phase. Weak interface bonding between phases could not effectively incorporate the contributions of each component while strong interface bonding could compromise each ingredient for thermal-expansion properties.

Molecular dynamics (MD) predictions [5] have shown that the axial CTEs of CNTs are negative in a wide low-temperature range, and vary nonlinearly with the temperature. These axial CTEs may become positive as the temperature increases. This indicates that the CTE could be significantly reduced by adding CNTs into a polymer matrix, provided good interfacial bonding is achieved.

Wei *et al.* [6] investigated the CTE of SWCNT–polyethylene composites through MD simulation shows that the CTE increases with nanotube loading, which is attributed to phonon modes and Brownian motions. Guo *et al.* [7] studied the properties of polyacrylonitrile/SWCNT composite films and the CTE of composite was 1.7×10^{-6} °C^{-1} at a weight loading of 40%. Xu *et al.* [8] found that the CTE of SWCNT/poly(vinylidene fluoride) composites decreased with increased SWCNT content. Wang *et al.* [9] studied the CTEs of nanocomposites reinforced by functionalized CNTs, and found that a reduction of 52% is observed below the glass transition temperature (T_g). However, the CTE above T_g increases significantly due to the contribution of phonon modes and Brownian motions. These studies show inconsistent results, which can be attributed to two competing mechanisms: (1) the high stiffness and low CTE will restrain the expansion of the matrix, causing the decrease in the CTE; (2) the phonon modes and Brownian motions increase the CTE. The resulting CTE is a combined effect of these two mechanisms, and highly dependent on the interfacial bonding.

Extensive investigations into the preparation and characterization of SWCNT–polymer composites have been reported [10,11]. However, the properties of the composites are not as expected because of poor dispersion and weak interfacial bonding. One possible solution to acquire high-performance nanocomposites is functionalization. Some recent experimental results indicate that the mechanical properties of SWCNT–polymer composites are significantly enhanced through functionalization [9,12,13], which demonstrates potential applications in structural and multifunctional materials.

Halpin–Tsai equation [10,14], Mori–Tanaka method [15–17], and finite element analysis [18] have been used for predicting the properties of composites. In this study, the stiffness and CTE of CNT reinforced polymer nanocomposites in various configurations, i.e., unidirectional, 2D random, and 3D random, are studied. The effects of CNT orientation, aspect ratio, and CNT volume content are investigated using Mori–Tanaka method. The distribution of CNT is characterized statistically and the corresponding stiffness and CTE are predicted. The results provide a practical insight into the reinforcing effects of CNT for polymers.

2. Approach

2.1. Mori-Tanaka method

The elastic and thermo-elastic properties of CNT composites are modeled using the method proposed by Taya *et al.* [19,20] based on Eshelby's inclusion theory [21] and Mori–Tanaka's mean field theory [15]. It is proved by Seidel and Lagoudas [16] that the Mori–Tanaka method provides accurate results. Using the Mori–Tanaka method, the effective stiffness is given by

$$\mathbf{C} = [(1 - V_\mathrm{f})\mathbf{C_m} + V_\mathrm{f}\mathbf{C_f A}][(1 - V_\mathrm{f})\mathbf{I} + V_\mathrm{f}\mathbf{A}]^{-1} \tag{1}$$

$$\mathbf{A} = \left[\mathbf{I} + \mathbf{S}\mathbf{C_m^{-1}}(\mathbf{C_f} - \mathbf{C_m})\right]^{-1} \tag{2}$$

For a general ellipsoidal inclusion, the components of Eshelby's tensor S_{ijkl} are dependent on the aspect ratio, and their expressions can be found in the literature [18]. For a fiber-like inclusion, e.g., CNT ($x_1 = x_2 < x_3$), the aspect ratio is given by $a = x_3/x_1 = l/d$, as shown in Figure 1.

In the stiffness tensor, the longitudinal and transverse stiffness components are C_{33} and C_{11}, respectively. For the purpose of comparison, the normalized longitudinal and transverse stiffness components are defined as

$$C_\mathrm{L}^* = C_{33}/C_\mathrm{m} \tag{3}$$

$$C_\mathrm{T}^* = C_{11}/C_\mathrm{m} \tag{4}$$

The effective CTE of the composite is given by

$$\boldsymbol{\alpha} = \boldsymbol{\alpha}_\mathbf{m} + V_\mathrm{f}\left\{\mathbf{C_m} + (\mathbf{C_f} - \mathbf{C_m})\left[(1 - V_\mathrm{f})\mathbf{S} + V_f\mathbf{I}\right]\right\}^{-1}\mathbf{C_f}(\boldsymbol{\alpha}_\mathbf{f} - \boldsymbol{\alpha}_\mathbf{m}) \tag{5}$$

The longitudinal and transverse CTEs are α_{33} and α_{11}, respectively. Similarly, the normalized longitudinal and transverse CTEs are defined as

$$\alpha_\mathrm{L}^* = \alpha_{33}/\alpha_\mathrm{m} \tag{6}$$

$$\alpha_\mathrm{T}^* = \alpha_{11}/\alpha_\mathrm{m} \tag{7}$$

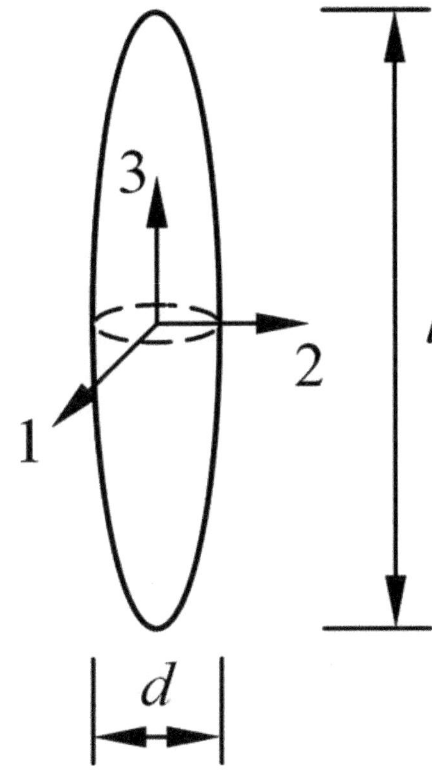

Figure 1. A fiber-like inclusion.

2.2. Transformation

When nanotubes are oriented at an arbitrary angle, the effective stiffness and CTE need to be transformed to the global coordinate system. This can be conveniently done using Euler angles, ϕ, θ, and β, which are referred to, respectively, as precession, nutation, and spin, as shown in Figure 2. For unidirectional composites, $\phi = \theta = \beta = 0$; for 2D random

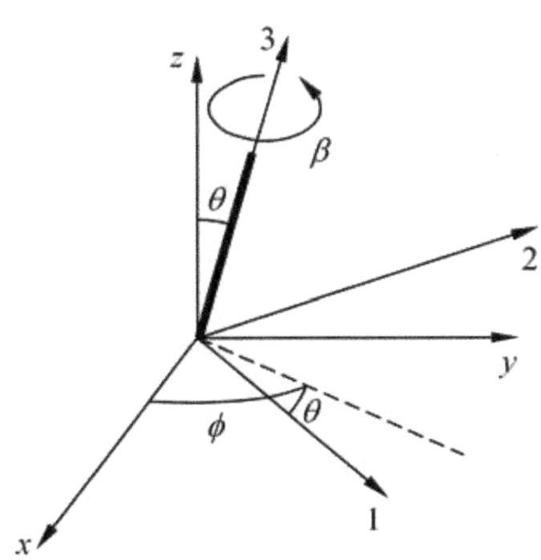

Figure 2. Euler angles.

composites, $\phi = \beta = 0$ and $0 \le \theta \le \pi$; and for 3D random composites, $\beta = 0,\ 0 \le \theta, \phi \le \pi$.

From the Euler angles, the directional cosines can be obtained, i.e.

$$\begin{array}{lll} l_1 = \cos\phi\cos\theta & m_1 = \sin\phi\cos\theta & n_1 = -\sin\theta \\ l_2 = -\sin\phi & m_2 = \cos\phi & n_2 = 0 \\ l_3 = \cos\phi\sin\theta & m_3 = \sin\phi\sin\theta & n_3 = \cos\theta \end{array} \tag{8}$$

The transformation matrix is given by

$$\mathbf{T} = \begin{bmatrix} l_1^2 & m_1^2 & n_1^2 & 2m_1n_1 & 2l_1n_1 & 2l_1m_1 \\ l_2^2 & m_2^2 & n_2^2 & 2m_2n_2 & 2l_2n_2 & 2l_2m_2 \\ l_3^2 & m_3^2 & n_3^2 & 2m_3n_3 & 2l_3n_3 & 2l_3m_3 \\ l_2l_3 & m_2m_3 & n_2n_3 & m_2n_3 + m_3n_2 & l_2n_3 + l_3n_2 & l_2m_3 + l_3m_2 \\ l_1l_3 & m_1m_3 & n_1n_3 & m_1n_3 + m_3n_1 & l_1n_3 + l_3n_1 & l_1m_3 + l_3m_1 \\ l_1l_2 & m_1m_2 & n_1n_2 & m_1n_2 + m_2n_1 & l_1n_2 + l_2n_1 & l_1m_2 + l_2m_1 \end{bmatrix} \tag{9}$$

The transformation of stress tensors is given with respect to Figure 3 by

$$\boldsymbol{\sigma} = \mathbf{T}\overline{\boldsymbol{\sigma}} \tag{10}$$

where $\overline{\boldsymbol{\sigma}}$ and $\boldsymbol{\sigma}$ are the stress tensors in the global (*oxy*) and local (*ox'y'*) coordinate systems, respectively.

The transformation of strain vectors is given by

$$\boldsymbol{\varepsilon} = \mathbf{T}^{\mathrm{T}}\overline{\boldsymbol{\varepsilon}} \tag{11}$$

where $\overline{\boldsymbol{\varepsilon}}$ and $\boldsymbol{\varepsilon}$ are the strain tensors in the global (*oxy*) and local (*ox'y'*) coordinate systems, respectively.

Therefore, the stiffness tensor in the global coordinate system is given by

$$\overline{\mathbf{C}} = \mathbf{T}^{-1}\mathbf{C}\left(\mathbf{T}^{-1}\right)^{\mathrm{T}} \tag{12}$$

The normalized effective stiffness as a function of orientation angle is defined as

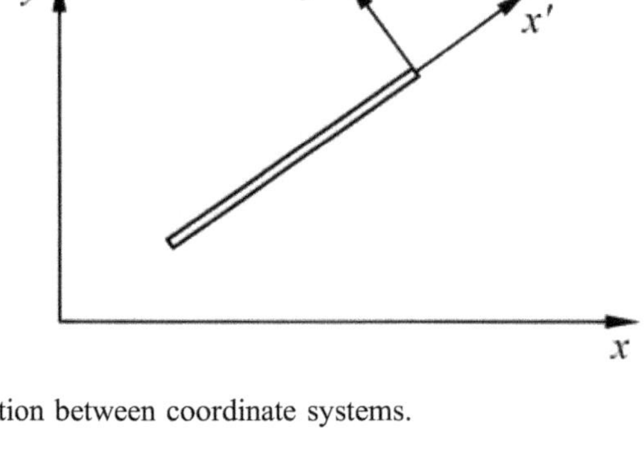

Figure 3. Transformation between coordinate systems.

$$\bar{C}^*(\theta) = \bar{C}_{33}/C_\mathrm{m} \tag{13}$$

The CTE in the global coordinate system is given by

$$\bar{\boldsymbol{\alpha}} = \mathbf{T}^\mathrm{T}\boldsymbol{\alpha} \tag{14}$$

where $\bar{\boldsymbol{\alpha}}$ and $\boldsymbol{\alpha}$ are the CTE in the global (*oxy*) and local (*ox'y'*) coordinate systems, respectively.

The normalized effective CTE as a function of orientation angle is defined as

$$\bar{\alpha}^*(\theta) = \bar{\alpha}_{33}/\alpha_\mathrm{m} \tag{15}$$

In addition to stiffness and CTE, Eshelby's tensors also need to be transformed in such a way that

$$\bar{\mathbf{S}} = \mathbf{T}^\mathrm{T}\mathbf{S}\mathbf{T} \tag{16}$$

where $\bar{\mathbf{S}}$ is the transformed Eshelby's tensor.

2.3. *Effects of parameters*

The effects of CNT orientation, CNT volume fraction on the effective properties of unidirectional, 2D random, and 3D random composites were studied. Based on the data from the literature, a SWCNT loading of 1% by weight corresponds to approximately 0.743% by volume.

In this study, the effective longitudinal and transverse moduli of CNT at the room temperature were taken to be 704 GPa and 346 GPa, respectively [16]. The longitudinal and transverse CTEs of SWCNTs at the room temperature were estimated to be -12×10^{-6} °C^{-1} and $-1.5 \pm 2 \times 10^{-6}$ °C^{-1}, respectively [22–24], which were further validated using X-ray measurements [25]. The matrix was epoxy. Its modulus and CTE are 3 GPa and 6.4×10^{-5} °C^{-1}, respectively.

There are numerous examples in the literature, which suggest that above a critical concentration of CNTs, the properties of either drop or the reinforcement efficiency decreases considerably [26,27]. Thus, in this study, only CNT loadings up to 5% (v/v) were considered.

3. Results

3.1. *Effects of orientation*

When the aspect ratio is 500 and the CNT volume fraction is 1%, the normalized effective stiffness vs. the orientation angle, θ, is shown in Figure 4. The longitudinal direction of the CNT is defined to be $\theta = 0$. From Equation (12), it can be derived that the stiffness is a function of $\cos^4\theta$. The relationship between the normalized stiffness and CNT orientation angle is given by

$$\bar{C}^*(\theta) = C_\mathrm{T}^* + \left(C_\mathrm{L}^* - C_\mathrm{T}^*\right)\cos^4\theta \tag{17}$$

Likewise, the normalized effective CTE vs. the orientation angle is shown in Figure 5. Similarly, the longitudinal direction of the CNT is defined to be $\theta = 0$. From Equation (14),

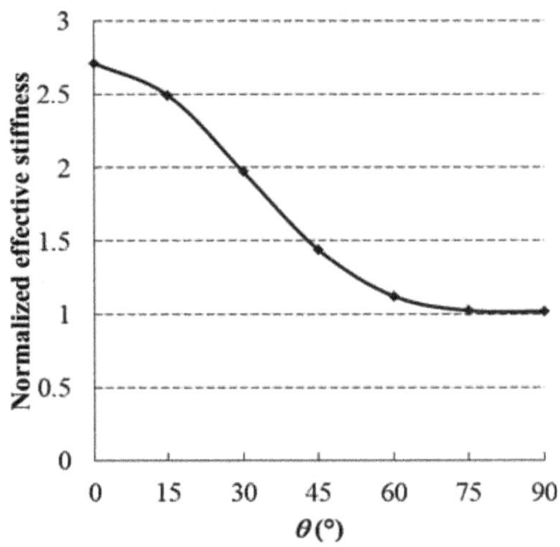

Figure 4. Normalized effective stiffness vs. CNT orientation for aspect ratio 500 and CNT volume fraction 1%.

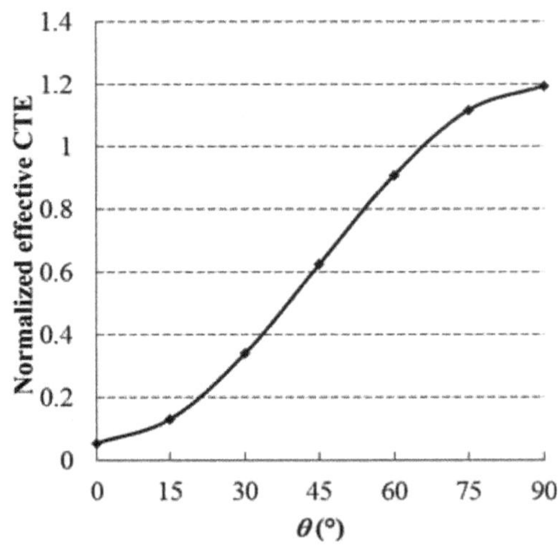

Figure 5. Normalized effective CTE vs. CNT orientation for aspect ratio 500 and CNT volume fraction 1%.

it can be derived that the effective CTE is a function of $\sin^2\theta$. The relationship between the normalized CTE and CNT orientation angle is given by

$$\bar{\alpha}^*(\theta) = \alpha_{\mathrm{L}}^* + \left(\alpha_{\mathrm{T}}^* - \alpha_{\mathrm{L}}^*\right)\sin^2\theta \qquad (18)$$

3.2. *Orientation average*

The effective modulus of a 2D random composite is given by taking the orientation average, i.e.

$$C_{2D}^{*} = \frac{\int_{0}^{\frac{\pi}{2}} \bar{C}^{*}(\theta)\,\mathrm{d}\theta}{\int_{0}^{\frac{\pi}{2}} \mathrm{d}\theta} = \frac{2}{\pi}\int_{0}^{\frac{\pi}{2}} \bar{C}^{*}(\theta)\,\mathrm{d}\theta \tag{19}$$

Substituting Equation (17) into Equation (19) yields

$$C_{2D}^{*} = \frac{3}{8}C_{L}^{*} + \frac{5}{8}C_{T}^{*} \tag{20}$$

The effective CTE of a 2D random composite is given by

$$\alpha_{2D}^{*} = \frac{\int_{0}^{\frac{\pi}{2}} \bar{C}^{*}(\theta)\bar{\alpha}^{*}(\theta)\,\mathrm{d}\theta}{\int_{0}^{\frac{\pi}{2}} \bar{C}^{*}(\theta)\,\mathrm{d}\theta} \tag{21}$$

Substituting Equations (17) and (18) into Equation (21) yields

$$\alpha_{2D}^{*} = \frac{(5C_{L}^{*} + 3C_{T}^{*})\alpha_{L}^{*} + (C_{L}^{*} + 7C_{T}^{*})\alpha_{T}^{*}}{2(3C_{L}^{*} + 5C_{T}^{*})} \tag{22}$$

Likewise, the effective stiffness of a 3D random composite is given by

$$C_{3D}^{*} = \frac{\int_{0}^{\frac{\pi}{2}} \bar{C}^{*}(\theta)\sin\theta\,\mathrm{d}\theta}{\int_{0}^{\frac{\pi}{2}} \sin\theta\,\mathrm{d}\theta} = \int_{0}^{\frac{\pi}{2}} \bar{C}^{*}(\theta)\sin\theta\,\mathrm{d}\theta \tag{23}$$

Substituting Equation (17) into Equation (23) yields

$$C_{3D}^{*} = \frac{1}{5}C_{L}^{*} + \frac{4}{5}C_{T}^{*} \tag{24}$$

Likewise, the effective CTE of a 3D random composite is given by

$$\alpha_{3D}^{*} = \frac{\int_{0}^{\frac{\pi}{2}} \bar{C}^{*}(\theta)\bar{\alpha}^{*}(\theta)\sin\theta\,\mathrm{d}\theta}{\int_{0}^{\frac{\pi}{2}} \bar{C}^{*}(\theta)\sin\theta\,\mathrm{d}\theta} \tag{25}$$

Substituting Equations (17) and (18) into Equation (25) yields

$$\alpha_{3D}^{*} = \frac{(15C_{L}^{*} + 20C_{T}^{*})\alpha_{L}^{*} + (6C_{L}^{*} + 64C_{T}^{*})\alpha_{T}^{*}}{21(C_{L}^{*} + 4C_{T}^{*})} \tag{26}$$

3.3. *Aspect ratio*

The effective longitudinal and transverse stiffness vs. the aspect ratio for the CNT reinforced composites at various CNT volume fractions from 1% to 5% is plotted in Figure 6. It is shown that the longitudinal stiffness increases with the aspect ratio while little change is seen for the transverse stiffness. When the aspect ratio is greater than 100, the change in stiffness becomes less.

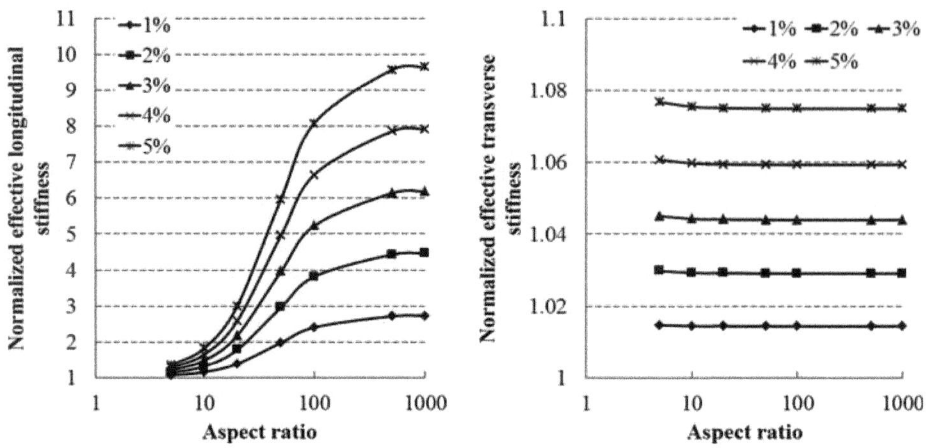

Figure 6. Normalized effective stiffness vs. aspect ratio.

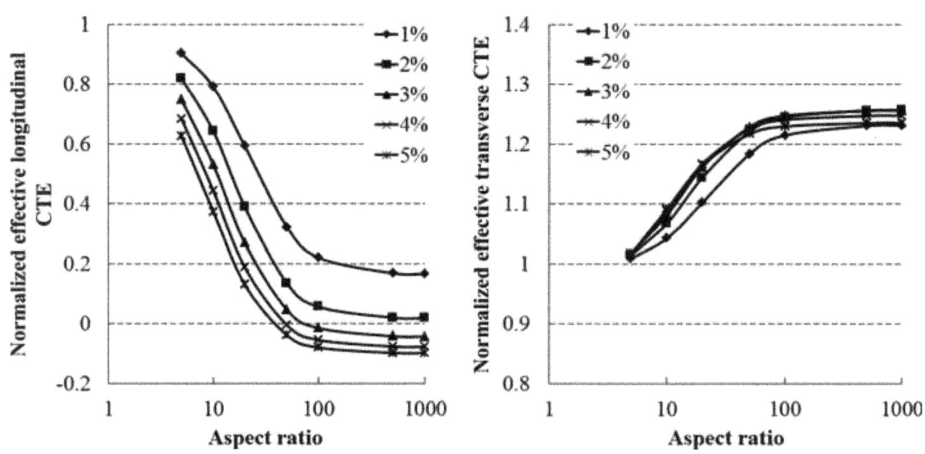

Figure 7. Normalized effective CTE vs. aspect ratio.

The normalized effective longitudinal and transverse CTEs vs. the aspect ratio for the CNT reinforced composites at various CNT volume fractions from 1% to 5% is plotted in Figure 7. It is shown that the longitudinal CTE decreases with the increasing aspect ratio while the transverse CTE increases with the decreasing aspect ratio. When the aspect ratio is greater than 100, the change in CTE becomes less.

3.4. CNT volume fraction

The normalized effective stiffness of the CNT composites vs. the CNT volume fraction is shown in Figure 8 for two aspect ratios, $a = 10$ and $a = 1000$, respectively. For both aspect ratios, when the CNT volume fraction is low (up to 5%), both the longitudinal stiffness and transverse stiffness of the CNT composites approximately linearly increases with the CNT volume fraction. The effective stiffness can be expressed by the following regression formulas.

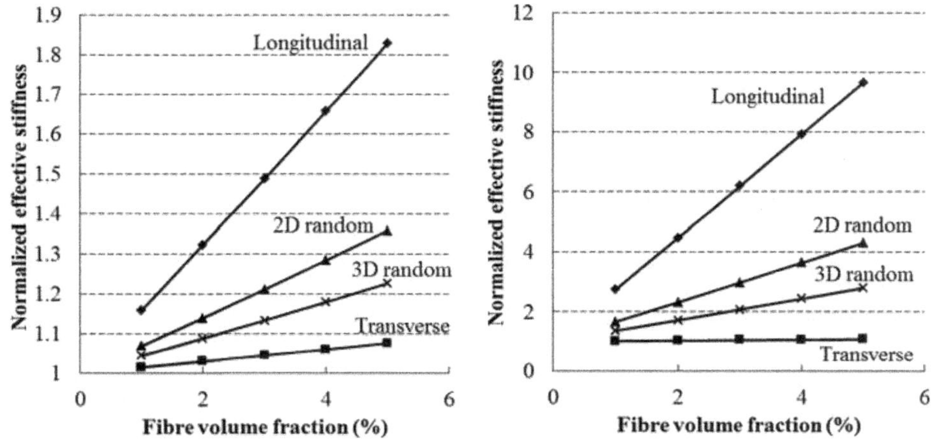

Figure 8. Normalized effective stiffness vs. CNT volume fraction for a = 10 (left) and a = 1000 (right).

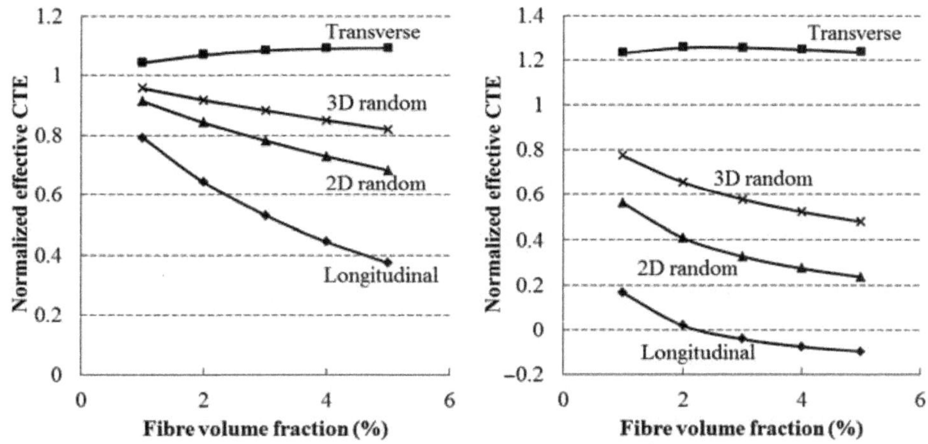

Figure 9. Normalized effective CTE vs. CNT volume fraction for a = 10 (left) and a = 1000 (right).

Likewise, the normalized effective CTE of the CNT composites vs. the CNT volume fraction is shown in Figure 9 for two aspect ratios, $a = 10$ and $a = 1000$, respectively. For both aspect ratios, the longitudinal CTE decreases with the increasing CNT volume fraction while the transverse CTE shows little change with the CNT volume fraction.

3.5. Regression formulas

It is shown in Figure 6 that the longitudinal stiffness vs. the aspect ratio shows a dose–response relationship, i.e.

$$C_{\mathrm{L}}^{*}(a) = c_1 + \frac{c_2 - c_1}{1 + \left(\frac{x_0}{a}\right)^{p}} \tag{27}$$

where c_1, c_2, x_0, and p are constants. It is shown from the data, as shown in Figure 6, that c_1 and c_2 are functions of V_f. The final regression formula for the longitudinal stiffness is then given by

$$C_L^*(V_f, a) = 1 + 0.03V_f + \frac{1.71V_f}{1 + \left(\frac{43.65}{a}\right)^{1.7}} \tag{28}$$

The effective transverse stiffness is only dependent on the CNT volume fraction, and the corresponding regression formula is

$$C_T^*(V_f) = 1 + 0.015V_f \tag{29}$$

In a similar way, the longitudinal and transverse CTEs can also be expressed by dose–response formulas, which are given by

$$\alpha_L^*(V_f, a) = -0.066 + 0.35V_f^{-0.66} + \frac{1.066 - 0.023V_f - 0.35V_f^{-0.66}}{1 + \left(\frac{a}{-6.52 + 28.03V_f^{-0.34}}\right)^{1.65}} \tag{30}$$

$$\alpha_T^*(V_f, a) = 1 + \frac{0.2}{1 + \left(\frac{14.29 + 9.81V_f^{-2.5}}{a}\right)^{1.65}} \tag{31}$$

4. Effects of statistical distributions

When there are CNTs of various aspect ratios in the range between a_1 and a_2, the normalized effective stiffness is derived by taking the aspect ratio average, which is given by

$$C_L^* = \int_{a_1}^{a_2} C_L^*(a) f(a)\, da \tag{32}$$

Likewise, the normalized effective CTE is given by

$$\alpha_L^* = \frac{\int_{a_1}^{a_2} \alpha_L^*(a) C_L^*(a) f(a)\, da}{\int_{a_1}^{a_2} C_L^*(a) f(a)\, da} \tag{33}$$

When the aspect ratios are uniformly distributed, the normalized effective longitudinal stiffness is derived by substituting Equation (28) into Equation (34), which is given by

$$C_L^* = 1 + 0.03V_f + \frac{1.71V_f}{a_2 - a_1} \left\{ \begin{array}{l} a_2 \left[{}_2F_1\left(1, -0.59; 0.41; -\left(\frac{43.65}{a_2}\right)^{1.7}\right)\right] \\ -a_1 \left[{}_2F_1\left(1, -0.59; 0.41; -\left(\frac{43.65}{a_1}\right)^{1.7}\right)\right] \end{array} \right\} \tag{34}$$

where ${}_2F_1(\underline{a}_1, \underline{a}_2; \underline{b}_1; z) = \sum_{n=0}^{\infty} \frac{\underline{a}_1^{(n)} \underline{a}_2^{(n)}}{\underline{b}_1^{(n)}} \frac{z^n}{n!}$ is the generalized hypergeometric function, and $x^{(n)} = x(x+1)(x+2)\cdots(x+n-1), \quad n \geq 1$ is the rising factorial or Pochhammer symbol.

For any other distributions, the normalized effective longitudinal stiffness can be derived via numerical integration.

For the normalized effective longitudinal and transverse CTEs, a closed form solution cannot be found, and numerical integration is needed. The effective CTE of 2D and 3D random composites can be obtained by substituting the longitudinal and transverse CTEs into Equations (22) and (26), respectively.

4.1. *Uniform distributions*

The stiffness and CTE for a number of uniform distributions of aspect ratios are shown in Table 1. It is shown that when the aspect ratio is greater than 100, both the stiffness and CTE are approximately constant.

4.2. *Normal distributions*

The stiffness and CTE for a number of normal distributions of aspect ratios are shown in Table 2. It is shown that if the aspect ratios are assumed to be normally distributed, the stiffness and CTE are mainly dependent on the mean.

4.3. *Weibull distributions*

The stiffness and CTE for a number of Weibull distributions of aspect ratios are shown in Table 3. It is shown that the stiffness and CTE are mainly dependent on the scale

Table 1. Stiffness and CTE for uniform distributions.

Range	C_L^*	α_L^*	α_T^*
5–50	1.5504	0.5750	1.1009
5–100	1.9102	0.4463	1.1383
50–100	2.6165	0.3661	1.1717
5–1000	2.2340	0.3045	1.1883
100–500	2.6484	0.2970	1.1954
100–1000	2.6911	0.2910	1.1975
500–1000	2.7253	0.2861	1.1993

Table 2. Stiffness and CTE for normal distributions.

Mean	Standard deviation	C_L^*	α_L^*	α_T^*
500	200	2.7003	0.2896	1.1980
500	100	2.7113	0.2881	1.1986
500	50	2.7129	0.2879	1.1986
150	50	2.5186	0.3163	1.1886
400	50	2.7003	0.2897	1.1980
600	50	2.7201	0.2869	1.1990
750	50	2.7264	0.2860	1.1993

Table 3. Stiffness and CTE for Weibull distributions.

Shape parameter	Scale parameter	C_L^*	α_L^*	α_T^*
1.5	500	2.6391	0.2987	1.1934
2	500	2.6708	0.2938	1.1964
2.5	500	2.6829	0.2920	1.1971
3	500	2.6914	0.2909	1.1976
1.5	200	2.5107	0.3194	1.1825
1.5	300	2.5791	0.3080	1.1887
1.5	400	2.6171	0.3020	1.1917
3	200	2.5873	0.3059	1.1924
3	300	2.6101	0.3025	1.1935
3	400	2.6679	0.2942	1.1964

parameter. When the scale parameter increases, the longitudinal stiffness increases but the longitudinal CTE decreases.

If the aspect ratio a and orientation angle θ are independent of each other, the effective stiffness of 2D and 3D random composites can be obtained by substituting Equations (29) and (36) into Equations (20) and (24), respectively.

5. Conclusions

The stiffness and CTE of CNT reinforced nanocomposites were studied with the aid of Mori–Tanaka method. The effects of CNT volume fraction, aspect ratio, and orientation were investigated. It is shown that the effective stiffness linearly increases with CNT volume fraction from 1% to 5%. The longitudinal CTE decreases with CNT volume fraction while the transverse CTE shows little change with CNT volume fraction. The longitudinal stiffness increases with the aspect ratio while little change is seen for the transverse stiffness. When the aspect ratio is greater than 100, the change in stiffness becomes less. The longitudinal CTE decreases with the increasing aspect ratio while the transverse CTE increases with the aspect ratio. When the aspect ratio is greater than 100, the change in CTE becomes less.

Regression formulas were developed to relate the stiffness and CTE to CNT volume fraction, aspect ratio, and orientation. Using these formulas, if the distribution of CNT is known, the stiffness and CTE can be conveniently obtained by numerical integration. A number of uniform, normal, and Weibull distributions were studied.

References

[1] S. Iijima, *Helical microtubules of graphitic carbon*, Nature 354 (1991), pp. 56–58. doi:10.1038/354056a0

[2] S. Iijima and T. Ichihashi, *Single-shell carbon nanotubes of 1-nm diameter*, Nature 363 (1993), pp. 603–605. doi:10.1038/363603a0

[3] D.S. Bethune, C.H. Klang, M.S. de Vries, G. Gorman, R. Savoy, J. Vazquez, and R. Beyers, *Cobalt-catalysed growth of carbon nanotubes with single-atomic-layer walls*, Nature 363 (1993), pp. 605–607. doi:10.1038/363605a0

[4] E.T. Thostenson and T.-W. Chou, *On the elastic properties of carbon nanotube-based composites: Modelling and characterization*, J. Phys. D: Appl. Phys. 36 (5) (2003), p. 573–582. doi:10.1088/0022-3727/36/5/323

[5] A. Alamusi, N. Hu, B. Jia, M. Arai, C. Yan, J. Li, Y. Liu, S. Atobe, and H. Fukunaga, *Prediction of thermal expansion properties of carbon nanotubes using molecular dynamics simulations*, Comput. Mater. Sci. 54 (2012), pp. 249–254. doi:10.1016/j.commatsci.2011.10.015

[6] C. Wei, D. Srivastava, and K. Cho, *Thermal expansion and diffusion coefficients of carbon nanotube–polymer composites*, Nano Lett. 2 (6) (2002), pp. 647–650. doi:10.1021/nl025554+

[7] H. Guo, T.V. Sreekumar, T. Liu, M. Minus, and S. Kumar, *Structure and properties of polyacrylonitrile/single wall carbon nanotube composite films*, Polymer 46 (9) (2005), pp. 3001–3005. doi:10.1016/j.polymer.2005.02.013

[8] Y. Xu, G. Ray, and B. Abdel-Magid, *Thermal behavior of single-walled carbon nanotube polymer–matrix composites*, Comp. Part A Appl. Sci. Manufact. 37 (1) (2006), pp. 114–121. doi:10.1016/j.compositesa.2005.04.009

[9] S. Wang, Z. Liang, P. Gonnet, Y. Liao, B. Wang, and C. Zhang, *Effect of nanotube functionalization on the coefficient of thermal expansion of nanocomposites*, Adv Funct. Mater. 17 (1) (2007), pp. 87–92. doi:10.1002/adfm.200600760

[10] M.T. Byrne and Y.K. Gun'ko, *Recent advances in research on carbon nanotube–polymer composites*, Adv. Mater. 22 (15) (2010), pp. 1672–1688. doi:10.1002/adma.200901545

[11] N.G. Sahoo, S. Rana, J.W. Cho, L. Li, and S.H. Chan, *Polymer nanocomposites based on functionalized carbon nanotubes*, Prog. Polym. Sci. 35 (7) (2010), pp. 837–867. doi:10.1016/j.progpolymsci.2010.03.002

[12] S. Wang, Z. Liang, T. Liu, B. Wang, and C. Zhang, *Effective amino-functionalization of carbon nanotubes for reinforcing epoxy polymer composites*, Nanotechnology 17 (6) (2006), pp. 1551–1557. doi:10.1088/0957-4484/17/6/003

[13] J. Zhu, J. Kim, H. Peng, J.L. Margrave, V.N. Khabashesku, and E.V. Barrera, *Improving the dispersion and integration of single-walled carbon nanotubes in epoxy composites through functionalization*, Nano Lett. 3 (8) (2003), pp. 1107–1113. doi:10.1021/nl0342489

[14] J.C. Halpin and N.J. Pagano, *The laminate approximation for randomly oriented fibrous composites*, J. Comp. Mater. 3 (4) (1969), pp. 720–724.

[15] T. Mori and K. Tanaka, *Average stress in matrix and average elastic energy of materials with misfitting inclusions*, Acta Metal. 21 (5) (1973), pp. 571–574. doi:10.1016/0001-6160(73)90064-3

[16] G.D. Seidel and D.C. Lagoudas, *Micromechanical analysis of the effective elastic properties of carbon nanotube reinforced composites*, Mech. Mater. 38 (8–10) (2006), pp. 884–907.

[17] V. Anumandla and R.F. Gibson, *A comprehensive closed form micromechanics model for estimating the elastic modulus of nanotube-reinforced composites*, Comp. Part A Appl. Sci. Manufact. 37 (12) (2006), pp. 2178–2185. doi:10.1016/j.compositesa.2005.09.016

[18] G.P. Tandon and G.J. Weng, *The effect of aspect ratio of inclusions on the elastic properties of unidirectionally aligned composites*, Polym. Comp. 5 (4) (1984), pp. 327–333. doi:10.1002/pc.750050413

[19] Y. Takao and M. Taya, *Thermal expansion coefficients and thermal stresses in an aligned short fiber composite with application to a short carbon fiber/aluminum*, ASME Trans. J. Appl. Mech. 52 (1985), pp. 806–810. doi:10.1115/1.3169150

[20] M. Taya and T.W. Chou, *On two kinds of ellipsoidal inhomogeneities in an infinite elastic body: An application to a hybrid composite*, Int. J. Solids Struct. 17 (6) (1981), pp. 553–563. doi:10.1016/0020-7683(81)90018-4

[21] J.D. Eshelby, *The determination of the elastic field of an ellipsoidal inclusion, and related problems*, Proc. R. Soc A Math. Phys. Eng. Sci. 241 (1226) (1957), pp. 376–396. doi:10.1098/rspa.1957.0133

[22] H. Jiang, B. Liu, Y. Huang, and K.C. Hwang, *Thermal expansion of single wall carbon nanotubes*, J. Eng. Mater. Technol. 126 (3) (2004), pp. 265–270. doi:10.1115/1.1752925

[23] Y.-K. Kwon, S. Berber, and D. Tománek, *Thermal contraction of carbon fullerenes and nanotubes*, Phys. Rev. Lett. 92 (1) (2004), 015901. doi:10.1103/PhysRevLett.92.015901

[24] R.B. Pipes and P. Hubert, *Helical carbon nanotube arrays: Thermal expansion*, Comp. Sci. Technol. 63 (11) (2003), pp. 1571–1579. doi:10.1016/S0266-3538(03)00075-7

[25] Y. Maniwa, R. Fujiwara, H. Kira, H. Tou, H. Kataura, S. Suzuki, Y. Achiba, E. Nishibori, M. Takata, M. Sakata, A. Fujiwara, and H. Suematsu, *Thermal expansion of single-walled carbon nanotube (SWNT) bundles: X-ray diffraction studies*, Phys. Rev B. 64 (24) (2001), 241402. doi:10.1103/PhysRevB.64.241402

[26] W. Jeong and M.R. Kessler, *Effect of functionalized MWCNTs on the thermo-mechanical properties of poly(5-ethylidene-2-norbornene) composites produced by ring-opening metathesis polymerization*, Carbon 47 (10) (2009), pp. 2406–2412. doi:10.1016/j.carbon.2009.04.042

[27] S. Kanagaraj, F.R. Varanda, T.V. Zhiltsova, M.S.A. Oliveira, and J.A.O. Simoes. *Mechanical properties of high density polyethylene/carbon nanotube composites*, Comp. Sci. Technol. 67 (15–16) (2007), pp. 3071–3077.

14

High electromechanical responses of ultra-high-density aligned nano-porous microwave exfoliated graphite oxide/polymer nano-composites ionic actuators

M. Ghaffari[a], Y. Zhou[b], M. Lin[c], Chong Min Koo[d] and Q.M. Zhang[a,b]*

[a]Department of Materials Science and Engineering, Materials Research Institute, The Pennsylvania State University, University Park, PA 16802, USA; [b]Department of Electrical Engineering, Materials Research Institute, The Pennsylvania State University, University Park, PA 16802, USA; [c]Materials Research Institute, The Pennsylvania State University, University Park, PA 16802, USA; [d]Center for Materials Architecturing, Institute for Multi-Disciplinary Convergence of Materials, Korea Institute of Science and Technology (KIST), Hwarangno 14-gil 5, Soengbuk-gu, Seoul 136-791, Republic of Korea

High elastic energy density and high-efficiency ionic electromechanical actuators were prepared from aligned activated microwave exfoliated graphite oxide (A-aMEGO)/ polymer nano-composites, and the electromechanical performance was characterized. The elastic modulus and elastic energy density of the ionic actuators can be tuned over a wide range by varying the polymer (poly (vinylidene fluoride/chlorotrifluoroethylene) [P(VDF-CTFE)]) concentration in the nano-composite actuators. The A-aMEGO/ P(VDF-CTFE) nano-composite actuators with 35 wt.% of polymer content exhibit an elastic energy density higher than 5 J/cm^3 and an electromechanical conversion efficiency higher than 3.5%, induced under 4 V. The results show the promise of high-density highly aligned graphene electrodes for high-performance ionic electromechanical transduction devices.

Keywords: nano-porous activated microwave exfoliated graphite oxide (aMEGO); actuator; self-assembly; elastic modulus; elastic energy density

1. Introduction

Materials that can generate large strain with a high force level under an applied electrical signal are of great importance for applications such as artificial muscles, precise motion and position control, and micro-electromechanical systems (MEMS) [1–9]. The significance of these materials is especially more evident when they can generate large electroactuation strain under a few volts, making them appealing for direct integration with advanced microelectronics and thus performing complex functions. In designing actuator devices from these materials, one has to pay special attention to not only the strain magnitude but also the force level the device can exert. In other words, generation of a large strain does not necessarily translate into high device performance. For instance, ionic polymer actuators such as gels and soft polymers that contain a high fraction of water can generate strains >50% under a few volts. Nevertheless, a major constraint of these gel

*Corresponding author. Email: qxz1@psu.edu

actuators for practical applications is their extremely low elastic modulus (<1 MPa) [10–12].

Our previous studies have revealed that a class of aligned activated 'nano-porous' microwave exfoliated graphite oxide (A-aMEGO) actuators with controlled nano-morphology can generate an electroactuation strain of more than 50% under a few volt electrical excitation [13]. The A-aMEGO actuators were fabricated by the aid of a modified self-assembly procedure and by using poly (tetrafluoroethylene) or poly (vinylidene fluoride/chlorotrifluoroethylene) (P(VDF-CTFE) as the polymer matrix in the system and 1-butyl-3-methylimidazolium tetrafluoroborate ([BMIM][BF_4]) solution in acetonitrile (AN) as the electrolyte. The principal actuation mechanism of these A-aMEGO actuators, as schematically illustrated in Figure 1, was believed to be due to the ingression and/or depletion of the mobile ions in the nano-pores to compensate the changes in the electronic charges on the A-aMEGO electrodes under the external voltage. Due to a very high degree of alignment (orientation) of the aMEGO sheets and nano-pore sizes that match the sizes of mobile ions, this process caused changes in the inter-sheet spacing that resulted in the unique large anisotropy of the electroactuation strain in A-aMEGO actuators. Namely, the large strain was induced along the direction of the average surface normal to the graphene sheets, while in the directions perpendicular to that, the strain was more than 10 times small.

As was investigated in the aforementioned study, the electromechanical performance of the A-aMEGO/polymer nano-composite devices depends on several parameters, including voltage, scan rate, electrolyte type, polymer type, and concentration. Specifically, to improve the force generation capability of these devices, the elastic modulus of the nano-composites has to be improved since following Hooke's law, the stress generated is proportional to the elastic modulus of the actuator material. It was shown that the nano-composite actuators with 10 wt.% P(VDF-CTFE) exhibited an elastic modulus of 67.5 MPa without actuation that reduced to about 9 MPa at 56.6% actuation strain under 4 V. As a result, the actuators showed an electromechanical conversion

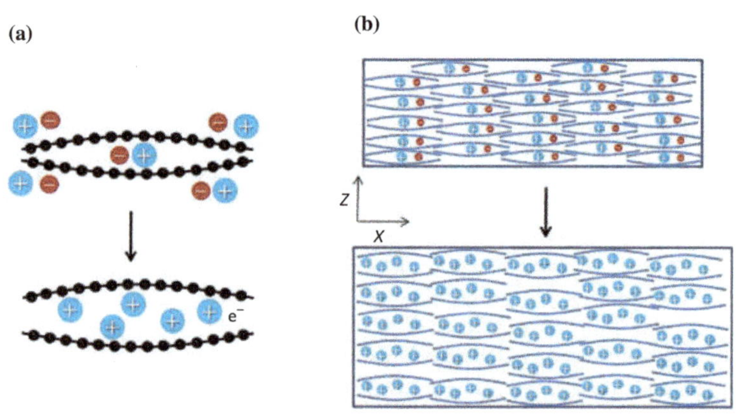

Figure 1. (a) An illustration of ionic electroactuation in graphene sheets where the ingression of the mobile ions (blue circles, cations here) into the interlayer space upon the application of an external voltage generates large strain when the interlayer spaces are comparable to the mobile ion sizes. e^- represents the electron charges on the cathode (graphene sheets). (b) Schematic of the nano-morphology of highly oriented aMEGO sheets of an ionic actuator employing the actuation mechanism in (a), generating large electroactuation strain along the *z*-direction ('thickness direction').

efficiency about 1.5%, not as high as expected considering the large electroactuation strain and reasonably high elastic modulus of the nano-composites in the non-actuated state.

In the present study, we investigate the influence of the polymer matrix concentration in the nano-composites on the electromechanical response of the actuators. By increasing the polymer matrix concentration (reducing the volume content of the voids, which were originally occupied by electrolyte in the micro-, meso-, and macro-pores), the elastic modulus of the A-aMEGO/polymer can be improved. Therefore, although the strain level would decrease by an increase in the polymer content, elastic energy density and electromechanical efficiency of the actuator device may be improved. For this purpose, 10, 20, 35 and 50 wt.% of P(VDF-CTFE) copolymer was incorporated into the A-aMEGO/polymer nano-composites and the corresponding electromechanical performance were characterized. The results show that the electroactuation and force generation capabilities depend critically on the polymer concentration, and the electromechanical performance of the composite actuator show an optimum state at 35 wt.% P(VDF-CTFE) with an electromechanical conversion efficiency of more than 3.5%, very high among ionic electroactive polymer actuators.

2. Experimental and characterization

The synthesis of 'nano-porous aMEGO' is presented in detail in Ref. [14]. A modified vacuum-assisted self-assembly method was employed to fabricate the highly aligned nano-porous aMEGO sheets [13,15,16]. During the filtration process, nano-porous aMEGO and P(VDF-CTFE) layers are pulled toward the Anodisc surface and stacked successively on top of each other. In this study, 10, 20, 35 and 50 wt.% polymer was incorporated into the nano-composites.

To fabricate the ionic actuators, the A-aMEGO/P(VDF-CTFE) composites were swelled by a 2 M solution of [BMIM][BF_4] (Iolitec) in AN. Square-shaped A-aMEGO/P(VDF-CTFE) (40–60 μm thick) were cut and placed in an electrolyte bath at room temperature. After the nano-composites were completely infiltrated by the electrolyte (~30 min), the electroactuation and cyclic voltammetry (CV) performance were characterized. The electroactuation strain in the ionic A-aMEGO/P(VDF-CTFE) nano-composites was characterized in a configuration reported in the previous study [13]. For the actuation characterization, two A-aMEGO/P(VDF-CTFE) nano-composites of 2 × 2-mm lateral size and 0.05 mm thick (along the *z*-direction, Figure 1(b)) were bonded to the bottom of a Teflon sample holder filled with [BMIM][BF_4]/AN electrolyte. The Teflon sample holder has two square dishes of 2.2-mm diameter and 1-mm depth connected by a 1-mm-wide channel for the ion conduction between the two nano-composite electrodes. The two nano-composites served as two electrodes, and two tungsten pins with flat heads of 1.5-mm diameter were used as the current collectors, which were pressed lightly to the top surfaces of the two nano-composites. The positions (both lateral and vertical) of the tungsten electrodes were precisely controlled by micro-manipulators. The electroactuation strains in both the cathode and the anode were characterized using a fiber optic displacement sensor (MTI 2100 Fotonic) that has a resolution of 25 nm. The capacitance of the ionic A-aMEGO/P(VDF-CTFE) nano-composite actuators was measured with a potentiostat (VersaSTAT 4, Princeton Applied Research).

The Young's moduli of the ionic A-aMEGO/ P(VDF-CTFE) nano-composites along the thickness (the *z*-direction in Figure 1(b)) under no applied voltage were measured using the Derjaguin–Muller–Toporov model from the data acquired with the nanomechanical characterization mode (peak force tapping) of a Bruker Icon atomic force microscope (AFM). The operation principle relies on the interaction of the modulated tip with the sample surface [17–19]. The elastic modulus of the nano-composite actuators under different applied voltages was characterized by measuring the change of the electroactuation strain under load along the *z*-direction (Figure 1). In this measurement, a load (0.025, 0.25, and 0.5 MPa) was applied to the composite sample and the thickness change due to the electroactuation was measured.

3. Experimental results and discussions

The electroactuation strain in the cathode of the ionic A-aMEGO/P(VDF-CTFE) nano-composites with [BMIM][BF_4]/AN electrolyte was characterized under different voltage at a 50-mV/s scan rate. The strain magnitude was calculated from the following:

$$\text{Strain} = \left(\frac{\text{Final thickness} - \text{Initial thickness}}{\text{Initial thickness}}\right) \times 100 \tag{1}$$

Figure 2(a) represents a typical thickness electroactuation strain hysteresis loop of the nano-composite actuator (along the *z*-direction Figure 1(b)) with 10 wt.% of P(VDF-CTFE), measured at the cathode under 4 V. Figure 2(b) summarizes the cyclic

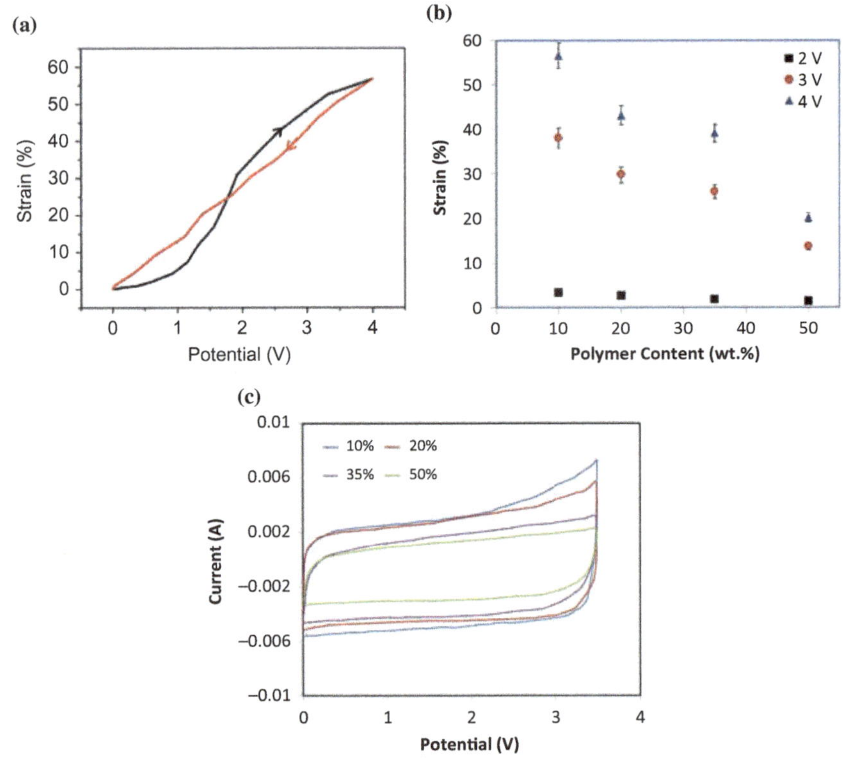

Figure 2. (a) An electroactuation cycle for A-aMEGO/P(VDF-CTFE) nano-composite containing 10 wt.% polymer as an example. Arrows indicate the strain direction. (b) Electroactuation strain on the cathode for the A-aMEGO/P(VDF-CTFE) nano-composites investigated at different applied peak voltages and 50-mV/s scan rate. (c) CV curves for A-aMEGO/P(VDF-CTFE) nano-composites at 50-mV/s scan rate at different polymer concentrations.

electroactuation thickness strain of the A-aMEGO/P(VDF-CTFE) nano-composites of different P(VDF-CTFE) wt.% in the composites, measured at the cathode under different voltages. Similar to previously observed trend, the strain value in the thickness direction increases almost linearly with the voltage and reaches the maximum value at 4 V. However, the strains perpendicular to the thickness direction were negligible and had a sign opposite to that of the thickness strain. Such large electroactuation strain anisotropy is caused by the unique nano-morphology of the highly oriented nano-pores of the aMEGO sheets. Another evident trend in Figure 2(b) is the reduction of the strain value with increased polymer concentration in the composites. For instance, under 4 V, the strain is 56.6% for the nano-composite containing 10 wt.% polymer, its value drops to 43.2% for the nano-composite with 20 wt.% of P(VDF-CTFE), and it further decreases to 20.3% for the composite containing 50 wt.% polymer. This is understandable since the elastic modulus of the composite film increases with the polymer content due to the replacement of voids filled with ionic liquid inside the pores in the original composite with a much stiffer polymer matrix. In addition, the increased polymer in the nano-composites causes a reduction of the mobile ion mobility in the composite matrix, which reduces the excess mobile ion concentration in the electrodes under a given voltage and scan rate as shown in the CV curves in Figure 2(c). Although the strain magnitude decreases by increase in the polymer content, nevertheless, the increased elastic modulus improves the device performance in terms of force generation capability, elastic energy density, and electromechanical conversion efficiency, as will be discussed in the following.

The charge/discharge behavior of the A-aMEGO/P(VDF-CTFE) nano-composites was characterized using CV experiments under 4-V peak voltage and 50-mV/s scan rate. As depicted in Figure 2(c), the CV curves for the nano-composites are nearly rectangular, which is the indication of non-Faradaic charge transfer in A-aMEGO-based electrodes similar to the previous observations [13]. The CV curves also reveal that the specific capacitance for the composites decreases with the polymer concentration. This is comprehendible since by addition of the polymer matrix, the mobility of the mobile ions is reduced. Hence, the amount of the excess mobile ions stored in the porous electrodes under a given voltage and scan rate decrease, which is reflected in the specific capacitance values.

As mentioned earlier, strain generation is not the sole parameter in defining the electromechanical performance of an actuator device; the elastic energy density and electromechanical conversion efficiency are also critical parameters. The elastic energy density is used to characterize an actuator material's strain and stress generation capability [1–3,20]. The strain by the electroactuation that changes the nano-composite thickness from the initial thickness l_0 to l under an applied voltage V can also be generated by applying a force F along the thickness direction ($F = \sigma A$, where σ is the stress and A is the nano-composite area that does not change with the applied voltage or stress due to the large elastic anisotropy of the A-aMEGO/polymer nano-composite). The total mechanical work W done by the external force, divided by the nano-composite initial volume ($\mathrm{Vol} = A \times l_0$), is the elastic energy density stored in the nano-composite actuator. For materials with large strain, the mechanical stress is $Y \times e$ where Y is the elastic modulus and e is true strain $e = \ln(l'/l_0)$. The work W done by the external force F is as follows:

$$W = \int_{l_0}^{l} F\,\mathrm{d}l' \tag{2}$$

Dividing W by Vol = $A \times l_0$ yields [2,21,22] the following:

$$U_m = \int_{l_0}^{l} Y(x) \ln(x) \, dx \tag{3}$$

which is the elastic energy density of the ionic A-aMEGO/P(VDF-CTFE) nano-composite actuators. The elastic energy density derived here is the total elastic energy density generated in the actuator, which is, in general, different from that delivered to an external load.

For the purpose of elastic energy density calculations, the elastic modulus Y of the A-aMEGO/P(VDF-CTFE) nano-composites along the z-direction when the composite actuators were not actuated was characterized by an AFM using a tapping method [21,23]. Afterwards, the elastic modulus of the nano-composite actuators under different applied voltages was characterized by measuring the reduction of the induced strain under load along the z-direction. In this measurement, a load (a fixed weight) was applied to the composite sample and the thickness change due to the electroactuation was measured. Using the relation that the strain under a load (stress) is

$$S(\text{strain}) = S_0 - \frac{\text{load}}{Y} \tag{4}$$

where S_0 is the strain generated under 4 V in the absence of the load and Y is the elastic modulus at strain S, the elastic modulus Y at S can thus be deduced when S is not far from S_0.

Figure 3(a) represents the strain data at the cathode side under 4 V for nano-composites with different polymer content under different applied loads. As expected, increasing the load value reduces the electroactuation strain. The data from this measurement were used to deduce elastic modulus (using Equation (4)) of the nano-composites in the actuation state under 4 V, which are presented in Figure 3(b). The results show that the elastic modulus of the nano-composites decreases with the actuation strain.

To deduce the elastic energy density of the nano-composite actuators from Equation (3), the dependence of the elastic modulus with the actuation strain ($Y(x)$) is deduced from the two data points in Figure 3(b) by assuming that the elastic modulus is a linear function of the

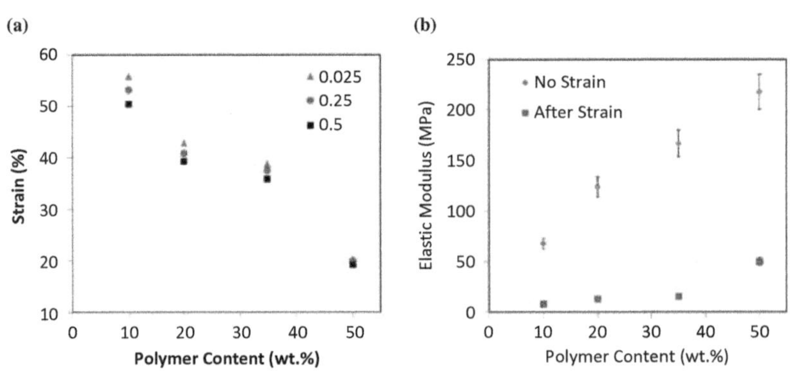

Figure 3. (a) Strain under different (0.025, 0.25, and 0.5 MPa) applied load. The error bars are not shown to have a clean graph. (b) Elastic modulus for A-aMEGO/P(VDF-CTFE) nano-composites at different polymer concentrations with and without electroactuation (at 4 V).

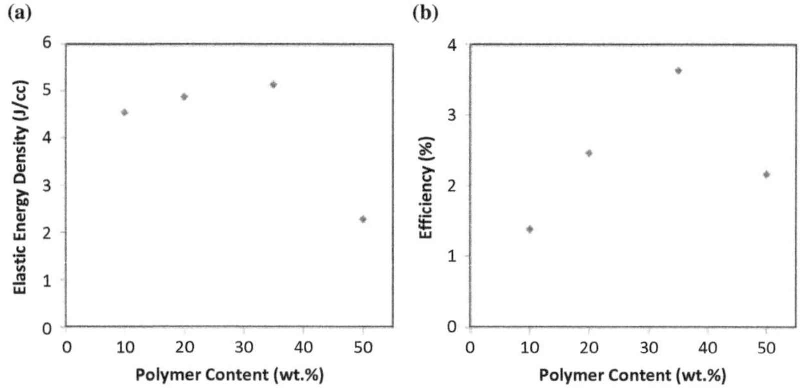

Figure 4. (a) Elastic energy density as a function of polymer content. (b) Efficiency for A-aMEGO/P(VDF-CTFE) nano-composites with different polymer concentrations.

strain ($Y = Y_0 - a \times Strain$), where a is a coefficient that is determined from the two data points in Figure 3(b). The elastic energy density of the nano-composite actuators with different polymer content thus derived is presented in Figure 4(a), which shows that the elastic energy density increased with the polymer content and reaches a maximum for the nano-composites with 35 wt.% P(VDF-CTFE), which is >5 J/cm^3. On the other hand, the nano-composites with 50 wt.% P(VDF-CTFE), due to the reduction of the strain, show an elastic energy density of 2.2 J/cm^3. The input electric energy density is deduced from the CV curves in Figure 2(c), and the ratio of elastic energy density and the input electric energy density yields the electromechanical conversion efficiency, which is presented in Figure 4(b). The nano-composites with 35 wt.% of P(VDF-CTFE) exhibited an electromechanical conversion efficiency of 3.7%, owing to the combined effect of a higher actuator elastic energy density and reduced input electric energy density compared with the nano-composites with 10 and 20 wt.% P(VDF-CTFE).

It should also be noted that the elastic energy density deduced in Figure 4(a) is the total stored electric energy density in the actuator device, which is different from that of the actuator delivers to an external load. For example, under 0.5-MPa load, the A-aMEGO/P(VDF-CTFE) with 10 wt.% polymer shows an electroactuation strain of 50%, which translates to a work density of 0.25 J/cm^3, which is different from the intrinsic elastic energy density presented in Figure 4(a). The result indicates that the work that an electroactuator delivers to a load will depend on the load condition.

As depicted in Figure 4(a), the elastic energy density increases by increasing polymer concentration. This can be attributed to the increase in the elastic modulus of the nano-composite with P(VDF-CTFE) content. Nevertheless, the increase in the energy density reaches a maximum value at 35 wt.% polymer and afterwards its value diminishes. Since the energy density is a function of both elastic modulus and strain, and given that usually an increase in one results in a decrease in the other, this observation is expected. Although the nano-composites with 50 wt.% polymer exhibit the highest modulus amongst all, however, this does not compensate the significantly reduced strain value for this specific composition and hence the elastic energy density is reduced. Very similar behavior was observed for the efficiency of the nano-composite, with the highest efficiency for the sample with 35 wt.% polymer. This again is related to the highest elastic energy exerted by this sample per unit value of the consumed electric energy, calculated form the CV curves. The elastic energy density and efficiency values for A-aMEGO/P(VDF-CTFE) nano-composites are among the highest for ionic actuators performing under voltages lower than 4 V. The high performance features of the A-aMEGO actuators render them potentially valuable systems for MEMS applications.

4. Conclusions

In this study, the electromechanical performance of A-aMEGO/P(VDF-CTFE) nano-composite ionic actuators was investigated and optimized by varying the polymer concentration in the composites. Under 4-V applied voltage, the nano-composites with 10 wt.% polymer exhibited the highest strain of 56.6% and the lowest elastic modulus of 67.5 MPa, whereas the nano-composites with 50 wt.% polymer showed the lowest strain of 20.3% and highest elastic modulus of 218.0 MPa. Similar trend was observed under 3 and 2 V. The actuation mechanism was due to ingression and/or depletion of excess mobile ions into/from the nanometer-sized pores of the aligned graphene sheets in response to the external applied voltage. The trade-off of the actuation strain and elastic modulus with the polymer content in the nano-composites results in a maximum elastic energy density for nano-composites with 35 wt.% polymer content, which also exhibits the highest electromechanical conversion efficiency of 3.7%, among the nano-composites studied.

Acknowledgments

The work was supported by NSF under grant number CMMI-1130437 and by a grant from Korean Institute of Science and Technology.

References

[1] Y. Bar-Cohen and Q.M. Zhang, *Electroactive polymer actuators and sensors*, MRS Bullet. 33 (2008), pp. 173–181. doi:10.1557/mrs2008.42.
[2] Q.M. Zhang, V. Bharti, and X. Zhao, *Giant electrostriction and relaxor ferroelectric behavior in electron-irradiated poly(vinylidene fluoride-trifluoroethylene) copolymer*, Science 280 (1998), pp. 2101–2104. doi:10.1126/science.280.5372.2101.
[3] R. Pelrine, R. Kornbluh, Q. Pei, and J. Joseph, *High-speed electrically actuated elastomers with strain greater than 100%*, Science 287 (2000), pp. 836–839. doi:10.1126/science.287.5454.836.
[4] W. Lu, A.G. Fadeev, B. Qi, E. Smela, B.R. Mattes, J. Ding, G.M. Spinks, J. Mazurkiewicz, D. Zhou, G.G. Wallace, D.R. MacFarlane, S.A. Forsyth, and M. Forsyth, *Use of ionic liquids for π-conjugated polymer electrochemical devices*, Science 297 (2002), pp. 983–987. doi:10.1126/science.1072651.
[5] M. Shahinpoor, Y. Bar-Cohen, J.O. Simpson, and J. Smith, *Ionic polymer-metal composites (IPMCs) as biomimetic sensors, actuators and artificial muscles – a review*, Smart Mater. Struct. 7 (1998), pp. R15–R30. doi:10.1088/0964-1726/7/6/001.
[6] R.H. Baughman, C. Cui, A.A. Zakhidov, Z. Iqbal, J.N. Barisci, G.M. Spinks, G.G. Wallace, A. Mazzoldi, D. De Rossi, A.G. Rinzler, O. Jaschinski, S. Roth, and M. Kertesz, *Carbon nanotube actuators*, Science 284 (1999), pp. 1340–1344.
[7] S. Liu, Y. Liu, H. Cebeci, R. Guzmán de Villoria, J.H. Lin, B.L. Wardle, and Q.M. Zhang, *High electromechanical response of ionic polymer actuators with controlled-morphology aligned carbon nanotube/nafion nano-composite electrodes*, Adv. Funct. Mater. 20 (2010), pp. 3266–3271. doi:10.1002/adfm.201000570.
[8] T. Thorsen, S.J. Maerkl, and S.R. Quarke, *Microfluidic large-scale integration*, Science 298 (2002), pp. 580–584. doi:10.1126/science.1076996.

[9] M.D. Lima, N. Li, M.J. de Andrade, S. Fang, J. Oh, G.M. Spinks, M.E. Kozlov, C.S. Haines, D. Suh, J. Foroughi, S.J. Kim, Y. Chen, T. Ware, M.K. Shin, L.D. Machado, A.F. Fonseca, J.D. W. Madden, W.E. Voit, D.S. Galvao, and R.H. Baughman, *Electrically, chemically, and photonically powered torsional and tensile actuation of hybrid carbon nanotube yarn muscles*, Science 338 (2012), pp. 928–932. doi:10.1126/science.1226762.
[10] P. Calvert, G. Sensors, and M.R.S.B. Actuators 33 (2008), pp. 207–212.
[11] M. Knoblauch, G.A. Noll, T. Müller, D. Prüfer, I. Schneider-Hüther, D. Scharner, A.J.E.V. Bel, and W.S. Peters, *ATP-independent contractile proteins from plants*, Nat. Mater. 2 (2003), pp. 600–603.
[12] D.J. Beebe, J.S. Moore, J.M. Bauer, Q. Yu, R.H. Liu, C. Devadoss, and B.H. Jo, *Functional hydrogel structures for autonomous flow control inside microfluidic channels*, Nature 404 (2000), pp. 588–590. doi:10.1038/35007047.
[13] M. Ghaffari, W. Kinsman, Y. Zhou, S. Murali, Q. Burlingame, M. Lin, R.S. Ruoff, and Q.M. Zhang, *Aligned nano-porous microwave exfoliated graphite oxide ionic actuators with high strain and elastic energy density*, Adv. Mater. 25 (2013), pp. 6277–6283.
[14] Y. Zhu, S. Murali, M.D. Stoller, K.J. Ganesh, W. Cai, P.J. Ferreira, A. Pirkle, R.M. Wallace, K.A. Cychosz, M. Thommes, D. Su, E.A. Stach, and R.S. Ruoff, *Carbon-based supercapacitors produced by activation of graphene*, Science 332 (2011), pp. 1537–1541. doi:10.1126/science.1200770.
[15] Q. Liang, X. Yao, W. Wang, Y. Liu, and C.P. Wong, *A three-dimensional vertically aligned functionalized multilayer graphene architecture: an approach for graphene-based thermal interfacial materials*, ACS Nano. 5 (2011), pp. 2392–2401. doi:10.1021/nn200181e.
[16] D.A. Dikin, S. Stankovich, E.J. Zimney, R.D. Piner, G.H.B. Dommett, G. Evmenenko, S.T. Nguyen, and R.S. Ruoff, *Preparation and characterization of graphene oxide paper*, Nature 448 (2007), pp. 457–460. doi:10.1038/nature06016.
[17] T. Ritter, X.C. Geng, K.K. Shung, P.D. Lopath, S.E. Park, and T.R. Shrout, *Single crystal PZN/PT polymer composites for ultrasound transducer applications*, IEEE Trans. Ultrason. Ferroelectr. Freq. Control 47 (2000), pp. 792–800. doi:10.1109/58.852060.
[18] B. Chu, X. Zhou, K. Ren, B. Neese, M. Lin, Q. Wang, F. Bauer, and Q.M. Zhang, *A dielectric polymer with high electric energy density and fast discharge speed*, Science 313 (2006), pp. 334–336. doi:10.1126/science.1127798.
[19] C.Z. Rosen, B.V. Hiremath, and R. Newnham, *Piezoelectricity*, American Institute of Physics, New York, 1992.
[20] V. Giurgiutiu and C.J. Rogers, *Dynamic power and energy capabilities of commercially-available electro-active induced-strain actuators*, J. Intell. Mater. Syst. Struct. 7 (1996), pp. 656–667. doi:10.1177/1045389X9600700605.
[21] D. Maugis, *Contact, Adhesion and Rupture of Elastic Solid*, Springer, New York, 2000.
[22] X. Zhou, X. Zhao, Z. Suo, C. Zou, J. Runt, S. Liu, S. Zhang, and Q.M. Zhang, *Electrical breakdown and ultrahigh electrical energy density in poly(vinylidene fluoride-hexafluoropropylene) copolymer*, Appl. Phys. Lett. 94 (2009), p. 162901.
[23] O. Sahin and N. Erina, *High-resolution and large dynamic range nanomechanical mapping in tapping-mode atomic force microscopy*, Nanotechnology 19 (2008), p. 445717. doi:10.1088/0957-4484/19/44/445717.

15

Research on automatic spraying of single-walled carbon nanotubes and detection of spraying effects

Jianwen Zhao, Jiangcheng Yu, Junyang Niu*, Yong Ge and Liwu Liu

HIT at Weihai, State Key Laboratory of Robotics and System, Song Jian Research Institute of Building 2, Wenhua Road (W), Weihai, Shandong, 264209, PR China

Single-walled carbon nanotubes (SWNTs) have been introduced as compliant electrodes for dielectric elastomers (DEs) due to fault tolerance. To acquire a better electrostrictive strain and longer lifetime, it is essential to obtain a certain and uniform width of the SWNT electrode. To ensure uniform width manually, a small flux and longer time are necessary. Moreover, it is difficult to control the width of the electrode for the randomness of manual spraying. Therefore, a new type of automatic spraying process is presented in this paper. The width and homogeneity of the electrode can be easily controlled by certain parameters of the process. Two methods for detecting the homogeneity of the electrode are introduced in this paper: Measurement of surface resistance and luminosity. The coefficient of variation (CV) values detected by the two methods are virtually equal and less than 8%, which shows the feasibility of the detection method and homogeneity of automatic spraying. The speed of automatic spraying is 102 mm^2/s, which is higher than that of manual spraying. The spraying process and the method used to detect homogeneity in this paper provide a reference for the relevant processes.

Keywords: single-wall carbon nanotube; automatic spraying; dielectric elastomer; homogeneity detection

1. Introduction

Acrylic elastomers have demonstrated actuated strains of over 200%, stresses of up to 7 MPa, and estimated energy densities of over 3 J/g, and such remarkable performance has attracted considerable attention [1–4]. To realize the electrostrictive strain of dielectric elastomer (DE) actuators, the sandwich structure of compliant electrode/DE /compliant electrode is applied [5]. Naturally, the research on compliant electrodes has stepped into the spotlight in regard to the characteristics of DEs. Progress has been made on different kinds of electrodes like graphite powder, carbon-filled grease, silver-filled grease [6], and single-walled carbon nanotubes (SWNTs) [7]. Among these, SWNTs are applied in the construction of DE actuators for exhibiting fault tolerance through the local degradation of SWNTs during dielectric breakdown [8,9]. To date, most researches have mainly focused on how to improve the electrostrictive strain and lifetime of DEs by the use of SWNTs [10]. Although the manufacturing process of the SWNT electrode has been studied [11], the manufacturing process of 3M VHB acrylic tapes' SWNT electrode is still carried out

*Corresponding author. Email: niujunyang0725@gmail.com

manually. Uncertain factors such as the homogeneity and thickness of the sprayed electrode are introduced by manual spraying, leading to longer spraying times, and thus the process cannot be controlled accurately. To provide a solution to these problems, a new type of automatic spraying process is introduced in this paper. With the prerequisite of zero damage to the SWNT structure, ultrasonic crushing equipment is used to improve the dispersibility of the SWNT. Sustainable and stable ejection from the SWNT sprinkler is achieved by this method. Finally, a continuous spraying and uniform electrode coating is realized and the area strain with fixed width is also measured in this paper.

2. Description of the automatic spraying system

The system consists of a controller (single-chip AT89C51), a powerful motor driver, and two mechanisms with translation movement, a pneumatic circuit, and a sprinkler (Figure 1).

A serpentine path is realized by the controller, stepping motor driver, and stepping motor. The sprinkler is installed on the ball screw by a clamping device. All units above are supported by standard sections and surrounded by a seal plate.

To enhance spraying, the unit describes a serpentine spraying path, with instant stopping and motion reset incorporated in the controller. Structure and design of the pneumatic circuit is shown in Figure 2.

The free end of the sprinkler is connected to the air compressor by an additional super-mist separator and solenoid valve. Similarly, the end in contact with liquid is connected to the sprinkling can, which is specifically designed according to the size of the sprinkler and the solution characteristics. The prototype is presented in Figure 3.

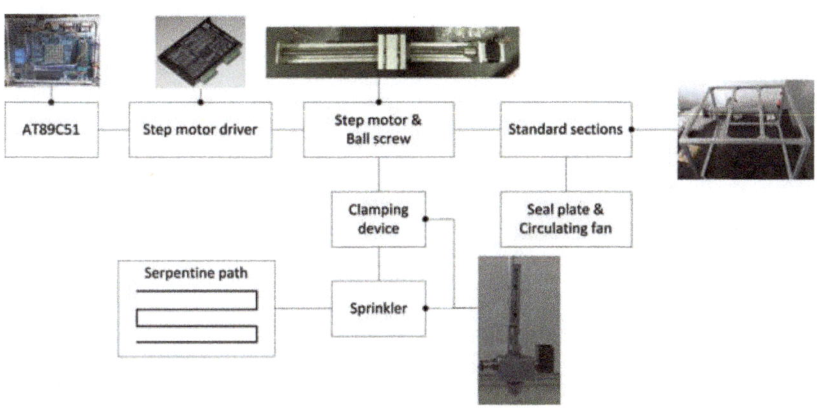

Figure 1. Mechanism of the new spraying machine.

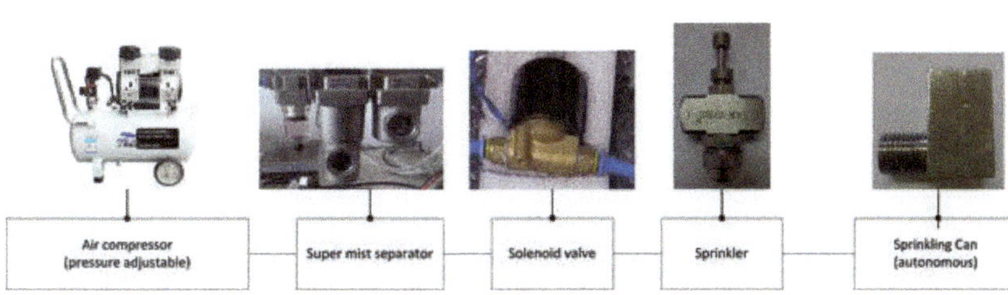

Figure 2. Structure and design of the pneumatic circuit.

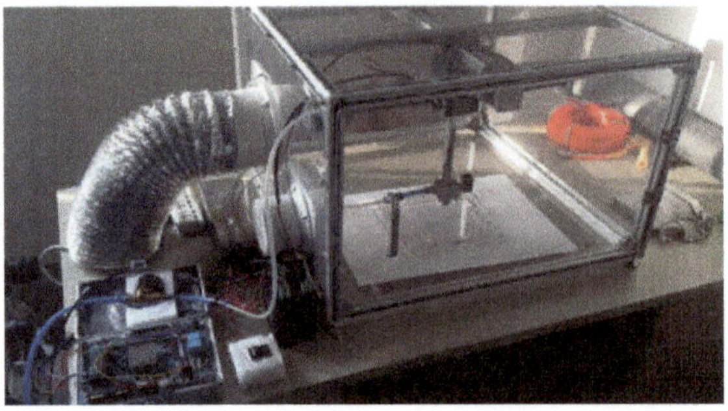

Figure 3. The prototype product.

3. Research on the dispersibility of SWNT by ultrasonic crushing

In solution, SWNTs readily become entangled and bunched together due to the van der Waals force acting among them. From the macro perspective, SWNT powders readily aggregate into larger particles and thus during the spraying process the sprinkler is easily jammed, affecting the fluency and uniformity of spraying. Therefore, the dispersibility of SWNT solutions needs to be improved. Three types of method are commonly used: physical, chemical, and radiation [12,13]. The physical method [14] is the most suitable in regard to DE actuators, by virtue of the absence of effect on SWNT conductivity [15]. However, the most widely used physical method today is ultrasonic cleaning, but energy transfer by this method is inefficient. The level of ultrasonic energy reduced gradually when passing through multimedia to reach the SWNT solution, and dispersibility is not ideal with this method. To enhance dispersibility, the concept of ultrasonic crushing is introduced in this article.

3.1. Enhanced dispersibility of SWNT solution by ultrasonic crushing

The increasing use of powders in SWNT solutions rapidly converts these into suspensions by precipitation when poorly dispersed. The degree of precipitation is used for comparison of dispersibility between ultrasonic crushing and ultrasonic cleaning. SWNTs (Chengdu Organic Chemicals Co. Ltd., Chinese Academy of Sciences) were dispersed in an aqueous solution (mixture of water and isopropanol 1:5 by volume) by either ultrasonic crushing or ultrasonic cleaning, yielding a solution concentration of 0.5 mg/ml. The ultrasonic cleaner was applied for 15 min under 480 W power, and ultrasonic crushing for the same time but with a pause of 3 min after running for 5 min oat200 W and 19.4 KHz. If the temperature of the solution becomes excessive, the isopropanol will evaporate rapidly. The solution was then placed in two test tubes and left to stand. Figure 4 shows dispersal of the respective solutions after having stood for 12 hours. As will be seen from this, ultrasonic crushing shows the better dispersibility.

3.2. Comparison of time to sprinkler blockage

Following addition of the dispersant, the SWNT solution will disperse efficiently but conductivity will be reduced [15]. During the spraying process, SWNT solution dispersed by physical methods with no addition of dispersant will eventually jam the sprinkler if its

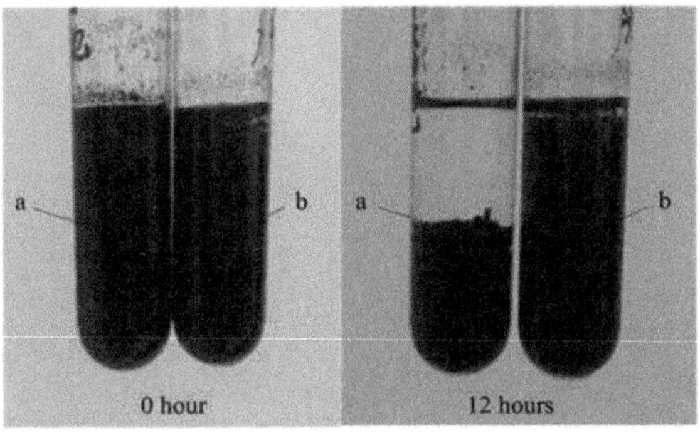

Figure 4. Degree of precipitation following dispersal by (a) ultrasonic cleaning and (b) ultrasonic crushing. After 12 hours of standing, the solution dispersed by ultrasonic cleaning has clearly precipitated.

inner diameter is 0.3 mm or less; time to blockage of a sprinkler is a key parameter in the spraying process.

The SWNT solution was prepared at a concentration of 1 mg/ml (water and isopropanol1:5 by volume). To clarify comparative results, the concentration used was higher than that previously and the solution divided in two. The first part was dispersed by ultrasonic crushing for 15 min (pausing for 3 min after running for 5 min), while the other was dispersed by ultrasonic cleaning for 30 min (pausing for 3 min after running for 15 min). The spraying pressure was 24 psi, and time to blockage was recorded from the start of the experiment. Three cases of failure occurred (discontinuous spraying, decreasing flux, and sprinkler clogging) during spraying, any of which could have affected the spraying results. The time when one of these failures first occurred was termed valid time, and we then calculated the average valid time. The results are shown in Table 1.

As will be seen from Table 1, the average time to blockage in ultrasonic crushing was greater than for ultrasonic cleaning, indicating that the solution dispersed by the former method can meet the demand for spraying of larger areas and wider electrodes.

Table 1. Time to sprinkler blockage during spraying.

	Discontinuity	Decrease	Blockage (s)	Minimum (s)	Average (s)
(a) Ultrasonic crushing					
1	/	/	475	475	
2	/	/	513	513	
3	/	/	432	432	478.2
4	/	/	502	502	
5	/	/	469	469	
(b) Ultrasonic cleaning					
1	100	120	150	100	
2	160	140	180	140	
3	256	215	280	215	131
4	100	/	116	100	
5	/	/	100	100	

Note: " / " denotes no occurrence.

3.3. Observation of SWNT structure

During the process of dispersion by ultrasonic crushing, the probe inserted directly into the SWNT solution leads to efficient energy transfer, but it may damage SWNT solution composition thereby affecting conductivity [16]. Thus it is necessary to observe SWNT solution composition to confirm whether it has been negatively impacted.

The SWNT solution (1 mg/ml) prepared previously was observed by scanning electron microscopy (SEM). The concentration of SWNT solution used as the electrode is generally no higher than 1 mg/ml. Images obtained by SEM are shown in Figure 5, where it will be seen that there is little variation in composition. On the other hand, the Raman spectra of the original and dispersed SWNTs were measured and compared (Figure 6). The black and red curves are Raman spectra of the original and dispersed SWNTs, respectively; the disordered carbon/graphitic carbon (D/G) peak coordinate values of the

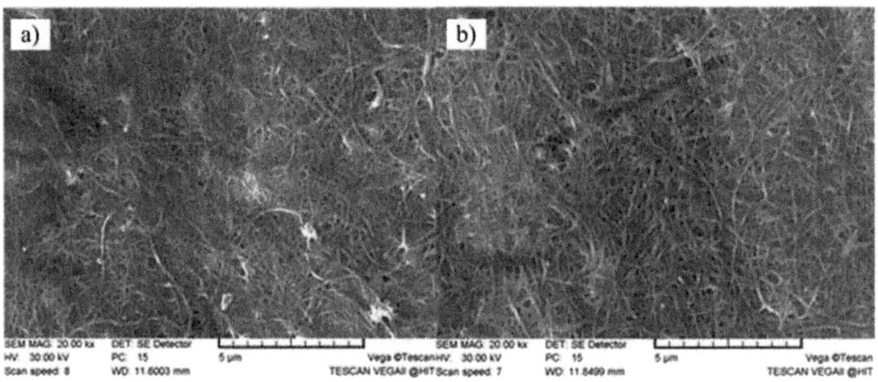

Figure 5. Structure of SWNT solutions under SEM. SWNT solution dispersed by (a) ultrasonic cleaner and (b) ultrasonic crushing.

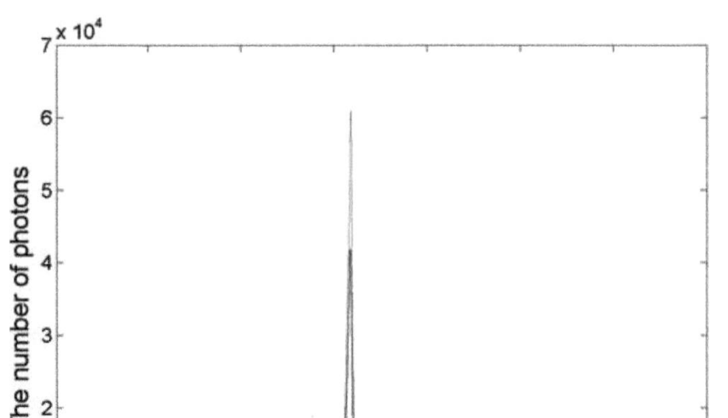

Figure 6. Raman spectra of original and dispersed SWNT solutions.

original are (1344.78, 2105.9) and (1588.62, 41,830), and those of the dispersed SWNTs are (1344.78, 2160.67) and (1591.37, 60,792.7). The D/G peak ratio of the latter is marginally less than that of the former. If the SWNT solution concentration had been reduced by ultrasonic crushing, D/G peak ratio would have been higher than the original, proof that this did not occur.

4. Comparison of the two methods in regard to electrode homogeneity detected

4.1. Detection of homogeneity by surface resistance

Homogeneity of SWNT electrodes cannot be guaranteed during manual spraying, hence the introduction of our spraying process to resolve this problem; the homogeneity of automatic SWNT electrode spraying was detected to confirm whether this process is feasible. The width, t of SWNT electrodes was estimated using $t = 1/R_s s$ [17], where R_s is surface resistance and s the SWNT conductivity; t has a proportional relation with R_s, so homogeneity can be analyzed by surface resistance.

SWNTs were dispersed in aqueous solution (mixture of water and isopropanol 1:5 by volume) with the aid of ultrasonic crushing for 15 min (halting for 3 min after operating for 5 min) at a concentration of 0.4 mg/ml. The operating parameters of ultrasonic crushing were 19.4 KHz, 200 W. The SWNTs were sprayed onto quadrille paper, either automatically or manually, ensuring consistency in spraying. The spraying parameters are summarized in Table 2. The parameter of pressure was optimized, with the original value ranging from 20 to 60 psi [18].

The schematic of the divided electrode during measurement of surface resistance is shown in Figure 7. The SWNT electrode was divided into five strips of 105 × 10 mm; a square 10 × 10 mm was then cut from a rectangle of 80 × 10 mm in the middle of each strip.

Table 2. Parameters of automatic and manual spraying.

Spray mode	Time (s)	Sprayed area (mm^2)	Pressure (psi)	Height (mm)	Flow rate (ml/min)
Manual/Automatic	90	105 × 70	24	100	1.0

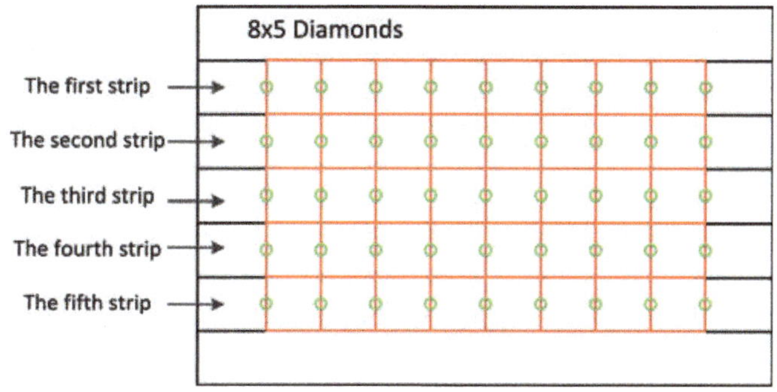

Figure 7. Schematic of the divided electrode.

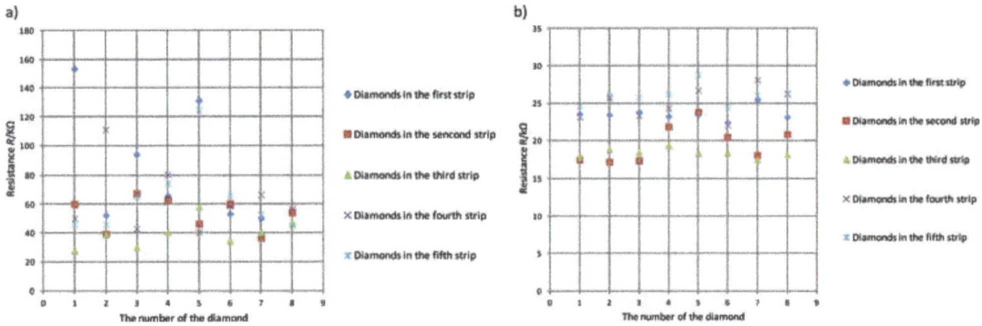

Figure 8. Results of manual (a) and automatic (b) spraying.

The surface resistence was measured at the spot of each square shown in Figure 7. Given that the edges of the spraying area may differ considerably from the central area, only the central 80 × 50 mm was measured. The results are shown in Figure 8.

As shown in Figure 8(a), there was no obvious distribution of manual spraying resistance, which is due to the lack of a specific motion path. The phenomenon of surface resistance in the third strip was lower than others in Figure 8(b) caused by the taper in the export of the sprinkler which sprays solution conically. The electrode coated in the third strip would be thicker because of the overlapping region between this and adjacent strips.

The characterization of thickness x is equal to $1/R$. The average value was calculated from the data in Figure 8, from which we derived $\bar{x}_{\text{automatic}} = 3.1549 \times 10^{-2}/K\Omega$, $\bar{x}_{\text{manual}} = 4.0291 \times 10^{-2}/\text{k}\Omega$.

According to the standard deviation (Equation (1)) and coefficient of variation (CV) formula (Equation (2)):

$$S = \sqrt{\frac{\sum_{i=1}^{n} (x_i - \bar{x})^2}{n-1}} \tag{1}$$

$$\text{CV} = \frac{S}{\bar{x}} \times 100\% \tag{2}$$

we derive $S_{\text{automatic}} = 2.2342 \times 10^{-3}$, $S_{\text{manual}} = 1.9952 \times 10^{-2}$, $\text{CV}_{\text{automatic}} = 7.08\%$, $\text{CV}_{\text{manual}} = 49.52\%$.

The result shows that the rate of change in surface resistance with the automatic process is better than that of the manual, because $\text{CV}_{\text{automatical}}$ is less than $\text{CV}_{\text{manual}}$. Thus the homogeneity of automatic electrode spraying is better.

4.2. Detection of homogeneity by luminosity

There is a quantitative relationship between the luminosity of the SWNT electrode and its width. Therefore, the homogeneity of SWNT electrodes can be confirmed by detection of luminosity. A square area of the electrode (10 × 10 cm) was sprayed onto the acrylic film automatically, using the parameters for solution and spraying described in Section 4.1. Figure 9 shows the set-up for detecting the luminosity of SWNT electrodes.

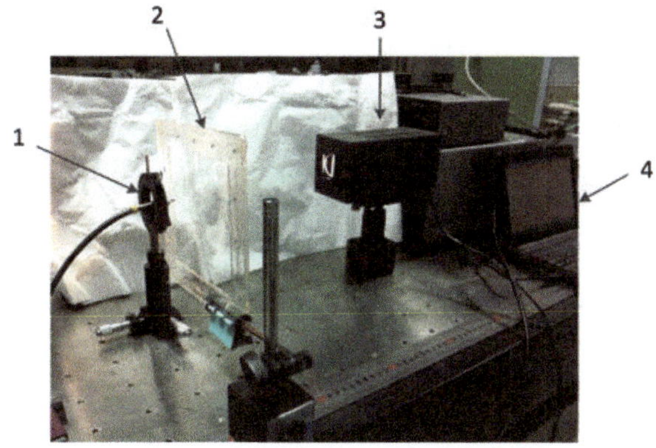

Figure 9. Set-up for light intensity measurement: (1) optical fiber; (2) DE with SWNT sample; (3) light source; (4) spectrograph.

The testing process comprises three steps: (1) measurement of the intensity of the light source; (2) measurement of the light intensity passing through the film without SWNTs; and (3) measurement the light intensity passing through the film with SWNTs. In the final step, measurement of the spraying area was divided into three parts, in each of which four measurement points were selected. The values obtained do not have units, because the intensity of the light measured by the instrument is a relative value. Luminosity, T is calculated by

$$T = \frac{I}{I_0} \times 100\% \tag{3}$$

where I and I_0 are the light intensity passing through the film and the original light intensity, respectively. The intensity of light passing through the medium is calculated by

$$I = I_0 e^{-\mu d} \tag{4}$$

where e, μ, and d are also the light intensity passing through the film. After performing Equation (4), d can be expressed as

$$d = -\frac{1}{\mu} \ln \frac{I}{I_0} = = -\frac{1}{\mu} \ln T \tag{5}$$

where d has a proportional relationship with $\ln T$ that can be derived from Equation (5). Thus the uniformity of film thickness (d) can be analyzed by the CV of $\ln T$.

During measurement, light intensity at different frequencies as obtained. Luminosity of the SWNT film was calculated by the average light intensity at all frequencies; the average original light intensity and that passing through the film without SWNTs was 1725 and 1564, respectively. Table 3.

Our calculations derived $\ln T$ and its mean, standard deviation, and CV: $\overline{\ln T} = -0.126$, $S_{\ln T} = 0.00955$, $CV_{\ln T} = 7.58\%$. The luminosity of SWNT film at 12 different measurement points varied marginally, which can be inferred from the value of $CV_{\ln T}$. Hence, the thickness at these measurement points is almost identical, indicating the homogeneity of SWNT electrodes sprayed automatically.

Table 3. Luminosity (I_0) calculated as 1564.

	Average light intensity passing through the film with SWNT, and results of analysis											
(a)	1	2	3	4	5	6	7	8	9	10	11	12
Light intensity	1401	1365	1356	1376	1384	1379	1395	1389	1370	1381	1365	1383
Luminosity T/%	89.6	87.3	86.7	88.1	88.5	88.2	89.2	88.8	87.6	88.3	87.3	88.4
$\ln T$	−0.110	−0.136	−0.143	−0.127	−0.122	−0.126	−0.114	−0.119	−0.132	−0.124	−0.136	−0.123

5. Validation of the automatic spraying technique

An important index of DE-coated SWNT electrodes is area strain, and therefore we measured this parameter and corresponding SWNT electrode width. Because the homogeneity of the electrode was validated above, knowing the width of the electrode when spraying automatically, and its relation to spraying parameters, is valuable. To this end, electrode width was measured by atomic force microscopy (AFM).

5.1. Width measurement

The factors affecting the width of the SWNT electrode are the distance between the sprinkler and the film, SWNT concentration, the solution flux (controlled by a throttle valve), air pressure, and sprinkler speed of the sprinkler (Table 4).

The width of the coated SWNT electrode was 69.05 nm, as measured by AFM at the step edges (Figure 10).

Table 4. The parameters of width measurement.

Parameter	Concentration (mg/ml)	Pressure (psi)	Flow (ml/min)	Height (mm)	Sprinkler speed (mm/s)
	0.4	24	0.59	100	13

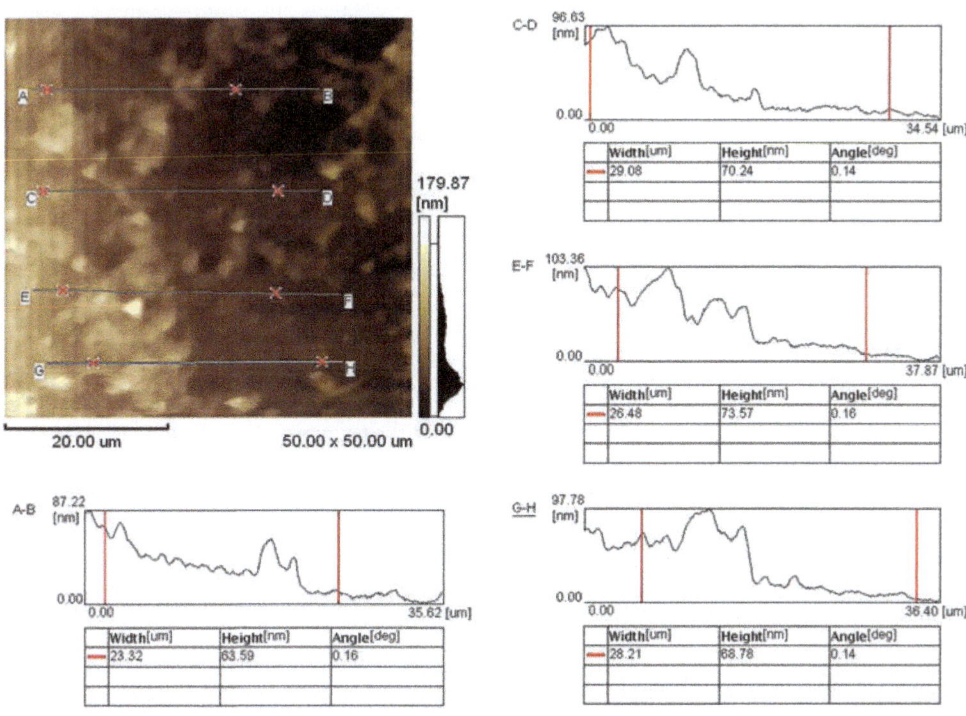

Figure 10. Scanning positions and measurement results.

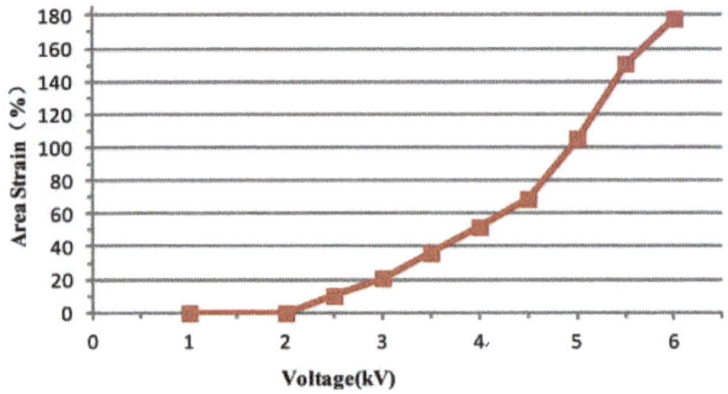

Figure 11. Area strain of the actuator with SWNT electrodes for applied voltages ranging from 0 to 6 kV. Maximum area strain is 177.8%.

5.2. Electrostrictive strain measurement

In this part of the experiment, acrylic film was biaxially prestrained by 300%. The SWNT solution was sprayed onto the film automatically and the film was then covered by a mask incorporating a hole of 30 mm diameter. The parameters of SWNT solution and spraying were as described in Section 5.1, as was the thickness. Area strain is shown in Figure 11 for voltages applied ranging from 0 to 6 kV. The maximum area strain was 177.8% at 6 KV.

Sixteen seconds was required for automatic spraying of an area 40 × 40 mm^2 at a spraying speed of 102 mm^2/s. To achieve the same thickness and homogeneity by manual spraying, the flux needs to be reduced and application time extended. Automatic spraying was shown to be more efficient than manual.

6. Conclusion

A novel technique of SWNT solution spraying, at a speed of 102 mm^2/s, is introduced in this paper. By calculating values of luminosity and surface resistance for different parts of SWNT electrodes, the CV was 7.08% and 7.58%, respectively. All the above parameters indicate that the application of this technique will yield a uniform SWNT electrode with high spraying speed.

Ultrasonic crushing is used to improve SWNT dispersibility, and such improvement is measured by standing and blockage experimentation. SEM was also used to prove that no damage had occurred to SWNT structure.

Two methods, measurement of luminosity and surface resistance, are described to detect the homogeneity of the SWNT electrode. The CV results from these two methods are virtually identical, which proves their feasibility.

Finally, with the application of ultrasonic crushing and spraying technique, the area strain on acrylic film (VHB4910, 300% biaxially prestrained) was measured at 177.8% at 6 KV, with electrode thickness of 69.05 nm. These results show that the technique can be applied in electrode spraying.

Negative aspects found were parameters such as distance between sprinkler and film, SWNT concentration, solution flux (controlled by the throttle valve), air pressure, and sprinkler speed; these still need to be optimized to enhance the spraying technique.

Acknowledgments

This work received financial support from the National Natural Science Foundation of China. We would like to thank Professor Qibing Pei for advice on the preparation of the SWNT solution, the School of Material Engineering for providing SEM, and the Department of Physics for guidance on the detection of homogeneity.

References

[1] R. Pelrine, R. Kornbluh, and G. Kofod, *High–strain actuator materials based on dielectric elastomers*, Adv. Mater. 12(16) (2000), pp. 1223–1225. doi:10.1002/1521–4095(200008)12:16<1223::AID–ADMA1223>3.0.CO;2–2.

[2] R. Pelrine, R. Kornbluh, Q. Pei, and J. Joseph, *High-speed electrically actuated elastomers with strain greater than 100%*, Science 287(5454) (2000), pp. 836–839. doi:10.1126/science.287.5454.836.

[3] Q. Pei, M.A. Rosenthal, R. Pelrine, S. Stanford, and R.D. Kornbluh, *Multifunctional electro-elastomer roll actuators and their application for biomimetic walking robots*, Smart Structures and Materials. International Society for Optics and Photonics, 2003: 281–290.

[4] J.S. Plante, L.M. Devita, and S. Dubowsky, *A road to practical dielectric elastomer actuators based robotics and mechatronics: Discrete actuation*. Proceedings of SPIE Electroactive Polymer Actuators and Devices, San Diego, CA. SPIE, Bellingham, WA, 2007: 652406–1.

[5] R.E. Pelrine, R.D. Kornbluh, and J.P. Joseph, *Electrostriction of polymer dielectrics with compliant electrodes as a means of actuation*, Sens. Actuators. A Phys. 64(1) (1998), pp. 77–85. doi:10.1016/S0924–4247(97)01657-9.

[6] P. Khodaparast, S.R. Ghaffarian, M.R. Khosroshahi, N. Yousefimehr, and D. Zamani, *Electrode structures in high strain actuator technology*, J. Optoelectron. Adv. Mater. 9(11) (2007), pp. 3585–3591.

[7] W. Yuan, T. Lam, J. Biggs, L. Hu, Z. Yu, S. Ha, D. Xi, M.K. Senesky, G. Grüner, and Q. Pei, *New electrode materials for dielectric elastomer actuators*, The 14th International Symposium on: Smart Structures and Materials & Nondestructive Evaluation and Health Monitoring. International Society for Optics and Photonics, 2007: 65240N-1–65240N-12.

[8] W. Yuan, L. Hu, S. Ha, T. Lam, G. Grüner, and Q. Pei, *Self-clearable carbon nanotube electrodes for improved performance of dielectric elastomer actuators*, The 15th International Symposium on: Smart Structures and Materials & Nondestructive Evaluation and Health Monitoring. International Society for Optics and Photonics, 2008: 69270P-1–69270P-12.

[9] P. Brochu and Q. Pei, *Advances in dielectric elastomers for actuators and artificial muscles*, Macromol. Rapid. Commun. 31(1) (2010), pp. 10–36. doi:10.1002/marc.200900425.

[10] W. Yuan, P. Brochu, S.M. Ha, and Q. Pei, *Dielectric oil coated single-walled carbon nanotube electrodes for stable, large-strain actuation with dielectric elastomers*, Sens. Actuators. A Phys. 155(2) (2009), pp. 278–284. doi:10.1016/j.sna.2009.09.003.

[11] Y. Zhou, L. Hu, and G. Gruner, *A method of printing carbon nanotube thin films*, Appl. Phys. Lett. 88(12) (2006), pp. 123109–123109-3. doi:10.1063/1.2187945.

[12] N. Pierard, A. Fonseca, Z. Konya, I. Willems, G. Van Tendeloo, and J.B. Nagy, *Production of short carbon nanotubes with open tips by ball milling*, Chem. Phys. Lett. 335(1–2) (2001), pp. 1–8. doi:10.1016/S0009-2614(01)00004-5.

[13] C.A. Dyke and J.M. Tour, *Covalent functionalization of single-walled carbon nanotubes for materials applications*, J. Phys. Chem. A 108(51) (2004), pp. 11151–11159. doi:10.1021/jp046274g.

[14] C. Ménard-Moyon, N. Izard, E. Doris, and C. Mioskowski, *Separation of semiconducting from metallic carbon nanotubes by selective functionalization with azomethine ylides*, J. Am. Chem. Soc. 128(20) (2006), pp. 6552–6553. doi:10.1021/ja060802f.

[15] E. Bekyarova, M.E. Itkis, N. Cabrera, B. Zhao, A. Yu, J. Gao, and R.C. Haddon, *Electronic properties of single-walled carbon nanotube networks*, J. Am. Chem. Soc. 127(16) (2005), pp. 5990–5995. doi:10.1021/ja043153l.

[16] T. Sina, Y. Xiaohong, and D. Jianwen, *The effects of both length and tube-diameter on the conductance of single-walled carbon nanotubes*, Acta. Phys. Sin. 54(1) (2005), pp. 333–337.

[17] W. Yuan, L.B. Hu, Z.B. Yu, T. Lam, J. Biggs, S. Ha, D. Xi, B. Chen, M. Senesky, G. Grüner, and Q. Pei, *Fault-tolerant dielectric elastomer actuators using single-walled carbon nanotube electrodes*, Adv. Mater. 20(3) (2008), pp. 621–625. doi:10.1002/adma.200701018.

[18] O. Assad, A.M. Leshansky, B. Wang, T. Stelzner, S. Christiansen, and H. Haick, *Spray-coating route for highly aligned and large-scale arrays of nanowires*, ACS Nano 6(6) (2012), pp. 4702–4712. doi:10.1021/nn204513y.

Permissions

All chapters in this book were first published in IJSNM, by Taylor & Francis Online; hereby published with permission under the Creative Commons Attribution License or equivalent. Every chapter published in this book has been scrutinized by our experts. Their significance has been extensively debated. The topics covered herein carry significant findings which will fuel the growth of the discipline. They may even be implemented as practical applications or may be referred to as a beginning point for another development.

The contributors of this book come from diverse backgrounds, making this book a truly international effort. This book will bring forth new frontiers with its revolutionizing research information and detailed analysis of the nascent developments around the world.

We would like to thank all the contributing authors for lending their expertise to make the book truly unique. They have played a crucial role in the development of this book. Without their invaluable contributions this book wouldn't have been possible. They have made vital efforts to compile up to date information on the varied aspects of this subject to make this book a valuable addition to the collection of many professionals and students.

This book was conceptualized with the vision of imparting up-to-date information and advanced data in this field. To ensure the same, a matchless editorial board was set up. Every individual on the board went through rigorous rounds of assessment to prove their worth. After which they invested a large part of their time researching and compiling the most relevant data for our readers.

The editorial board has been involved in producing this book since its inception. They have spent rigorous hours researching and exploring the diverse topics which have resulted in the successful publishing of this book. They have passed on their knowledge of decades through this book. To expedite this challenging task, the publisher supported the team at every step. A small team of assistant editors was also appointed to further simplify the editing procedure and attain best results for the readers.

Apart from the editorial board, the designing team has also invested a significant amount of their time in understanding the subject and creating the most relevant covers. They scrutinized every image to scout for the most suitable representation of the subject and create an appropriate cover for the book.

The publishing team has been an ardent support to the editorial, designing and production team. Their endless efforts to recruit the best for this project, has resulted in the accomplishment of this book. They are a veteran in the field of academics and their pool of knowledge is as vast as their experience in printing. Their expertise and guidance has proved useful at every step. Their uncompromising quality standards have made this book an exceptional effort. Their encouragement from time to time has been an inspiration for everyone.

The publisher and the editorial board hope that this book will prove to be a valuable piece of knowledge for researchers, students, practitioners and scholars across the globe.

List of Contributors

Maria Cinefra and Stefano Valvano
Department of Mechanical and Aerospace Engineering, Politecnico di Torino, Torino, Italy

Erasmo Carrera
Department of Mechanical and Aerospace Engineering, Politecnico di Torino, Torino, Italy
School of Aerospace, Mechanical and Manufacturing Engineering College, RMIT University, Melbourne, Australia
Department of Mathematics, Faculty of Science, King Abdulaziz University, P.O. Box 80203, Jeddah 21589, Saudi Arabia

Nanying Ning, Liqun Zhang and Ming Tian
State Key Lab of Organic-Inorganic Composites, Beijing University of Chemical Technology, Beijing 100029, China
Laboratory of Carbon Fiber and Functional Polymers, Ministry of Education, Beijing University of Chemical Technology, Beijing 100029, China

S. Mantry, B.B. Jha and B.K. Mishra
CSIR-Institute of Minerals & Materials Technology, Bhubaneswar 751013, Odisha, India

A. Mandal and M. Chakraborty
School of Minerals, Metallurgical and Materials Engineering, Indian Institute of Technology Bhubaneswar, Bhubaneswar 751007, Odisha, India

D.K. Mishra
Department of Physics, Institute of Technical Education and Research, Siksha 'O'Anusandhan University, Khandagiri Square, Bhubaneswar 751030, Odisha, India
Department of Physics, College of Science, Engineering and Technology, University of South Africa, Johannesburg, South Africa

R. Zhu and F.G. Yuan
Department of Mechanical and Aerospace Engineering, North Carolina State University, Raleigh, NC 27695, USA

X.N. Liu and G.K. Hu
School of Aerospace Engineering, Beijing Institute of Technology, Beijing 100081, People's Republic of China

G.L. Huang
Department of Mechanical and Aerospace Engineering, University of Missouri, Columbia, MO 65211, USA

Shouhu Xuan, Yangguang Xu, Taixiang Liu and Xinglong Gong
CAS Key Laboratory of Mechanical Behavior and Design of Materials, Department of Modern Mechanics, University of Science and Technology of China, Hefei 230026, P. R. China

Qinghua Zhu and Mengyan Shen
Department of Physics and Applied Physics, and Nanomanufacturing Center, University of Massachusetts Lowell, 1 University Avenue, Lowell, MA 01854, USA

Xuhui Zhang, Tengfei Lin, Zhenghai Tang, Kailun Sun and Baochun Guo
Department of Polymer Materials and Engineering, South China University of Technology, Guangzhou 510640, P. R. China

Xiuying Zhao and Yonglai Lu
Key Laboratory of Beijing City on Preparation and Processing of Novel Polymer Materials, Beijing University of Chemical Technology, Beijing 100029, P. R. China Beijing Engineering Research Center of Advanced Elastomers, Beijing University of Chemical Technology, Beijing 100029, P. R. China

Jingna Yang, Detao Zhao and Liqun Zhang
Key Laboratory of Beijing City on Preparation and Processing of Novel Polymer Materials, Beijing University of Chemical Technology, Beijing 100029, P. R. China

Wencai Wang
Beijing Engineering Research Center of Advanced Elastomers, Beijing University of Chemical Technology, Beijing 100029, P. R. China

Toshio Nishi
Department of Organic and Polymeric Materials, Graduate School of Science and Engineering, Tokyo Institute of Technology, 2-12-1 O-okayama, Meguro-ku, Tokyo 152-8552, Japan

Qingzhao Kong
Department of Mechanical Engineering, University of Houston, Houston, TX 77204, USA

Qian Feng
Hubei Key Laboratory of Earthquake Early Warning, Institute of Seismology, China Earthquake Administration, Wuhan 430071, China
Wuhan Institute of Earthquake Engineering, Wuhan 430071, China

Gangbing Song
Department of Mechanical Engineering, University of Houston, Houston, TX 77204, USA
School of Civil Engineering, Dalian University of Technology, Dalian, Liaoning, China

Leonid A. Kaledin, Fred Tepper and Tatiana G. Kaledin
Argonide Corporation, Nanomaterial & Filtration Technologies, 291 Power Court, Sanford, FL 32771, USA

Chi Li, Yuhan Xie, Guorui Li, Xuxu Yang, Yongbin Jin and Tiefeng Li
Institute of Applied Mechanics, Zhejiang University, 38 Zheda Road, Hangzhou 310027, China

Sara Filippi and Giovanni Polacco
Department of Civil and Industrial Engineering, University of Pisa, Largo Lucio Lazzarino, 1, 56126 Pisa, Italy

Chensong Dong
Department of Mechanical Engineering, Curtin University, GPO Box 1987, Perth, WA 6845, Australia

Q.M. Zhang
Department of Materials Science and Engineering, Materials Research Institute, The Pennsylvania State University, University Park, PA 16802, USA
Department of Electrical Engineering, Materials Research Institute, The Pennsylvania State University, University Park, PA 16802, USA

M. Ghaffari
Department of Materials Science and Engineering, Materials Research Institute, The Pennsylvania State University, University Park, PA 16802, USA

Y. Zhou
Department of Electrical Engineering, Materials Research Institute, The Pennsylvania State University, University Park, PA 16802, USA

M. Lin
Materials Research Institute, The Pennsylvania State University, University Park, PA 16802, USA

Chong Min Koo
Center for Materials Architecturing, Institute for Multi-Disciplinary Convergence of Materials, Korea Institute of Science and Technology (KIST), Hwarangno 14-gil 5, Soengbuk-gu, Seoul 136-791, Republic of Korea

Jianwen Zhao, Jiangcheng Yu, Junyang Niu, Yong Ge and Liwu Liu
HIT at Weihai, State Key Laboratory of Robotics and System, Song Jian Research Institute of Building 2, Wenhua Road (W), Weihai, Shandong, 264209, PR China

Index